T0186240

Practical
Robot Design
Game Playing Robots

Practical
Robot Design
Game Playing Robots

Jagannathan Kanniah

M. Fikret Ercan

Carlos A. Acosta Calderon

CRC Press
Taylor & Francis Group
Boca Raton London New York

CRC Press is an imprint of the
Taylor & Francis Group, an **informa** business

MIX
Paper from
responsible sources
FSC® C014174
www.fsc.org

CRC Press
Taylor & Francis Group
6000 Broken Sound Parkway NW, Suite 300
Boca Raton, FL 33487-2742

© 2014 by Taylor & Francis Group, LLC
CRC Press is an imprint of Taylor & Francis Group, an Informa business

No claim to original U.S. Government works

Printed on acid-free paper
Version Date: 20130612

International Standard Book Number-13: 978-1-4398-1033-0 (Hardback)

Library of Congress Cataloging-in-Publication Data

Kanniah, Jagannathan.
 Practical robot design : game playing robots / Jagannathan Kanniah, M. Fikret Ercan, Carlos A. Acosta Calderon.
  pages cm
 Includes bibliographical references and index.
 ISBN 978-1-4398-1033-0 (hardback)
 1. Robots--Design and construction. 2. Robotics. I. Title.

TJ211.K355 2013
629.8'92--dc23   2013017526

Visit the Taylor & Francis Web site at
http://www.taylorandfrancis.com

and the CRC Press Web site at
http://www.crcpress.com

Contents

Preface.. xiii
Acknowledgments... xv
Authors...xvii

Chapter 1   Game Robotics 1

    1.1   Introduction 1
    1.2   Robotics Games and Engineering Education 1
    1.3   Robotic Games in Singapore 2
        1.3.1   Pole-Balancing Robot Race............................. 2
        1.3.2   Wall-Climbing Robot Race 3
        1.3.3   Robot Colonies 3
        1.3.4   Humanoid Robot Competition 4
        1.3.5   Other Competitions and Open Category 4
    1.4   Robotic Games around the World 6
    1.5   Overview of the Book.. 11
    References 12

Chapter 2   Basic Robotics 13

    2.1   Introduction to Robotic Systems 13
        2.1.1   Terminology Used in Robotics........................ 13
    2.2   Coordinate Transformations and Finding Position of
        Moving Objects in Space... 14
        2.2.1   Composite Rotations 17
        2.2.2   Homogeneous Transformation Matrix 19
        2.2.3   Composite Transformations 20
            2.2.3.1   Matrix Multiplication Order in
                    Composite Transformations....................... 20
        2.2.4   Mathematical Description of Objects....................23
    2.3   Wheel Drive in Mobile Robots...................................... 28
        2.3.1   Differential Drive 32
        2.3.2   Ackermann Steering (Car-Like Drive).................... 34
        2.3.3   Track Drive... 34
        2.3.4   Omniwheel Drive 35
        2.3.5   Odometry.. 36
        2.3.6   Case Study of Odometry for a Differential
            Drive Robot 37
    2.4   Robotic Arms 39
        2.4.1   Forward Kinematic Solutions.......................... 45
        2.4.2   Inverse Kinematics 46

          2.4.3   Case Study: Three-Link Articulated Robot Arm 48
     References 52

Chapter 3   Sensors 53

     3.1   Sensors Used in Game Robotics................................... 54
          3.1.1   Measuring Robot Speed 54
          3.1.2   Measuring Robot Heading and Inclination 56
          3.1.3   Measuring Range 57
          3.1.4   Detecting Color 63
     References 64

Chapter 4   Robot Vision.. 65

     4.1   Introduction 65
     4.2   Camera Systems for Robotics....................................... 66
     4.3   Image Formation 67
     4.4   Digital Image-Processing Basics.................................. 71
          4.4.1   Color and Color Models 71
     4.5   Basic Image-Processing Operations............................. 73
          4.5.1   Convolution 74
          4.5.2   Smoothing Filters 76
     4.6   Algorithms for Feature Extraction 77
          4.6.1   Thresholding 78
          4.6.2   Edge Detection 81
          4.6.3   Color Detection 83
     4.7   Symbolic Feature Extraction Methods.......................... 86
          4.7.1   Hough Transform 86
          4.7.2   Connected Component Labeling........................... 92
     4.8   Case Study: Tracking a Colored Ball 98
     4.9   Summary 100
     References 101

Chapter 5   Basic Theory of Electrical Machines and Drive Systems................ 103

     5.1   Actuators for Robots.. 103
     5.2   Electrical Actuators 104
          5.2.1   Fundamental Concepts of Generating and
                   Motoring... 104
          5.2.2   DC Machines.. 106
          5.2.3   AC Motor Drives 112
     5.3   Specific Needs of Robotics Drives 113
          5.3.1   DC Permanent Magnet Motors 114
          5.3.2   Servo Motors 114
          5.3.3   Stepper Motors 116
          5.3.4   Brushless DC Motors 122
     5.4   Drive Systems... 126
          5.4.1   DC Motor Control 126

5.4.2   Stepper Motors Drivers 128
          5.4.2.1   Sequence Generator 129
          5.4.2.2   Operating Modes of L297 130
          5.4.2.3   Applications 131
5.4.3   Brushless DC Motor Drive 133
          5.4.3.1   Back EMF Sensing-Based Switching 134
          5.4.3.2   Sensor-Based Switching 136
5.5   Conclusion 137
References 137

Chapter 6   Motor Power Selection and Gear Ratio Design for Mobile Robots 139

6.1   Gear Ratio for a Mobile Robot 139
6.2   Power Requirement of the Drive Motor 141
        6.2.1   Role of Motor Inertia and Friction 143
6.3   Typical Motor Characteristics Data Sheet 145
6.4   Friction Measurement in a Linear Motion System 146
6.5   First Approach: Gear Ratio Design 148
6.6   Second Approach: System Performance
        as a Function of Gear Ratio 153
6.7   Gear Ratio Design for Stepper Motors 156
6.8   Design Procedures for Mobile Robot That Are Not
        Ground Based 158
6.9   Conclusion 164
References 164

Chapter 7   Control Fundamentals 165

7.1   Control Theory for Robotics 165
7.2   Types of Plants 166
        7.2.1   Linear versus Nonlinear Plants 166
        7.2.2   Time-Invariant versus Time-Variant Plants 167
7.3   Classification Based on Control System 167
        7.3.1   Analog versus Digital Systems 168
        7.3.2   Open-Loop versus Closed-Loop Systems 168
7.4   Need for Intelligent Robot Structure 169
7.5   A Typical Robot Control System 170
7.6   Trends in Control 171
7.7   Conclusion 171
References 172

Chapter 8   Review of Mathematical Modeling, Transfer Functions, State
                Equations, and Controllers 173

8.1   Introduction 173
8.2   Importance of Modeling 174
8.3   Transfer Function Models 174
        8.3.1   Different Forms of Transfer Functions 175

8.4   Steps in Modeling 176
8.5   Some Basic Components Often Encountered in Control
       Systems 177
       8.5.1   Electrical Components 177
       8.5.2   Mechanical Components 178
8.6   Block Diagram Concepts 179
       8.6.1   Block Diagram Reductions 180
8.7   Some System Examples 180
8.8   State Equations 191
       8.8.1   Basic Concepts of State Equations from
               Differential Equations 191
       8.8.2   State Equations from Plant Knowledge 193
       8.8.3   State Equations Directly from Transfer Functions 195
8.9   Time Domain Solutions Using Transfer
       Functions Approach 203
       8.9.1   Analytical Solution for Mass, Spring, and
               Damper System in Closed Loop 204
       8.9.2   Simulation Solution for Mass, Spring, and
               Damper System in Closed Loop 205
       8.9.3   PID Controller Response 207
8.10   Time Domain Solutions of State Equations 209
       8.10.1 Time Domain Solutions Using Analytical
               Methods 209
8.11   Regulator and Servo Controllers 214
8.12   Conclusion 216
References 216

Chapter 9   Digital Control Fundamentals and Controller Design 217

9.1   Introduction 217
9.2   Digital Control Overview 217
       9.2.1   Signal Sampler 218
       9.2.2   Digital Controller 219
       9.2.3   Zero-Order Hold 219
9.3   Signal Representation in Digital Systems 221
       9.3.1   Sampling Process 221
               9.3.1.1   Sampling for Reconstruction 223
               9.3.1.2   Sampling for the Purpose of Control 223
       9.3.2   Z-Transform of Signals 223
               9.3.2.1   Z-Transform of Continuous Signals 224
               9.3.2.2   Z-Transform of Signals Represented
                         Only as Sample Count, k 226
9.4   Plant Representation in Digital Systems 228
       9.4.1   Transfer Function of ZOH 229
       9.4.2   Z-Transform of Plant Fed from ZOH 230
       9.4.3   Tustin's Approximation 230

9.5   Closed-Loop System Transfer Functions 231
        9.5.1   Systems with Digital Instrumentation...................... 232
9.6   Response of Discrete Time Systems, Inverse
        Z-Transforms... 234
        9.6.1   Partial Fraction Technique 234
        9.6.2   Difference Equation Techniques 234
        9.6.3   Time Domain Solution by MATLAB® 235
9.7   Typical Controller Software Implementation 237
        9.7.1   Integral Calculations 240
        9.7.2   Derivative Calculations 240
        9.7.3   Implementation of a Digital Controller 240
9.8   Discrete State Space Systems.. 241
        9.8.1   Discrete State Space System from Discrete
                 Transfer Functions... 241
        9.8.2   Discrete State Space Model from Continuous
                 State Space Model.. 242
                 9.8.2.1   Analytical Method.................................. 242
                 9.8.2.2   MATLAB Approach............................... 245
        9.8.3   Time Domain Solution of Discrete State Space
                 Systems.. 245
                 9.8.3.1   Computer Calculations 246
                 9.8.3.2   Z-Transform Approach 246
9.9   Discrete State Feedback Controllers 250
        9.9.1   Concept of State Controllability............................ 250
        9.9.2   Concept of State Observability 251
        9.9.3   Common Condition for Controllability and
                 Observability of Sampled Data Systems 252
        9.9.4   Design of Pole Placement Regulators Using
                 State Feedback.. 253
                 9.9.4.1   Comparison of Coefficients Method........ 254
                 9.9.4.2   MATLAB Method of Pole Placement 256
                 9.9.4.3   MATLAB Simulation of the Controller
                            Performance... 256
        9.9.5   Steady-State Quadratic Optimal Control 258
                 9.9.5.1   Use of MATLAB in LQC Design 259
        9.9.6   A Simple Servo Controller 260
9.10   Typical Hardware Implementation of Controllers................ 264
9.11   Conclusion 266
References 266

Chapter 10 Case Study with Pole-Balancing and Wall-Climbing Robots.......... 267

10.1   Introduction 267
10.2   Pole-Balancing Robot.. 268
        10.2.1   Mathematical Modeling 269
        10.2.2   Transfer Function for Pole Angle Control................ 278

10.2.3   Pole-Balancing Robot State Model 278
10.2.4   State Model for the Pole-Balancing Robot from
         Robot and Motor Data.. 282
10.2.5   Pole Placement Controller with Servo Input
         Used as Offset 283
10.2.6   LQC Controller with Servo Input Used as Offset.... 286
         10.2.6.1   Effect of a Change in Q Matrix 287
10.2.7   Implementation of the Pole-Balancing Robot
         Controller Using DSP Processor 291
         10.2.7.1   Hardware Setup 291
         10.2.7.2   Software for the Robot............................. 293
10.2.8   Two-Degree-Freedom Pole-Balancing Robot 299
         10.2.8.1   Control Philosophy 299
10.2.9   Estimation of Angular Friction Term b Used in
         PBR from Experiment... 300
10.3   Wall-Climbing Robots.. 306
       10.3.1   Flipper Wall-Climbing Robot 306
                10.3.1.1   Overall System Configuration of
                           Flipper WCR.. 308
                10.3.1.2   Control of Suction Pad Arms and
                           Cruise Motor.. 308
                10.3.1.3   Operation Sequence of the Flipper WCR 310
       10.3.2   Design of a Wall-Climbing Robot Using
                Dynamic Suction... 312
                10.3.2.1   Dynamic Suction Principle....................... 313
                10.3.2.2   Operation of the WCR Using
                           Bernoulli's Principle 314
10.4   Conclusion 315
References 315

Chapter 11   Mapping, Navigation, and Path Planning....................................... 317

11.1   Introduction 317
11.2   Perception 317
       11.2.1   From Sensor Measurements to Knowledge Models 318
       11.2.2   Map Representation.. 321
       11.2.3   Metric Map... 322
                11.2.3.1   Case Study 325
       11.2.4   Topological Map... 328
                11.2.4.1   Case Study of Topological Map................ 328
11.3   Navigation... 330
       11.3.1   Wall Following 331
       11.3.2   Obstacle Avoidance with Vector Force Histogram....333
                11.3.2.1   Case Study of Obstacle Avoidance with
                           Vector Force Histogram............................. 335

11.4 Path Planning 341
      11.4.1 Wavefront Planner 341
      11.4.2 Path Planning Using Potential Fields 343
              11.4.2.1 Case Study of Path Planning Using
                        Potential Fields 344
      11.4.3 Path Planning Using Topological Maps 350
              11.4.3.1 Case Study of Path Planning Using
                        Topological Maps 350
      References 356

Chapter 12 Robot Autonomy, Decision-Making, and Learning 357

      12.1 Introduction 357
      12.2 Robot Autonomy 357
      12.3 Decision-Making 358
            12.3.1 Classical Decision-Making 359
            12.3.2 Reactive Decision-Making 360
                    12.3.2.1 Case Study on Reactive Decision-
                              Making 361
            12.3.3 Hybrid Decision-Making 365
      12.4 Robot Learning 366
            12.4.1 Artificial Neural Networks 368
                    12.4.1.1 Perceptron 369
                    12.4.1.2 Case Study on Perceptron with Learning ...372
                    12.4.1.3 Multilayer Perceptron 377
            12.4.2 Q-Learning 379
                    12.4.2.1 Case Study Q-Learning 380
      12.5 Conclusion 384
      References 386

Index 387

Preface

Robotic games and competitions are spawned from mainstream robotics, and they are very popular among the engineering students, robotics enthusiasts, and hobbyists. Over the last decade, hundreds of robotic competitions have been organized in different parts of the world. The interest in robotic games has also reached greater heights with the availability of many affordable parts and components that can be acquired easily over the Internet. Game robotics is a passion and provides great fun and learning experiences.

As in every field of engineering, progress is also inevitable for robotic games. The complexity of the games during the last decade has increased tremendously. Robots developed to compete in such games are becoming more and more sophisticated. Consequently, this makes robotic games not only entertaining, but also a great way of learning engineering concepts and establishing the link between theory and practice. Needless to say, robotics is a multidisciplinary subject. It expands to various engineering and scientific disciplines such as electrical engineering, mechanical engineering, computing, and many more. It is even a unifying platform for different courses taught in one discipline. For instance, electronics, microprocessors, electrical machines, and control theory are all distinct fields taught in electrical engineering. Each of these courses has vast course materials and research opportunities individually. Robotics is an application platform where all these fields converge naturally. However, for students and robotics fans who are designing robots for games and competitions, such a vast sources of material can be overwhelming. Our primary objective in this book is to provide a starting point and immediate knowledge needed for game robotics.

There are many good journals, workshops, books, and online resources for hobby robotics, and they provide many creative ideas. The current state of robotic games is reasonably advanced, as the mentioned competitions are becoming more and more complex. The knowledge and experience required for designing robots for such games also demand good understanding of engineering concepts. Robotic applications such as soccer-playing humanoids or wall-climbing robots not only require expertise in robot intelligence and programming, but also require designing robots well so that they can perform their actions and motions appropriately. Therefore, in this book, we present some of the fundamental concepts and show how they benefit the design process. In particular, we discuss the necessary basics to make the right choices for gears and actuators as well as modeling and low-level controlling of robot motions in Chapters 5 through 9. We present the application of these concepts in game robotics with some case studies in Chapter 10.

The authors of this book have been involved in robotic games and have designed many robots together with their students and colleagues for more than a decade. The book resulted from our earlier notes prepared for a summer course for those students taking up robotic games as their final year project. We hope that this book will empower undergraduate students in terms of the necessary background as well as the understanding of how various engineering fields are amalgamated in robotics. We hope that students and robot enthusiasts will benefit from this book in their endeavor to build cool robots while having fun with robotic games.

MATLAB® is a registered trademark of The MathWorks, Inc. For product information, please contact:

The MathWorks, Inc.
3 Apple Hill Drive
Natick, MA 01760-2098 USA
Tel: 508 647 7000
Fax: 508-647-7001
E-mail: info@mathworks.com
Web: www.mathworks.com

Acknowledgments

This book is the result of many years of research and development activities in the area of game robotics carried out at Singapore Polytechnic. We are grateful to all our students who spend long hours to labs to design, build, and tune robots for competitions. Their passion and drive naturally got us involved even more, and we share the fun of robotics with them. We are thankful to Jacqueline Oh, Lius Partawijiya, Mohd Zakaria, and Zar Ni Lwin for their interest and expertise in robot design and all the technical support they have provided over the years.

Authors

Jagannathan Kanniah received his BE and MSc degrees in electrical engineering from Annamalai University, India, in 1969 and 1971, respectively. He received his PhD from the University of Calgary, Canada in 1983. He is a senior member of IEEE, a member of IET, and a chartered engineer. He served as an academic staff in various institutions in India from 1971 to 1978. He worked as postdoctoral fellow at the University of Calgary from 1982 to 1983. After leaving Canada, he joined Singapore Polytechnic and rose to the level of principal lecturer. During his service at Singapore Polytechnic, he went as a visiting scientist to Lund Institute of Technology, Sweden, for three months during 1992 and to Massachusetts Institute of Technology, for 3 months during 1999. He has been the section head of the robotics and automation group since 1994 and a technology leader and the manager of the Singapore Robotics Games center since 1996 at Singapore Polytechnic until he retired in 2007. He continued to work until 2011 at the SRG (Singapore Robotic Games) center. His research interests are in the area of power systems, adaptive control, instrumentation, and robotics. He has more than 35 publications, including many journal papers. He has supervised many student groups working on robotics that took part and won many awards in the Singapore robotics game events over the years.

M. Fikret Ercan received his BSc and MSc degrees in electronics and communications engineering from Dokuz Eylul University, Turkey, in 1987 and 1991, respectively. He received his PhD from the Hong Kong Polytechnic University in 1998. He is a senior member of IEEE and a member of IEEE Ocean Engineering Society. His research interests are in the field of image processing, robotics, and computing. He has written one book (*Digital Signal Processing Basics*, Pearson, 2009) and two book chapters, and has more than 80 publications, including journal papers. Prior to his academic carrier, he worked in the electronic and computer industries primarily as a research and development engineer in various countries, including Turkey, Taiwan, and Hong Kong. He is currently a senior lecturer at Singapore Polytechnic. In addition to his research in image processing and computing, he has been actively involved in robotic games since 2000. The student teams that he had led participated in competitions both locally and overseas.

Carlos A. Acosta Calderon received his BEng degree in computer systems from the Pachuca Institute of Technology, Mexico, in 2000, his MSc degree in computer science (robotics and intelligent machines) from the University of Essex, United Kingdom, in 2001, and his PhD degree in computer science from the University of Essex, United Kingdom, in 2006. Currently, he is a lecturer at the School of Electrical and Electronic Engineering at Singapore Polytechnic. His research interests include social robots, coordination of multirobot systems, learning by imitation, and humanoid robots. He has published two book chapters and over 50 papers in journals and conferences. He also serves as a member of the Technical Committee of the Humanoid League at the RoboCup Competition, and as a member of the Organizing Committee of the RoboCup Singapore Open. He has been involved in robotics games since 2006 and guides student teams in local and international robotics competitions. They are NJRC, WRO, RoboCup, and SRG.

1 Game Robotics

1.1 INTRODUCTION

Robotics is a fast-developing and highly popular field of engineering. It encompasses a wide range of disciplines such as electrical engineering, mechanical engineering, computer science, biology, sociology, and so on. A great deal of developments in robotics was due to its applications in manufacturing. The need for more and more automation in assembly lines was the main driving force for it. Robots can do repetitive and mundane jobs a lot faster, more accurately, and cheaper than human beings. Their use in industry naturally increases productivity and makes it more flexible. Therefore, for a long period of time, robotics remained popular in manufacturing and industry. However, during the last decade, robotics found applications in many fields other than manufacturing such as service robotics, medicine, entertainment, and education. Advances made in computer technology, sensor technology, semiconductor technology, and artificial intelligence were also instrumental. We are now seeing interesting examples such as human-like robots interacting with people, robots dancing to a tune, robots playing musical instruments, robots playing football, or robots assisting surgeons in operating theaters (Baltes et al. 2010; Gao et al. 2010; Kaneko et al. 2009; Ogura et al. 2006; Taylor and Stoianovici 2003).

A distinct class of robotics, namely game robotics, emerged recently, due to the demand from academic institutions. It is also called edutainment robotics since it combines education with entertainment. Game robotics makes learning more fun and entertaining for the students. In this book, we are particularly interested in game-playing robots. We try to provide a reference book for senior students doing their projects in robotics or a guidebook for a robot enthusiast who wants to have a higher level of understanding of robotics. We emphasize mainly practical aspects of robotics and try to show how it is linked to conventional subjects learned from engineering textbooks.

1.2 ROBOTICS GAMES AND ENGINEERING EDUCATION

Game robotics is essentially entertainment robotics and also serves very well in engineering education (Malec 2001). The thrill of taking part in an "Olympics-like competition for robots" generates more interest than what can be achieved by a mundane project otherwise (Martin 2001). Needless to say, robotics easily captures the interest of young people. Even the older generation, who grew up watching robots in sci-fi movies, may find building robots interesting. However, building robots requires an understanding of various aspects of engineering and science such as mechanics, analog and digital circuits, programming, microcontrollers, and control theory just to name a few. Our experiences over the years show that the students who are engaged in building robots, unsurprisingly, are motivated to learn all these fields and more. They demonstrated a better understanding of linking theory with practice. These observations are also backed by many studies published on engineering education. For instance, a study conducted by Pisciotta et al. (2010) shows that students who are engaged in robotics projects perform better in math–science, electronics, and logic. Our experience over the years, working with students taking up robotic projects showed that there are three major changes in students' behavior. First, they become self-learners and autonomous. Second, their confidence in their engineering skills improves significantly. Third, they learn to collaborate and to be team players because robotic game projects are usually team projects. These are highly sought after traits in the industry.

1.3   ROBOTIC GAMES IN SINGAPORE

During the last decade, robotic games became very popular and spread all over the world. There are a vast number of robotic games, festivals, and competitions held in various parts of the world. In 1991, the first Singapore robotics festival was organized to create awareness about robotics. Later, the event was renamed as Singapore Robotic Games (SRG). The first SRG competition was held in 1993. Since then, the competition is held annually and it draws a lot of attention (SRG 2012). The competitions are open to public and tertiary institutions. At the beginning, the main events were few; today, competition have grown to more than 15 categories and are evolving continuously in their complexity year after year. Typically, game rules are revised every 3 years to accommodate the latest advances in technology and to make the games more challenging. For example, the pole-balancing race used to be a game where a mobile robot needs to balance a free-falling pole while moving from one point to another. Challenge has evolved, a robot now has to move on a platform with variable slopes and negotiate randomly placed obstacles on its path while balancing a pole. At present, there are 14 categories: pole-balancing robot, intelligent robot, robot colonies, wall-climbing robot, robot sumo, and legged robot race just to name a few. More details about the Singapore Robotic Games can be found on the competition web site (SRG 2012). In the following sections, we will give a brief description of technically challenging games, some of which are also presented as case studies in this book.

1.3.1   POLE-BALANCING ROBOT RACE

This game is inspired from a well-known control theory problem, which is balancing an inverted pendulum. A robot supports an inverted pendulum, which is free to swing around the horizontal axis, and balances it vertically by moving the point of support. The competition platform consists of a horizontal wooden surface with a dimension of 3 by 1.5 m. A robot is required to vertically balance the pole at the starting zone, then move toward the other end of the platform and go back to the starting zone while negotiating all the slopes and obstacles along the way. The above cycle is repeated and the robots are ranked based on the number of successful cycles within 5 min of time. Figure 1.1 shows a snapshot of a pole-balancing robot in action.

FIGURE 1.1   Pole-balancing competition.

1.3.2 WALL-CLIMBING ROBOT RACE

The objective of this competition is to demonstrate vertical and horizontal surface-climbing abilities of robots. A competition platform is made of a wooden plank forming the floor, wall, and ceiling sections, all of which are 2 m long. During the competition, a robot starts from the frontmost part of the floor, moves toward the wall, climbs the wall, reaches the ceiling, travels toward the edge of the ceiling, and finally travels back to the starting point. Robots are ranked based on their completion time of this task. The competition platform is nonmagnetic, which makes this game more challenging. Most of the robots in this competition employ pneumatic principles with a variety of creative techniques to accomplish the task in the shortest possible time. Figure 1.2 shows a snapshot of such a robot moving along the ceiling.

1.3.3 ROBOT COLONIES

The objective of the competition is to build a pair of autonomous and cooperative mobile robots. Their task is to search, detect, and collect colored pellets and deposit them in a designated container. Each container is reserved for one color, and they are

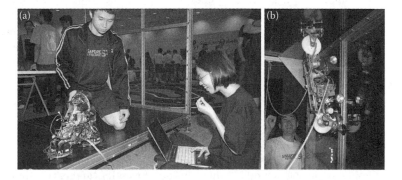

FIGURE 1.2   Wall-climbing robot. (a) Fine tuning a wall-climbing robot, (b) robot performing a climb.

FIGURE 1.3   Robot colony game.

located at the opposite sides of the platform. There are two different colored pellets used on the competition platform, and they are randomly placed. The goal is to collect and deposit an equal number of pellets of the two different colors. The major constraint for the robots is that they have to operate within their dedicated zones. Each robot is allowed to deposit one designated color at the collection point, which implies that at some point in time robots have to swap the pellets that they have collected to complete the task. For instance, a robot assigned to collect blue pellets will also collect green pellets that fell in its zone. However, it needs to transfer green pellets to its partner, which is in charge of green pellets in a dedicated zone. The center part of the platform is allocated for this purpose where two robots are allowed to be at the same time. This game induces the principles of autonomous and mobile robotics as well as instills an understanding of multirobotic collaboration, coordination, and communication. Figure 1.3 shows a snapshot of the robot colony competition.

1.3.4   Humanoid Robot Competition

The primary objective of humanoid robot competition is to encourage technological advances in humanoid robot technology so that robots can walk and run like human beings. The competition is between bipedal robots, and there is no predefined race arena for this game. The participating robots compete on the natural floor surface, which can be carpet, concrete, parquet, and so on. However, a race track is created using white reflective tape. A robot that covers the track from the starting point to the end in a shortest period of time is the winner. Figure 1.4 shows an instance from this competition.

1.3.5   Other Competitions and Open Category

In addition to the aforementioned categories, there are many more interesting games, which form more than 15 categories of games organized by the Singapore Robotic Games society. Each one of these games specially targets certain technical challenges involved in robotics. For instance, the intelligent robot game targets robot autonomy, object recognition, and handling aspects. Each team is required

FIGURE 1.4   Humanoid robot game. (a) Humanoid robot following a line, (b) tracking a ball.

to design and build either one or more autonomous robots to collect objects of various shapes and colors scattered in the competition arena. The collected objects are to be delivered to three different goal containers according to their respective colors within 6 min. Also, underwater robot competition aims to raise interest in marine engineering. Compared to land robots, designing underwater robots presents totally new challenges such as controlling robot buoyancy, autonomy, sensing, and maneuvering in water. During the competition, teams try to complete the given task with their robot either in the remote operative vehicle (ROV) category or the autonomous underwater vehicle (AUV) category. In the robot sumo competition, participants build mobile robots that can push an opponent out of the ring. This game requires an understanding of dynamics, friction, power, and motor control concepts. Additionally, an open category allows participants to show off their creativity and technical skills. Participants demonstrate interesting tasks that their robots can perform. Figure 1.5 shows some snapshots of these competitions.

FIGURE 1.5   Other interesting games in Singapore Robotic Games. (a) Open category robots at display, (b) sumo robot competition, (c) open category robot in action, (d) intelligent robot competition, (e) schools robotic competition-robo can collector.

1.4　ROBOTIC GAMES AROUND THE WORLD

Robotics competitions are appearing all over the world, each with its own set of unique objectives and rules. Some of these competitions may be started as national or regional events, but soon turned into an international event. It is impossible to list them all in this section; however, we will briefly mention those well-established and popular competitions.

Micromouse: This is perhaps one of the earliest robotics competitions. In this event, a robot mouse tries to solve a maze made of 16×16 cells. The technical challenge involves finding an optimum path and reaching the goal in the shortest time. Competitions are held worldwide. This game is also part of the Singapore Robotics Games. A description of the competition and its rules can be found in SRG (2012).

FIRA: This is one of the most established competitions around the world. It began in South Korea in 1995; since then it is held annually in different venues. The Federation of International Robot-Soccer Association (FIRA) was founded in June 1997 (FIRA 2012). This initiative gives a good platform for research on multiagents while two robot teams play soccer. The participants deal with problems such as cooperation, distributed control, effective communication, adaptation, and reliability. There are seven leagues in FIRA, each league focuses on a different type of robot and problem: HuroSot (humanoid robots), AmireSot (fully autonomous onboard robot), MicroSot (each team consists of three robots with dimensions $7.5 \times 7.5 \times 7.5$ cm), NanoSot (each team consists of five robots with dimensions $4 \times 4 \times 5.5$ cm), AndroSot (team of three robots, which are remotely controlled, with dimensions up to 50 cm), RoboSot (team of three robots fully autonomous or semi-autonomous, with dimensions of 20×20 cm $\times$ no limit in height), and SimuroSot (Simulation server for games of 5 vs. 5 and 11 vs. 11 games).

RoboCup: RoboCup is an international initiative to promote robotics and artificial intelligence research by providing a standard platform through which a wide range of technologies can be integrated and examined (RoboCup 2012). By 2050, the RoboCup Federation aims to develop autonomous humanoid robots advanced enough to compete against the human World Cup champions. If robots are able to play soccer and beat the champion human team, the technology developed in this team of robots will be good enough to provide robots that can help in any task.

The first RoboCup competition was held in 1997 in Nagoya, Japan. Since then, the competition has traveled all over the world to cities including Osaka, Bremen, Atlanta, Melbourne, and Singapore. Today, RoboCup is one of the largest robotics events in the world; thousands of participants from more than 40 countries are taking part in this annual event. It has grown so big that its influence can be seen with the amount of participants in the regional and country level events (called RoboCup Opens). These events are mainly used as a qualification stage for teams that seek a place in the international competition.

RoboCup events consist of competition, exhibition, and symposium. The competition is mainly divided into two great categories, the Junior Competitions for kids and teenagers up to the age of 19 years, and the Senior Competitions with no restriction on age, but mainly captivated by colleges and universities. The junior event includes four competitions: RoboCupJunior Soccer, RoboCupJunior Rescue, RoboCupJunior

Dance, and CoSpace. RoboCupJunior is a new and exciting way for young engineers to understand science and technology through hands-on experience with electronics, hardware, and software. It also offers opportunities to learn about teamwork while sharing ideas with friends. The development of study materials and innovative teaching methods are among the objectives of RoboCupJunior. It primarily focuses on education and comprises four challenges:

- *RoboCupJunior Soccer* is a challenge whereby teams are required to design and program two robots to play a game with an opposing pair of robots by kicking an infrared-transmitting ball into their designated goal. There are two different leagues to separate students from primary and secondary schools.
- *RoboCupJunior Dance* involves real team effort where participants are required to create dancing robots and program them to dance to music. Besides choreographing the motions of the robots, students are also expected to participate in the performance. Robots and students perform on a white-floored stage, which is bounded by black lines forming a square. Robots are not allowed to cross these boundaries. The judges evaluate the entertainment factor of the performance as well as the technical design of the robots used for the dance.
- *RoboCupJunior Rescue* is a challenge, in which robots need to complete the rescue mission by following a winding line or rooms to a designated area. This whole process is timed. The rescue robot will start by following a line and travel into different rooms. When inside the rooms, the robot will continue following the line without colliding with the obstacles or losing track of the line that might be disrupted. In addition, there are victims in the floor marked as human shapes with colors that the robots could recognize. Once the rescue robot passes over the victim, it blinks its LEDs, which indicates that the victim has been rescued by the robot. Robots encounter their final challenge when they proceed to the second level. The slope that connects that level is a hard challenge for the small motors of most robots. For those robots that manage to reach the second level, the challenge becomes even tougher since there is no line to follow at this stage. They are expected to rescue the victim and find their way back to the slope. The task is completed when the robot returns to its starting point.
- *CoSpace Dance/CoSpace Rescue Challenge* is an educational initiative to promote knowledge and skills in engineering design, programming, electronic control, and the world of 3D simulation through robotics for young minds. Using virtual environments provides great flexibility to manipulate the environment where the researchers can develop and experiment with new algorithms. It is not limited to a single robot; it can be a multirobot system too. Each robot in the system can be controlled in different ways, autonomous, semiautonomous, and manual. In each case, a multirobot system consists of a mix of different control forms. The main advantage of this platform, in contrast to other available simulators, becomes evident when the work requires cooperation between real robots and their virtual

counterparts. The CoSpace Development Platform eases the job by providing virtual environments and robots for the students. It also provides a large set of options to be controlled and monitored, such as time, number of teams, number of robots per team, obstacles, goals, and so on.

RoboCup senior competitions comprise: RoboCupSoccer, RoboCupRescue, and RoboCup@Home. There are also sponsor competitions like the Festo Logistics Competition (Festo 2012) and demo leagues like the Virtual Reality Competition.

RoboCupSoccer: The main focus of RoboCupSoccer is two teams of robots competing on a designated soccer field to score against the opposing team as many times as possible. The research outcome is mainly focused on multiagent cooperation and coordination, kinematics, and dynamics of robots. There are five leagues in this event: Small Size League, Middle Size League, Humanoid League, Standard Platform League, and Simulation League.

RoboCupRescue: This aims to promote research and development in disaster rescue, with robots exploring a simulated disaster site to locate and identify signs of life and produce a map of the site to safely perform a rescue. The competition aims to develop intelligent agents and robots to respond to disasters. There are two leagues here: Rescue Robot and Rescue Simulation.

RoboCup@Home: This aims to develop service and assistive robot technology for personal domestic applications. The focus is on developing robotic applications in human–machine interaction to enrich daily living. Participants compete in an environment simulating current societal issues such as aging, urbanization, healthcare, and assisted living.

RoboCup Singapore Open: RoboCup Singapore Open is a national-level robotics competition mainly for the RoboCupJunior competitions for students up to the age of 19. Participants compete in four challenges, RoboCupJunior Soccer, RoboCupJunior Rescue, RoboCupJunior Dance and CoSpace; the shortlisted teams advance to represent Singapore in the RoboCupJunior league during the international RoboCup event.

The objective of RoboCup Singapore Open is strongly educational, allowing local students the opportunity to participate, interact, share, and learn from their international peers. By competing in various leagues, the participants learn more than just artificial intelligence and mechatronics, but also creativity and human endeavor. It is truly a unique learning journey for students who have the opportunity to combine creativity with scientific knowledge in a project-oriented activity. Figure 1.6 shows some snapshots from RoboCup events.

MATES' ROV competition: Another well-known competition to robotics enthusiast is coordinated by the Marine Advance Technology and Education Center (MATE) (MATE 2012). This is a competition of ROVs built by students. The competitions take place across the United States, Canada, Hong Kong, and Scotland. Student teams from middle school to university levels participate in these events under different categories with different levels of sophistication of ROVs and mission requirements. The objective of the competition is not only to develop problem solving, critical thinking, and teamwork skills of students, but also to connect them with employers and working professionals from marine industries and introduce marine-related career opportunities.

FIGURE 1.6 Snapshots from RoboCup event. (a) Humanoid adult size, (b) humanoid kid size, (c) RoboCupJunior dance, (d) RoboCupJunior participants, (e) participants fine tuning robots before the competition.

Robotic sumo wrestling: It was introduced in Japan by Dr. Mato Hattori; it became very popular and it is adopted by other robotic events such as Singapore robotic games, Seattle Robotics Society, and many others. It involves two contestants who operate their robots in the sumo ring (Miles 2002).

World Robotic Sailing Championship: This is relatively a new competition started in Austria. World robotic sailing championship is a competition of fully autonomous and unmanned sailing boats (WRSC 2012).

International Aerial Vehicles Competition: This is a competition of autonomous flying robots, which is sponsored by the Association for Unmanned Vehicle Systems International (AUVSI 2012). The complexity of the competition increases with different missions defined over the years.

FIRST (For Inspiration and Recognition of Science and Technology): This competition was founded in 1989 in the United States to inspire interest in science and technology among young people (FIRST 2012). There are various categories of competition targeting different age groups. In junior FIRST LEGO League, children in the age group of 6–9 design and build a model using LEGO components. In FIRST LEGO League, children are exposed to real-world science and technology problems, and they develop their own solutions to these problems using autonomous LEGO robots. FIRST Tech Challenge is a higher level and designed for high school students. Teams of students design, build, and program robots and compete against other teams. FIRST Robotics Competition is for the age group of 14–18. Students taking part in this competition work under strict rules with limited resources and tight deadlines, which expose them to real world engineering problems. They design and build a robot that can perform a set of prescribed tasks against other competitors.

A common factor for all these robotic games is that they provide avenues to young engineers to develop their engineering skills. Hopefully, students will understand scientific concepts better, apply engineering principles into practice, and follow the current technological developments. As we mentioned, it is impossible to list every robotics competition here. However, we can classify these robotic games based on the technology involved as shown in Figure 1.7. Basically, two main categories of robot competitions take place: autonomous and remotely operated. A large number of robotic games falls into the autonomous category since challenges that can be posed upon these robots are vast. Among the autonomous robots, we find games designed either for a single robot or a team of robots. For a single robot, the technical complexity involves object handling, navigation, intelligence, and other well-known aspects of autonomy. However, basically two types of challenges are imposed. Some games target technical complexity involved in certain robot motions such as climbing, balancing a pole, bipedal or hexapod walking, and so on. Robots designed for these competitions require a good understanding of control theory and system dynamics. The other type of technical challenge is in robot intelligence. For instance, micromouse and RoboCup rescue games require robots to navigate autonomously, understand their environment, and find an optimum path to target. These games require a good understanding of higher level control of robots such as map generation, decision making, path planning, and so on. In the autonomous category, many exciting games are also designed for a team of robots. They instill principles of multirobotics such as robot collaboration, communication, and collective problem solving. Robot soccer and robot colony are the examples of such games.

In contrast to autonomous robots, we find a number of competitions that require robots to be controlled and operated by participants. Needless to say entertainment aspect is higher in these competitions. The users' direct control of robot actions significantly increases the level of entertainment and fun factor in such games. For instance, a remotely controlled robot sumo game counts not only the robot's design, but also the operating skills of its user. On the other hand, the objective in

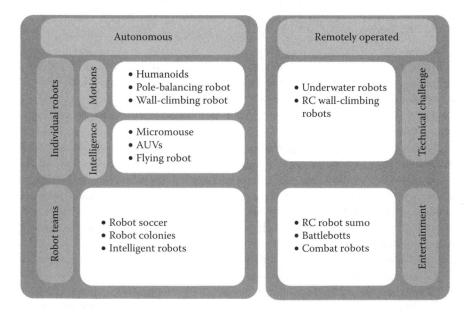

FIGURE 1.7   A classification of robotic games.

ROV competitions is to promote awareness for marine engineering where ROVs are widely used. Therefore, remote operation has practical needs and uses.

1.5 OVERVIEW OF THE BOOK

This chapter briefly introduced robotic games and some of the popular national and international competitions.

Chapter 2 introduces some of the necessary fundamental knowledge in robotics that will benefit the design process. Here, we consider mobile robots and relevant principles such as forward and inverse kinematics.

Chapter 3 discusses the available sensor technology that can be utilized in game robots. Sensing is needed for controlling robot actions, or detecting opponents, identifying objects to manipulate, and so on.

Chapter 4 discusses utilizing camera and image-processing techniques as a sensory unit for robots. In some of the games, such as robot soccer or robot colonies, having visual capability provides significant advantage. Image processing is a vast research field, though this chapter summarizes the immediate knowledge needed to incorporate a vision unit to robot design. The algorithms are presented with MATLAB® code explicitly for better understanding, although many of them are already available as built-in functions in MATLAB itself.

Chapter 5 discusses actuators. Actuators determine robot actions and play a key role in successful design. In this chapter, various actuators available to robotics and their operation principles are discussed. It is a topic worthy of a textbook alone; however, the objective is to give enough preliminary knowledge for understanding and utilizing actuators in robot design.

Chapter 6 discusses some of the basic calculations needed before starting to build a robot, which is often overlooked by robotic enthusiasts. It is important to make an appropriate choice such as required power, gear ratio, and so on when selecting actuators at the beginning of the design stage to avoid the trial and error method.

Chapter 7 introduces control principles and their relevance to robotics.

Chapter 8 discusses the concepts of mathematical modeling, state equations, and transfer functions. These are primarily analytical tools and help to understand the basic system dynamics to design a suitable controller.

Chapter 9 discusses discrete time-control concepts and their implementation in robotics. Robot actions are typically controlled by microprocessors or computers, which are inherently discrete. In Chapters 8 and 9, we give a number of examples and demonstrate how MATLAB can be used in these studies. We aim to illustrate thought processes involved in the design of low-level control of robot motions.

Chapter 10 presents various robot designs such as pole-balancing robot, wall-climbing robot, as case studies overtly showing how theory discussed in previous chapters are put into practice. However, the application of control principles discussed in previous chapters is vast; it can be used for basic speed control of an autonomous mobile robot or to control motions of an autonomous flying machine.

Chapter 11 discusses robot map building and navigation, which are typically used in every mobile robot application.

Chapter 12 discusses robot autonomy, decision making, and learning, in other words, robot intelligence. This can be perceived as the higher-level control of robots. These techniques are important for robotic games such as soccer-playing robots, robot rescue, and humanoids as they require robots to deal with a dynamic world and make decisions autonomously.

Throughout the text, MATLAB is used as the main tool to programming and algorithm examples as well as concepts and simulation studies.

REFERENCES

AUVSI-Association for Unmanned Vehicle Systems International. 2012. http://www.auvsi. org/Home/.

Baltes, J., Lagoudakis, M.G., Naruse, T., and Shiry, S. 2010. *RoboCup 2009: Robot Soccer World Cup XIII*. Series: Lecture Notes in Artificial Intelligence, Vol. 5949. Heidelberg: Springer-Verlag.

Festo. 2012. http://www.festo-didactic.com/int-en/.

FIRA-The Federation of International Robot-soccer Association. 2012. http://fira.net/.

FIRST (For Inspiration and Recognition of Science and Technology) website.2012. http:// www.usfirst.org/.

Gao, Q., Zhang, B., Wu, X., Cheng, Z., Ou, Y., and Xu, Y. 2010. A music dancing robot based on beat tracking of musical signal. *IEEE International Conference on Robotics and Biomimetics (ROBIO)*, Tianjin, China, 1536–1541.

Kaneko, K., Kanehiro, F., Morisawa, M., Miura, K., Nakaoka, S., and Kajita, S. 2009. Cybernetic human HRP-4C. *9th IEEE-RAS International Conference on Humanoids*, Paris, France, 7–14.

Malec, J. 2001. Some thoughts on robotics for education. *Proceedings of the 2001 AAAI Spring Symposium on Robotics and Education*. Palo Alto, California.

Martin, F. 2001. *Robotic Explorations, A Hands-On Introduction to Engineering*. Upper Saddle River, New Jersey: Prentice-Hall.

MATE-Marine Advance Technology and Education Center. 2012. http://www.marinetech.org/ rov_competition/.

Miles, P. 2002. *Robot Sumo: The Official Guide*. Berkley: McGraw-Hill.

Ogura, Y., Aikawa, H., Shimomura, K., et al. 2006. Development of a humanoid robot WABIAN-2. *Proceedings of the IEEE International Conference on Robotics and Automation*, Orlando, Florida, 76–81.

Pisciotta, M., Vello, B., Bordo, C., and Morgavi, G. 2010. Robotic competition: A classroom experience in a vocational school. *6th WSEAS/IASME International Conference on Educational Technologies (EDUTE'10)*, Sousse, Tunisia, 151–156.

RoboCup Federation. 2012. http://www.robocup.org.

SRG-Singapore Robotics Games. 2012. http://guppy.mpe.nus.edu.sg/srg.

Taylor, R.H., and Stoianovici, D. 2003. Medical robotics in computer-integrated surgery. *IEEE Transactions on Robotics and Automation* 19:765–781.

WRSC-World Robotic Sailing Championship. 2012. http://www.roboticsailing.org/.

2 Basic Robotics

2.1 INTRODUCTION TO ROBOTIC SYSTEMS

In the past, our encounters with robots were mainly as an automation tool for speeding up the manufacturing process. This is evident with the early definition of robotics. For instance, The British Robot Association (BRA) defines a robot as "a reprogrammable device with a minimum of four degrees of freedom designed to both manipulate and transport parts, tools or specialised manufacturing implements through variable programmed motions for the performance of the specific manufacturing task." Similarly, International Standards Organisation (ISO) defines a robot as "an automatically controlled, reprogrammable, multipurpose, manipulative machine with several degrees of freedom, which may be either fixed in place or mobile for use in industrial automation applications." Nowadays, robotics finds its place in many diverse areas from medicine to planetary explorations. For instance, two landmark examples are the da Vinci surgical robot used in surgeries demanding great care and precision and the Curiosity Rover sent to explore Mars.

2.1.1 TERMINOLOGY USED IN ROBOTICS

Robots can be designed for various applications; nevertheless, terms used to describe their features and capabilities are common. This is also valid for robots designed for games and competitions as well. Some of these terms are briefly discussed here:

> *Degree of freedom (DOF) and degree of mobility (DOM):* The term "degree of freedom" describes the number of independent movements that an object can perform in a three-dimensional space. If an object is moving freely in space, it has six DOFs. Three of them are about its location in 3D space and three of them are for its orientation as illustrated in Figure 2.1. The location of an object can be defined by translations along the x, y, and z axes. Similarly, its orientation can be defined with three rotations around the x, y, and z axes. Their combination can define the position of an object in 3D space entirely.
>
> To have six DOFs, a robot should have at least six joints, each acting upon one of the motions. Robots with fewer than six joints obviously have constrained motion, and many game robots do not need six DOFs. On the other hand, humanoid robots have more than six joints; these surplus joints enhance performance, providing human-like motions. In other words, each joint contributes a DOM. A joint that provides translation or rotation adds to the DOM, but not necessarily to the DOFs.

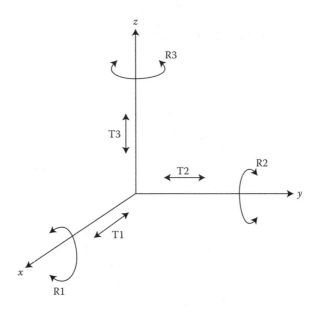

FIGURE 2.1   Representation of the six degrees of freedom.

Work envelope (work space): The work envelope implies all the points in space that a robot can reach. For instance, a robot arm fixed on a workbench can reach to only a limited geometry. Depending on robot type and configuration, its work envelope will have different shape.

Autonomous robots: Autonomous robots perform their tasks in unstructured environments without human interference. There are various degrees of autonomy. A fully autonomous robot is capable of making its decisions and takes action upon them. Autonomous robots are highly complex and many robotic games are intended for autonomous robots. Robot colonies, intelligent robot game, and autonomous sumo are some of the examples.

Remotely operated robots: A remotely operated robot (also known as a tele-robot) takes instructions from a human operator from a distance. The human operator performs live actions in a distant environment and through the sensors can measure the consequences of robot actions. Robot sumo is an example of remotely operated robots in game robotics. However, in practice, tele-robots have a wide range of use such as explosive disposal and surgery.

2.2   COORDINATE TRANSFORMATIONS AND FINDING POSITION OF MOVING OBJECTS IN SPACE

An important part of robotics study is forward kinematics, which concerns the position and orientation of a robot and its end effectors (such as robot gripper). In this section, we will not consider the details of the robot, its sources of motion, and so on.

We will simply assume a rigid object freely moving in 3D space. As mentioned earlier, there are two possible motions of a rigid object in space: rotation and translation. Provided that geometrical representation of an object is given, it will be enough to define the position and orientation of the coordinate system for reconstructing the object at arbitrary places.

We now consider a point P in x, y plane as shown in Figure 2.2a and assume that point P is rotated about θ degrees along the z axis. We can calculate the new coordinates of point P using trigonometry. The coordinates of point P before the rotation can be written as

$$P_x = r\cos\phi \quad \text{and} \quad P_y = r\sin\phi \tag{2.1}$$

After a rotation about θ degrees, P' defines the new coordinates of point P and it can be calculated as follows:

$$P'_x = r\cos(\phi + \theta) \quad \text{and} \quad P'_y = r\sin(\phi + \theta) \tag{2.2}$$

Using trigonometric identities, we obtain

$$P'_x = r(\cos\phi \times \cos\theta) - r(\sin\phi \times \sin\theta)$$

$$P'_y = r(\sin\phi \times \cos\theta) + r(\cos\phi \times \sin\theta) \tag{2.3}$$

And by using Equation 2.1 in Equation 2.3, we get

$$P'_x = P_x \cos\theta - P_y \sin\theta$$

$$P'_y = P_y \cos\theta + P_x \sin\theta \tag{2.4}$$

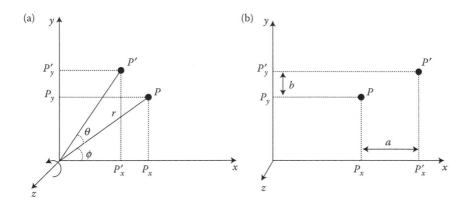

FIGURE 2.2 (a) Rotation along the z axis and (b) translation in the xy plane.

In matrix form

$$\begin{bmatrix} P'_x \\ P'_y \end{bmatrix} = \begin{bmatrix} \cos\theta & -\sin\theta \\ \sin\theta & \cos\theta \end{bmatrix} \begin{bmatrix} P_x \\ P_y \end{bmatrix} \tag{2.5}$$

Equation 2.5 defines a rotation of θ angle about the z axis in matrix form. Equation 2.5 operates on x and y coordinates of point P. Normally, a point in three-dimensional space is defined with its three components x, y, and z. By considering this, a rotation matrix can be defined as a 3×3 matrix. Thus, Equation 2.5 is rewritten as

$$\begin{bmatrix} P'_x \\ P'_y \\ P'_z \end{bmatrix} = \begin{bmatrix} \cos\theta & -\sin\theta & 0 \\ \sin\theta & \cos\theta & 0 \\ 0 & 0 & 1 \end{bmatrix} \begin{bmatrix} P_x \\ P_y \\ P_z \end{bmatrix} \tag{2.6}$$

We now have a rotation matrix, which represents a rotation of θ angle along the z axis

$$R_z(\theta) = \begin{bmatrix} \cos\theta & -\sin\theta & 0 \\ \sin\theta & \cos\theta & 0 \\ 0 & 0 & 1 \end{bmatrix} \tag{2.7}$$

Similarly, a rotation around y axis is defined as

$$R_y(\alpha) = \begin{bmatrix} \cos\alpha & 0 & \sin\alpha \\ 0 & 1 & 0 \\ -\sin\alpha & 0 & \cos\alpha \end{bmatrix} \tag{2.8}$$

and a rotation around x axis is defined as

$$R_x(\gamma) = \begin{bmatrix} 1 & 0 & 0 \\ 0 & \cos\gamma & -\sin\gamma \\ 0 & \sin\gamma & \cos\gamma \end{bmatrix} \tag{2.9}$$

Rotation along z axis is called roll, rotation along y is called pitch, and rotation along x is called yaw.

We now consider the linear translations shown in Figure 2.2b. New coordinates of point P after the linear translations will be

$$P'_x = P_x + a$$

$$P'_y = P_y + b \tag{2.10}$$

$$P'_z = 0$$

We can organize these equations in a matrix form by taking into consideration x, y, and z coordinates and obtain

$$\begin{bmatrix} P'_x \\ P'_y \\ P'_z \\ 1 \end{bmatrix} = \begin{bmatrix} 1 & 0 & 0 & a \\ 0 & 1 & 0 & b \\ 0 & 0 & 1 & 0 \\ 0 & 0 & 0 & 1 \end{bmatrix} \begin{bmatrix} P_x \\ P_y \\ P_z \\ 1 \end{bmatrix} \tag{2.11}$$

We can easily derive a generic equation representing translations in all three axes as follows:

$$\begin{bmatrix} P'_x \\ P'_y \\ P'_z \\ 1 \end{bmatrix} = \begin{bmatrix} 1 & 0 & 0 & k_x \\ 0 & 1 & 0 & k_y \\ 0 & 0 & 1 & k_z \\ 0 & 0 & 0 & 1 \end{bmatrix} \begin{bmatrix} P_x \\ P_y \\ P_z \\ 1 \end{bmatrix} \tag{2.12}$$

Here, k_x, k_y, and k_z are the displacements along the x, y, and z coordinates. It is important to note that the resulting translation matrix is now 4×4 and a point is now needed to be defined with 4 components instead of 3.

2.2.1 COMPOSITE ROTATIONS

An object in space may perform more than one rotation. This makes the calculation of its final position complicated. The solution becomes easier by assuming that a separate coordinate system is attached to the object as shown in Figure 2.3a. Let us assume that point P represents the object and a coordinate frame, which is coincident with the reference frame is firmly attached to it. We can find a transformation matrix by decomposing individual motions. At first, a coordinate frame attached to point P is rotated along the z axis about 90° as shown in Figure 2.3b. It is important to note that a clockwise rotation is considered negative and a counterclockwise rotation is considered positive. Transformation after this rotation is represented with coordinate

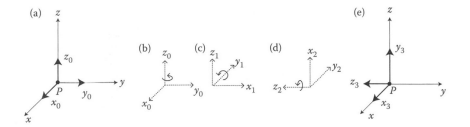

FIGURE 2.3 (a) Reference and object frames, (b) first rotation along the z axis, (c) second rotation along the y axis, (d) final rotation along the z axis, and (e) final coordinate frame compared to the reference frame.

frame x_1, y_1, z_1. The following motion is a rotation along y_1 axis about $-90°$ (see Figure 2.3c) and the resulting coordinate frame is x_2, y_2, z_2. The final rotation is $-90°$ along z_2 (Figure 2.3d) resulting in coordinate frame x_3, y_3, z_3 as shown in Figure 2.3e. The sequence of motions in this example is roll, pitch, and roll. The number of rotations and the angles are not limited; however, for the convenience of illustration, in this example, we chose right angles only.

In matrix form, the first motion is defined as

$$R_z(\theta) = \begin{bmatrix} \cos\theta & -\sin\theta & 0 \\ \sin\theta & \cos\theta & 0 \\ 0 & 0 & 1 \end{bmatrix} \quad \theta = 90 \tag{2.11}$$

Rotation matrix for the second motion is

$$R_y(\alpha) = \begin{bmatrix} \cos\alpha & 0 & \sin\alpha \\ 0 & 1 & 0 \\ -\sin\alpha & 0 & \cos\alpha \end{bmatrix} \quad \alpha = -90 \tag{2.12}$$

Rotation matrix for the last motion is

$$R_z(\phi) = \begin{bmatrix} \cos\phi & -\sin\phi & 0 \\ \sin\phi & \cos\phi & 0 \\ 0 & 0 & 1 \end{bmatrix} \quad \phi = 90 \tag{2.13}$$

Now using the rotation matrices for each motion as given earlier, we can obtain a complex rotation matrix from coordinate frame x_0, y_0, z_0 to x_2, y_2, z_2 using the post-multiplication rule:

$$R(\text{total}) = R_z(\theta)R_y(\alpha)R_z(\phi) \tag{2.14}$$

$$R(\text{total}) = \begin{bmatrix} \cos(90) & -\sin(90) & 0 \\ \sin(90) & \cos(90) & 0 \\ 0 & 0 & 1 \end{bmatrix} \begin{bmatrix} \cos(-90) & 0 & \sin(-90) \\ 0 & 1 & 0 \\ -\sin(-90) & 0 & \cos(-90) \end{bmatrix}$$
$$\begin{bmatrix} \cos(90) & -\sin(90) & 0 \\ \sin(90) & \cos(90) & 0 \\ 0 & 0 & 1 \end{bmatrix}$$

$$R(\text{total}) = \begin{bmatrix} -1 & 0 & 0 \\ 0 & 0 & -1 \\ 0 & -1 & 0 \end{bmatrix} \tag{2.15}$$

2.2.2 Homogeneous Transformation Matrix

The transformation matrices discussed earlier can represent rotation type of motions, but not the translations. It is possible to combine rotation and translation into a single transformation matrix. Let us assume that after the rotation transformations, the coordinate frame shown earlier in Figure 2.3e is now translated to x_4, y_4, z_4 as illustrated in Figure 2.4. The overall transformation can be disassembled as a rotation, which transforms the frame x_0, y_0, z_0 to x_3, y_3, z_3 and a translation, which brings x_3, y_3, z_3 frame to x_4, y_4, z_4. These rotation and translation motions can be shown in a compact form as a 4×4 matrix, which is known as homogeneous transformation matrix. It maps a position vector from one coordinate system to another.

$$H = \begin{bmatrix} R & & & K \\ 0 & 0 & 0 & 1 \end{bmatrix} \tag{2.16}$$

Here R is a 3×3 rotation matrix and K is a 3×1 translation vector. A homogeneous transformation matrix combines the position vector K with a rotation matrix R to provide a complete description of the position and orientation of a second coordinate system with respect to the base frame. By adding a fourth row, consisting of three "zeros" and a "one," a homogeneous transformation matrix is constructed.

$$H = \begin{bmatrix} r_{11} & r_{12} & r_{13} & k_x \\ r_{21} & r_{22} & r_{23} & k_y \\ r_{31} & r_{32} & r_{33} & k_z \\ 0 & 0 & 0 & 1 \end{bmatrix} \tag{2.17}$$

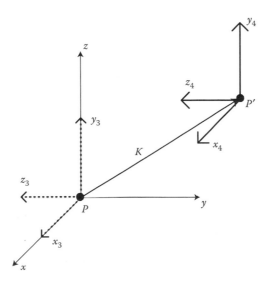

FIGURE 2.4   Transformation operation comprising both rotation and translation.

2.2.3   Composite Transformations

In practice, motions of a rigid body may be composed of a sequence of homogeneous rotations and homogeneous translations. A composite homogeneous transformation matrix, which represents the entire sequence of rotations and translations, can be obtained by multiplying together all these transformation matrices (Fu et al. 1987). However, it is important to do the right order of the matrix multiplication.

Assume that x_r, y_r, z_r represents a reference coordinate frame and x_0, y_0, z_0 represents the coordinate frame attached to the object and they both are initially coincident. Assume that two types of motions are defined on the coordinate frame x_0, y_0, z_0. One of them is a translation type of motion along y axis by B units. The corresponding homogeneous translation matrix will be

$$H_{TR} = \begin{bmatrix} 1 & 0 & 0 & 0 \\ 0 & 1 & 0 & B \\ 0 & 0 & 1 & 0 \\ 0 & 0 & 0 & 1 \end{bmatrix} \tag{2.18}$$

The second motion is θ angle rotation of x_0, y_0, z_0 coordinate frame about z_0 axis (or z_r axis since these coordinates are initially coincident), the homogeneous rotation matrix is

$$H_{ROT} = \begin{bmatrix} \cos\theta & -\sin\theta & 0 & 0 \\ \sin\theta & \cos\theta & 0 & 0 \\ 0 & 0 & 1 & 0 \\ 0 & 0 & 0 & 1 \end{bmatrix} \tag{2.19}$$

We now multiply the translation and rotation matrices to obtain the homogeneous transformation matrix. If the sequence of multiplication is done as follows

$$H = H_{TR} H_{ROT} \tag{2.20}$$

then the homogeneous transformation matrix in Equation 2.20 defines a motion, which is B units translation of x_0, y_0, z_0 coordinate frame along y_0 axis, followed by a rotation of θ angle along z_0 axis. On the other hand, the opposite sequence of multiplication

$$H = H_{ROT} H_{TR} \tag{2.21}$$

means x_0, y_0, z_0 coordinate frame is rotated θ angle along z_0 axis followed with B units of translation along the y_0 axis. The difference between these two motions can be seen clearly in Figure 2.5.

2.2.3.1   Matrix Multiplication Order in Composite Transformations

There are two types of 4×4 matrices used in robot motion calculations. The first type describes the transformations of a given coordinate system relative to the base

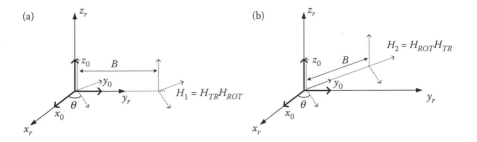

FIGURE 2.5 Composite transformations.

coordinate frame. In this fixed coordinate frame approach, all the successive transformations are defined relative to the original world coordinate frame (or static reference frame). Let us assume that a sequence of transformations is applied to a local frame. These transformations, relative to a global frame, are given as $H_1, H_2, \ldots, H_n$, where transformation H_1 is applied first and transformation H_n is applied last. In this case, we premultiply the transformation matrices.

$$H = H_n \times \cdots \times H_2 \times H_1 \qquad (2.22)$$

The second type describes the relationship between any two coordinate systems in a chain of moving local frame transformations. In this case, we assume that every successive transformation is defined based on the moving local frame. Let us assume that a sequence of transformations is applied to a local frame and transformations, relative to a moving local frame, are given as $H_1, H_2, \ldots, H_n$, where transformation H_1 is applied first and transformation H_n is applied last. In this case, we postmultiply the transformation matrices as follows

$$H = H_1 \times H_2 \times \cdots \times H_n \qquad (2.23)$$

Visualizing transformations based on a fixed reference coordinate frame is rather difficult. In composite transformations, we will consider motions of a robot or a rigid body in space as a sequence of homogeneous coordinate transformations based on a moving local frame. We will assume that the moving object approaches the target point step by step from the origin of the reference frame. We assign a separate coordinate system to each step of the transformation to visualize these motions easily. The transformation matrix $H_{i,n}$ describes the position and orientation of the n-th coordinate system relative to the i-th one as follows

$$H_{i,n} = \prod_{i}^{n-1} H_{i,i+1} \; = \; H_{i,i+1} \times H_{i+1,i+2} \times \cdots \times H_{n-1,n} \qquad (2.24)$$

where $i = 0, 1, \ldots, n-1$, which may start from any number less than n. The resultant transformation matrix $H_{i,n}$ describes the state of the nth coordinate system relative to any i-th component's coordinate frame. Thus, the coordinates of the target point P_t in the n-th coordinate system relative to any i-th component's coordinate frame can be expressed as

$$p_t^{(i)} = H_{i,n} \times p_t^{(n)} \tag{2.25}$$

$$\begin{bmatrix} x_t^{(i)} \\ y_t^{(i)} \\ z_t^{(i)} \\ 1 \end{bmatrix} = H_{i,n} \times \begin{bmatrix} x_t^{(n)} \\ y_t^{(n)} \\ z_t^{(n)} \\ 1 \end{bmatrix} \tag{2.26}$$

EXAMPLE 2.1

We consider a series of transformations shown in Figure 2.6. The figure on the left shows the final position of coordinate frame x', y', z', which is initially coincident with the reference frame x, y, z. We can visualize the transformation made by the frame x', y', z' by breaking it down to individual motions and by assigning a separate frame for each step of motion. These sequences of motions are illustrated in Figure 2.6. Assuming moving frames, the first motion is a rotation of 180° about the z_0 axis, which translates frame x_0, y_0, z_0 to x_1, y_1, z_1. The second motion transforms x_1, y_1, z_1 to x_2, y_2, z_2 by a rotation about 90° along y_1 axis. The final motion transforms x_2, y_2, z_2 to x_3, y_3, z_3 by a translation of 10 units along z_2 axis. The coordinate frame x_3, y_3, z_3 shows the final position of frame x', y', z'. Based on the moving frames method, we postmultiply the transformation matrices starting from the first motion to the last, as follows

$$H = H_1 \times H_2 \times H_3$$

$$H = \begin{bmatrix} \cos(180) & -\sin(180) & 0 & 0 \\ \sin(180) & \cos(180) & 0 & 0 \\ 0 & 0 & 1 & 0 \\ 0 & 0 & 0 & 1 \end{bmatrix} \times \begin{bmatrix} \cos(90) & 0 & \sin(90) & 0 \\ 0 & 1 & 0 & 0 \\ -\sin(90) & 0 & \cos(90) & 0 \\ 0 & 0 & 0 & 1 \end{bmatrix}$$

$$\times \begin{bmatrix} 1 & 0 & 0 & 0 \\ 0 & 1 & 0 & 0 \\ 0 & 0 & 1 & 10 \\ 0 & 0 & 0 & 1 \end{bmatrix}$$

$$H = \begin{bmatrix} 0 & 0 & -1 & 10 \\ 0 & -1 & 0 & 0 \\ -1 & 0 & 1 & 0 \\ 0 & 0 & 0 & 1 \end{bmatrix} \tag{2.27}$$

We now visualize the same transformations made by the frame x', y', z' by breaking it down to individual transformations based on reference frame x, y, z. In this case, the first motion is a rotation of 180° about z axis, which translates frame

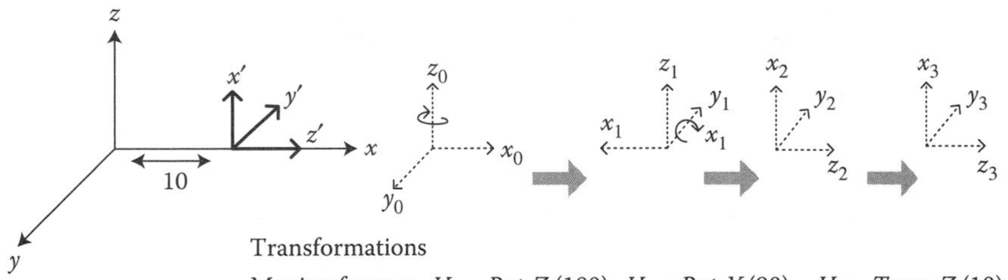

Transformations

Moving frames: $H_1 = Rot\text{-}Z\,(180)$ $H_2 = Rot\text{-}Y\,(90)$ $H_3 = Trans\text{-}Z\,(10)$

Fixed frame: $H_1 = Rot\text{-}Z\,(180)$ $H_2 = Rot\text{-}Y\,(-90)$ $H_3 = Trans\text{-}X\,(10)$

FIGURE 2.6   Series of transformations regarding Example 2.1.

x_0, y_0, z_0 to x_1, y_1, z_1. The second motion transforms x_1, y_1, z_1 to x_2, y_2, z_2 by a rotation about $-90°$ along the y axis. The final motion transforms x_2, y_2, z_2 to x_3, y_3, z_3 by a translation of 10 units along the x axis. In the case of the fixed frame definition of motions, we premultiply transformation matrices starting from the last motion toward the first one, as follows

$$H = H_3 \times H_2 \times H_1$$

$$H = \begin{bmatrix} 1 & 0 & 0 & 10 \\ 0 & 1 & 0 & 0 \\ 0 & 0 & 1 & 0 \\ 0 & 0 & 0 & 1 \end{bmatrix} \times \begin{bmatrix} \cos(-90) & 0 & \sin(-90) & 0 \\ 0 & 1 & 0 & 0 \\ -\sin(-90) & 0 & \cos(-90) & 0 \\ 0 & 0 & 0 & 1 \end{bmatrix}$$

$$\times \begin{bmatrix} \cos(180) & -\sin(180) & 0 & 0 \\ \sin(180) & \cos(180) & 0 & 0 \\ 0 & 0 & 1 & 0 \\ 0 & 0 & 0 & 1 \end{bmatrix}$$

$$H = \begin{bmatrix} 0 & 0 & -1 & 10 \\ 0 & -1 & 0 & 0 \\ -1 & 0 & 1 & 0 \\ 0 & 0 & 0 & 1 \end{bmatrix} \tag{2.28}$$

By comparing, Equations 2.27 and 2.28, we can see that we get the same homogeneous transformation matrix for the entire transformation.

2.2.4   MATHEMATICAL DESCRIPTION OF OBJECTS

To use a homogeneous transformation matrix to determine the position of an object (manufactured part, robot manipulator, or a mobile robot itself) after its motions, we need to represent the objects mathematically. We assume that the object of interest is surrounded by planar surfaces, and it is described as a $4 \times N$ matrix. Here, N indicates the number of vertices of the object chosen to represent the object. There are a couple of ways to represent the object. We can consider that the origin of an object's

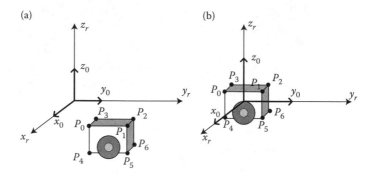

FIGURE 2.7 (a) Origin of the object's coordinate frame x_0, y_0, z_0 is in an arbitrary position in space. (b) Object's center of gravity is aligned with the origin of its coordinate frame.

coordinate system is positioned arbitrarily in the space (as shown in Figure 2.7a) and describes a general matrix presentation of the object with N vertices as follows:

$$M_{object} = \begin{bmatrix} x_0 & x_1 & \cdots & x_{N-1} \\ y_0 & y_1 & \cdots & y_{N-1} \\ z_0 & z_1 & \cdots & z_{N-1} \\ 1 & 1 & \cdots & 1 \end{bmatrix} \tag{2.29}$$

Alternatively, we can consider that the origin of the object coordinate system is fixed to one of its features (typically to its center of gravity) and derive object description matrix accordingly (see Figure 2.7b). It is important to remember that in either case during the initial state of homogeneous transformations, the reference coordinate system and the object coordinate frame are coincident.

Let us consider the robot-like object shown in Figure 2.7b represented by its eight vertices $[P_0, P_1, \ldots, P_7]$ in Cartesian coordinates. The origin of the fixed coordinate system is chosen at the center of gravity of the object. Assume that this object is a cube with dimension A. The corresponding columns of the object description matrix for the vertex P_0 will be $[A/2 - A/2 \quad A/2 \quad 1]^T$, for the vertex P_1 will be $[A/2 \quad A/2 \quad A/2 \quad 1]^T$, and so on. We can write a description matrix of this object as follows:

$$M_{obj} = \begin{bmatrix} A/2 & A/2 & -A/2 & -A/2 & A/2 & A/2 & -A/2 & -A/2 \\ -A/2 & A/2 & A/2 & -A/2 & -A/2 & A/2 & A/2 & -A/2 \\ A/2 & A/2 & A/2 & A/2 & -A/2 & -A/2 & -A/2 & -A/2 \\ 1 & 1 & 1 & 1 & 1 & 1 & 1 & 1 \\ P_0 & P_1 & P_2 & P_3 & P_4 & P_5 & P_6 & P_7 \end{bmatrix} \begin{matrix} x \\ y \\ z \\ \\ \end{matrix} \tag{2.30}$$

Let us now perform a translation and a rotation on the object just described. These motions will be described by a 4×4 transformation matrix H. The relation between the starting and final positions of an object is

$$M_{obj\_new} = H \times M_{obj\_start} \tag{2.31}$$

Here, M_{obj_start} is the description matrix of the moving object at the starting position, and M_{obj_new} is the new description matrix after the transformation. The expanded form of the equation becomes

$$
\begin{bmatrix}
x_0' & x_1' & \cdots & x_{N-1}' \\
y_0' & y_1' & \cdots & y_{N-1}' \\
z_0' & z_1' & \cdots & z_{N-1}' \\
1 & 1 & \cdots & 1
\end{bmatrix}
=
\begin{bmatrix}
 & R & & K \\
0 & 0 & 0 & 1
\end{bmatrix}
\times
\begin{bmatrix}
x_0 & x_1 & \cdots & x_{N-1} \\
y_0 & y_1 & \cdots & y_{N-1} \\
z_0 & z_1 & \cdots & z_{N-1} \\
1 & 1 & \cdots & 1
\end{bmatrix}
\tag{2.32}
$$

The matrix on the left-hand side represents the vertices of the object in their new position after transformation. Matrices on the right-hand side represent the transformation matrix and the vertices of the object in the starting coordinate frame, respectively.

Case Study

In this case study, we will study a more realistic scenario using the robot shown in Figure 2.8a. It is a mobile land robot, hence its motions are limited to the x,y plane. However, this does not affect the principles of our calculations. The robot has eight ultrasonic range sensors and a digital compass and their positions on the robot body are illustrated in Figure 2.8b. The maximum range of the ultrasonic range sensors is about 3 m and their resolution is in centimeters. Robot captures ultrasonic and compass readings regularly. On board, compass provides robot orientation with 1° resolution. However, as shown in Figure 2.9, the angle received from the compass shows robot's orientation based on the earth's magnetic field, and it needs to be converted to Cartesian coordinates to use in our calculations. More details on range sensors will be discussed in the next chapter. At this point, the information we need to do a realistic computation is distance traveled by the robot, its orientation, and the readings from its range sensors.

(a)   (b)

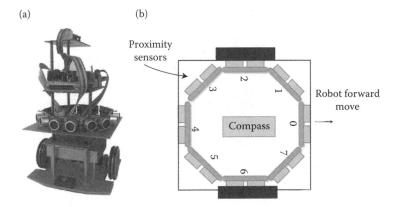

FIGURE 2.8   (a) Mobile robot used in experiments. (b) Position of ultrasonic sensors.

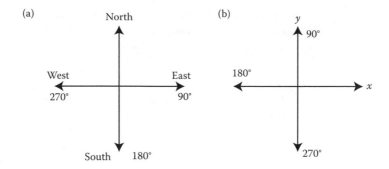

FIGURE 2.9 (a) Angle readings from digital compass. (b) Its correspondence in Cartesian coordinates.

Let us assume that the robot shown in Figure 2.8 has a local coordinate frame attached to its center of gravity as shown in Figure 2.10. Four corners of the robot are used in the object description matrix, as well as coordinates of the sensors on the robot based on this local frame at robot's center of gravity. In the figure, distance d indicates the range reading of a sensor and θ is the angle of the sensor. All the units are in centimeters. Based on the four corner points selected, an object description matrix for the robot can be written as

$$M_{obj} = \begin{bmatrix} 7 & 7 & -7 & -7 \\ 7 & -7 & -7 & -7 \\ 0 & 0 & 0 & 0 \\ 1 & 1 & 1 & 1 \\ P_0 & P_1 & P_2 & P_3 \end{bmatrix} \begin{matrix} x \\ y \\ z \\ \\ \end{matrix} \qquad (2.33)$$

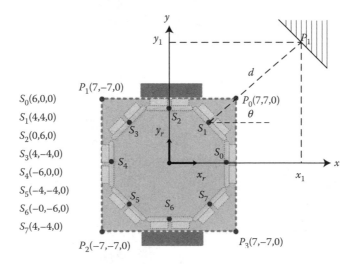

FIGURE 2.10 Local coordinate frame and selected vertices to describe the robot.

We now consider a number of motions performed by the robot. Robot in this example is capable of making a turn and moving forward or backward by adjusting the speed and the direction of its wheels. For simplicity, we assume that robot performs either rotation or translation type of move at any given time. At this point, we do not consider other parameters involved in robot motion such as speed, odometry errors, and so on, which will be discussed later. Let us assume that the robot travels 100 cm along the x axis and then makes a rotation to face $0°$ north ($90°$ rotation about z axis) followed with 30 cm translation along the x axis, followed with a rotation to face $300°$ south (another $120°$ rotation about z axis). These motions in matrix form can be written as follows:

$$T_{TR_1} = \begin{bmatrix} 1 & 0 & 0 & 100 \\ 0 & 1 & 0 & 0 \\ 0 & 0 & 1 & 0 \\ 0 & 0 & 0 & 1 \end{bmatrix}$$

$$T_{ROT_1} = \begin{bmatrix} \cos(90) & -\sin(90) & 0 & 0 \\ \sin(90) & \cos(90) & 0 & 0 \\ 0 & 0 & 1 & 0 \\ 0 & 0 & 0 & 1 \end{bmatrix} = \begin{bmatrix} 0 & -1 & 0 & 0 \\ 1 & 0 & 0 & 0 \\ 0 & 0 & 1 & 0 \\ 0 & 0 & 0 & 1 \end{bmatrix}$$

$$T_{TR_2} = \begin{bmatrix} 1 & 0 & 0 & 30 \\ 0 & 1 & 0 & 0 \\ 0 & 0 & 1 & 0 \\ 0 & 0 & 0 & 1 \end{bmatrix}$$

$$T_{ROT_2} = \begin{bmatrix} \cos(120) & -\sin(120) & 0 & 0 \\ \sin(120) & \cos(120) & 0 & 0 \\ 0 & 0 & 1 & 0 \\ 0 & 0 & 0 & 1 \end{bmatrix} = \begin{bmatrix} -0.5 & -0.866 & 0 & 0 \\ 0.866 & -0.5 & 0 & 0 \\ 0 & 0 & 1 & 0 \\ 0 & 0 & 0 & 1 \end{bmatrix} \quad (2.34)$$

The overall transformation matrix is then

$$H = T_{TR_1} \times T_{ROT_1} \times T_{TR_2} \times T_{ROT_2}$$

$$H = \begin{bmatrix} 1 & 0 & 0 & 100 \\ 0 & 1 & 0 & 0 \\ 0 & 0 & 1 & 0 \\ 0 & 0 & 0 & 1 \end{bmatrix} \times \begin{bmatrix} 0 & -1 & 0 & 0 \\ 1 & 0 & 0 & 0 \\ 0 & 0 & 1 & 0 \\ 0 & 0 & 0 & 1 \end{bmatrix} \times \begin{bmatrix} 1 & 0 & 0 & 30 \\ 0 & 1 & 0 & 0 \\ 0 & 0 & 1 & 0 \\ 0 & 0 & 0 & 1 \end{bmatrix} \times \begin{bmatrix} -0.5 & -0.866 & 0 & 0 \\ 0.866 & -0.5 & 0 & 0 \\ 0 & 0 & 1 & 0 \\ 0 & 0 & 0 & 1 \end{bmatrix}$$

$$H = \begin{bmatrix} -0.866 & 0.5 & 0 & 100 \\ -0.5 & -0.866 & 0 & 30 \\ 0 & 0 & 1 & 0 \\ 0 & 0 & 0 & 1 \end{bmatrix} \quad (2.35)$$

Multiplying with the object description matrix, we obtain positions of these four points in the reference frame after the transformations as

$$
M_{\text{obj\_new}} = \begin{bmatrix} -0.866 & 0.5 & 0 & 100 \\ -0.5 & -0.866 & 0 & 30 \\ 0 & 0 & 1 & 0 \\ 0 & 0 & 0 & 1 \end{bmatrix} \times \begin{bmatrix} 7 & 7 & -7 & 7 \\ 7 & -7 & -7 & -7 \\ 0 & 0 & 0 & 0 \\ 1 & 1 & 1 & 1 \end{bmatrix}
$$

$$
M_{\text{obj\_new}} = \begin{bmatrix} 97.43 & 90.43 & 102.56 & 90.43 \\ 20.43 & 32.56 & 39.56 & 32.56 \\ 0 & 0 & 0 & 0 \\ 1 & 1 & 1 & 1 \end{bmatrix} \tag{2.36}
$$

We can expand the object description matrix by including range sensors and their readings. In this case, we are not only concerned with the position of the robot, but also calculate the coordinates of the obstacles detected by the range sensors and mark them in the reference frame. For instance, referring to Figure 2.10, range sensor reading S_1 for the point P_1 can be translated into robot coordinate frame as $[4 + d\cos\theta, 4 + d\sin\theta, 0, 1]^T$. Since we know the sensor angle θ and its position in local frame, we can easily calculate coordinates of point P_1. In this example, we will assume that the robot makes a 30° turn toward north initially and then travels along its x axis about 100 cm. It is important to note that we assume the local coordinate frame is fixed to the center of the robot as shown in Figure 2.10, therefore all the forward movements (translations) will be along the x axis of this robot. The example code given in Figure 2.11 computes the final position of the robot together with its sensor readings, which are generated randomly in this case. A plot of these computation results is shown in Figure 2.12.

2.3 WHEEL DRIVE IN MOBILE ROBOTS

Technological advances enable us to envision robots taking more substantial roles in our daily lives. Robots will be interacting with human beings and operating in the same environment. An important feature of robots operating in such environment is their mobility. A wheel driver system enables a robot to gain mobility by the use of wheels, and this is one of the simplest methods to achieve mobility. However, the type of wheels, size, and their placement may increase or decrease the performance of the robot for different types of tasks. It is important to understand the different types of wheels and their advantages and disadvantages so that we can select a proper wheel drive system when designing a mobile robot. A mobile robot design also includes the control of these wheels to guide the robot properly and recognize its whereabouts after its motions. The kinematic analysis of the wheel drives help us understand how the wheel driver is going to move about under different parameters.

A common wheel drive system consists of the following components: motors, gearboxes, motor wheels, and caster wheels (Fred 2001). Motors provide the rotational motion with certain torque and speed, and the gearbox is used to amplify or reduce this motor torque/speed. Motor wheels provide the actual mobility to the robot, and they are driven by the motor/gearbox system. The caster wheels are

```
%-----------------------------------

% RSensor(1)      -> Robot ID

% RSensor(2)~(9) -> Proximity sensor readings

% RSensor(10)     -> Compass


RSensor=[1 15 55 53 59 59 50 57 55 45];% Random values of proximity sensors


% Proximity sensor orientations on the robot body (in radian)

Theta=[0  0.7854 1.5708 2.3562 3.1416 3.9270 4.7124 5.4978];


%Object matrix

%Position of four corner points

%        s0 s1 s2 s3 s4 s5 s6 s7 p0 p1 p2 p3

Rob1_Pr=[0 0 0 0 0 0 0 0 7 -7 -7 7;

         0 0 0 0 0 0 0 0 7 7 -7 -7;

         0 0 0 0 0 0 0 0 0 0 0 0;

         1 1 1 1 1 1 1 1 1 1 1 1];


Offset=[ 6 4 0 -4 -6 -4 0 4 0 0 0 0;  %Position of sensors on robot body

         0 4 6 4  0 -4 -6 -4 0 0 0 0;

         0 0 0 0 0 0 0 0 0 0 0 0;

         0 0 0 0 0 0 0 0 0 0 0 0];


Rob1_Prnew=[0 0 0 0 0 0 0 0 0 0 0 0;

            0 0 0 0 0 0 0 0 0 0 0 0;

            0 0 0 0 0 0 0 0 0 0 0 0;

            1 1 1 1 1 1 1 1 1 1 1 1];


T_rot=[1 0 0 0;   %Blank rotation matrix
```

FIGURE 2.11   MATLAB® code for demonstrating homogeneous transformations.

```
        0 1 0 0;

        0 0 1 0;

        0 0 0 1];


T_trans=T_rot;%Blank translation matrix

RSensor(1,10)=RSensor(1,10);

A1=(RSensor(1,2:9));                         %get proximity sensor readings

cmp=RSensor(1,10);                           %get compass reading and translate

ang1=2*pi*((-1*cmp)+90)/360;                 %to XY frame


%angle and position of sensor are known

%convert distance readings from sensors to x-y coordinates

fori=1:8

    Rob1_Pr(1,i)=(A1(i)*cos(Theta(i)));

    Rob1_Pr(2,i)=(A1(i)*sin(Theta(i)));

end

    Rob1_Pr=Rob1_Pr+Offset;

% Rotation matrix

T_rot(1,1)=cos(ang1); T_rot(1,2)=-sin(ang1);

T_rot(2,1)=sin(ang1); T_rot(2,2)=cos(ang1);


% Translation matrix

T_trans(1,4)=100;


Rob1_Prnew=(T_rot*T_trans)*Rob1_Pr;


%....... Display robot and sensor readings

% create square for robot

Robfrm1=[Rob1_Prnew(1,9:12),Rob1_Prnew(1,9:9); Rob1_Prnew(2,9:12),

Rob1_Prnew(2,9:9)];

%show sensor readings

plot(Rob1_Prnew(1,1:8),Rob1_Prnew(2,1:8),'og',Rob1_Prnew(1,9:12),Rob1_Prnew
```

FIGURE 2.11 (continued) MATLAB code for demonstrating homogeneous transformations.

```
(2,9:12),'sr');

% draw robot and draw its compass reading

CompX1=30*cos(ang1);CompY1=30*sin(ang1);

Rob1OX=sum(Rob1_Prnew(1,9:12))/4; Rob1OY=sum(Rob1_Prnew(2,9:12))/4;

line(Robfrm1(1,:),Robfrm1(2,:));

line([Rob1OX Rob1OX+CompX1],[Rob1OY

Rob1OY+CompY1],'Marker','d','LineStyle','-');

% draw platform

axis([-50 400 -50 400]); whitebg([0.2 0.2 0.2]);

line([0 0 350 350 0],[0 350 350 0 0],'Color','yellow','LineStyle','-');

title('Area Map (West)'); xlabel('X (east)'); ylabel('Y (south)');
```

FIGURE 2.11   (continued) MATLAB code for demonstrating homogeneous transformations.

mainly used to balance the robot chassis when it is in motion and they are not driven by the motor. Let us discuss further the different types of wheels.

There are a few types of wheels used in robotics, and they are shown in Figure 2.13. As mentioned, rigid caster wheels are attached to robots as a support mechanism. This is the most basic type of wheels, and it is widely used. Swivel caster wheels function the same way; however, they are able to rotate 360° due to the swivel joint provided in their design. This type of caster wheel will align itself to the direction of robot motion by design. Steerable caster wheels are rigid wheels, and they are

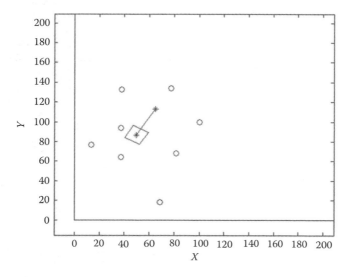

FIGURE 2.12   Position of the robot after transformations and distance readings from its six range sensors marked with circles.

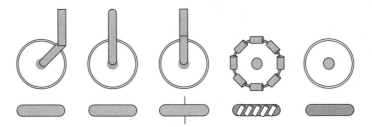

FIGURE 2.13 Different type of wheels: (a) swivel caster, (b) rigid caster, (c) steerable caster, (d) omniwheel, and (e) motor wheel.

driven by a motor to rotate around the z axis. They are similar to the swivel caster with a difference that they not only do more than just support the robot chassis, but also affect their motions.

Most motor wheels used in mobile robots are steerable. They are directly driven by the motor to provide forward or backward movement for the robot. Omniwheel is a special type of wheel that has rollers on the wheel itself. These kinds of wheels are also called mecanum wheels. There are two types of mecanum wheels, mecanum wheels with roller at 45°, and mecanum wheels with rollers at 90°. These two types have a totally different look and design. When an omniwheel is driven, the rollers will be in contact with the ground and cause a rotation. The motion effect from all the rollers on the wheel will result in different motion directions. In other words, the motions provided by these wheels are not limited to backward and forward only; they can move in any direction.

There are different types of wheel drive configurations for mobile robots, and new configurations can be created by adding more wheels to the systems. Let us discuss the four basic configurations and their control.

2.3.1 DIFFERENTIAL DRIVE

This type of wheel drive system consists of two independent motor-driven wheels. It can have one or two caster wheels to support the robot. By controlling the velocity of each wheel, robot motions can be controlled. Figure 2.14 illustrates the motion of the robot with different velocities of the wheels. A major challenge in differential drive systems is that the wheel speed has to be controlled properly and precisely to achieve desired robot motions (Siegwart and Nourbakhsh 2004). For instance, both

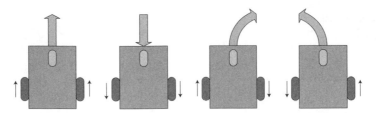

FIGURE 2.14 Analysis of motion of a differential drive mobile robot for different left and right velocities.

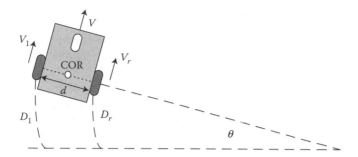

FIGURE 2.15 The parameters and variables involved in the motion of a differential drive mobile robot.

wheels have to be driven at the same speed to move the robot forward. If one wheel is slightly faster than the other one, the faster wheel will overtake the slower one and cause a turning motion. A greater difference in the speed of the two wheels will determine how sharp the turn will be. If one wheel stops and the other wheel is still turning, the robot will spin around itself with the stopped wheel being the center point of this motion. When both wheels are driven in the opposite directions at the same speed, then the robot makes a turn on the spot. This turn will be around a point that is in the middle of the two motor wheels, which is also known as the center of rotation (COR).

The wheels of the mobile robot shown in Figure 2.15 are traveling at different velocities, causing a rotation toward the right-hand side. By referring to the snapshot of this traveling robot, it is possible to calculate the kinematic equations that describe the position of the mobile robot at a particular time. The robot speed V is defined as the average of the left and right wheels

$$V = \frac{V_l + V_r}{2} \tag{2.37}$$

and the speed on both axes will be

$$\dot{x} = V \cos\theta \tag{2.38}$$

$$\dot{y} = V \sin\theta \tag{2.39}$$

The angular speed $\dot{\theta}$ can be given as

$$\dot{\theta} = \frac{V_l - V_r}{d} \tag{2.40}$$

where d is the distance between the wheels of the robot. As it is observed from above, the motion of this type of drive is a function of two variables, the left and right velocities. By controlling these two velocities, we are able to control the motion of any robot using differential drive systems.

2.3.2   ACKERMANN STEERING (CAR-LIKE DRIVE)

This is a commonly used steering mechanism in cars, buses, and other land vehicles. It is also used in mobile robots. Basically, it is a four wheel drive system in which there are two steerable wheels and two motor-driven rear wheels. The front wheels are steerable to control the direction of the robot or the vehicle and the motor-driven rear wheels provide the forward or backward moves as illustrated in Figure 2.16.

Let us assume that the robot shown in Figure 2.16 has the velocity of V and heading orientation of θ. The steering angle of the robot is defined as α, the radius of its wheels as R, and the wheelbase as B. Assume that the robot has the front-driven speed of ω (radian/s), then the kinematic equations of this drive can be defined as follows (Siegwart and Nourbakhsh 2004):

$$V = R\omega \cos \alpha \tag{2.41}$$

$$\dot{x} = V \cos \theta \tag{2.42}$$

$$\dot{y} = V \sin \theta \tag{2.43}$$

$$\dot{\theta} = \frac{R\omega}{B} \sin \alpha \tag{2.44}$$

2.3.3   TRACK DRIVE

The track drive is primarily a differential drive system. The main difference is that it uses tracks instead of wheels (see Figure 2.17). There is no need for caster wheels in this case since the tracks cover a large surface and keep the robot in balance. To move forward, both motors have to drive the tracks at the same speed. Similar to differential drive system, in track drive systems, a robot can make turns by stopping one motor or by slowing one motor than the other. It can also make turns on the spot by driving the tracks in the opposite direction and at the same speed. The

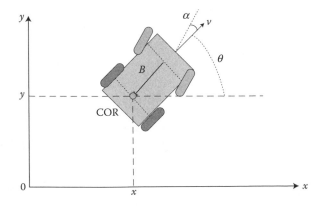

FIGURE 2.16   The parameters and variables involved in the motion of an Ackermann drive mobile robot.

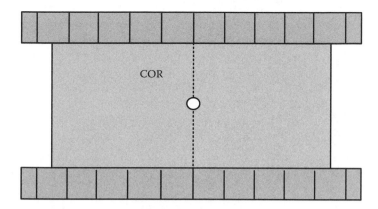

FIGURE 2.17 Track drive system.

track drive system is slower compared to the wheel drive system since there is more friction. However, it has certain advantages, which wheels cannot accomplish such as climbing up slopes, stairs, or going over obstacles and potholes (Mataric 2007). Robots that operate in unstructured terrains such as disaster zones and wheelchairs that climb stairs are some application examples of their use.

The kinematic equations of track drive systems are the same as the differential drive. However, it is important to note that these systems will suffer greater error compared to wheeled systems when the kinematic equations are used to calculate the robot position. This is mainly due to the skid-steering operation of the track drive. Figure 2.17 shows the COF of a track drive, and it is always located at the center of the robot.

2.3.4 OMNIWHEEL DRIVE

The omniwheel drive is a unique drive system when compared to the other wheeled drive systems we have discussed so far. A common feature of the systems discussed is that not all the wheels are motor driven or are used to steer the robot. In omni-wheel drive systems, all the wheels are motor driven and they are collectively used to steer the robot. Another major difference is that each omniwheel has rollers assembled on the wheel itself. The rollers act when the wheel is driven. The minor effects from all the rollers result in an ability to move different directions instantly. Owing to this special feature, an omniwheeled robot is able to move in any direction at all times, unlike the other drive systems, which all have some limitations. An omnidirectional robot comes with either three or four wheels. A major disadvantage of this drive system is the difficulty in controlling them. They have poor efficiency since not all the wheels are making the move in the direction of the robot motion. Furthermore, they have high slip; therefore, position control based on the motor encoders will not be accurate.

Figure 2.18 shows an example of a four wheel omniwheel robot and the resultant robot motions when each wheel is driven. To move forward, all four wheels have to be driven forward or backward with the same speed to achieve forward and backward moves. To glide the robot left, wheels 1 and 4 are driven backwards and wheels 2 and 3 are driven forward. Similarly, to slide right, wheels 1 and 4 are

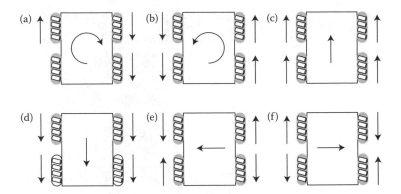

FIGURE 2.18   Different motions for a four omniwheel drive mobile robot. (a) Right turn motion, (b) left turn motion, (c) forward move, (d) reverse move, (e) robot sliding left, and (f) robot sliding right.

driven forward and wheels 2 and 3 are driven backwards. To accomplish a clockwise turn on the spot, wheels 1 and 3 are driven forward and wheels 2 and 4 are driven backwards. A counterclockwise turn is done by driving wheels 1 and 3 backwards and wheels 2 and 4 forward. RoboCupRescue robots and two degree freedom pole, balancing robots are examples of robots with omniwheels in competitions.

2.3.5 ODOMETRY

By using the kinematic equations presented earlier, it is possible to calculate the location of a mobile robot. Odometry is a mathematical procedure to determine the current location of a robot by discrete-time integration of its previous positions and velocity information over a given length of time (Siegwart and Nourbakhsh 2004). The general equations for the odometry are defined as:

$$x(k) = x(k - 1) + \Delta x \tag{2.45}$$

$$y(k) = y(k - 1) + \Delta y \tag{2.46}$$

$$\theta(k) = \theta(k - 1) + \Delta\theta \tag{2.47}$$

The aforementioned equations give the current position of the robot by simply adding the amount of displacement that has been made from its previous position. This difference in position is calculated by using the kinematic equation depending on the type of wheel drive the system robot is using. To calculate this difference in position, the system reads the encoder values from the motor wheels, and the kinematic equations tell us how to use these values to calculate the difference in position.

Odometry is subject to cumulative errors due to the inaccuracy of the encoders, unevenness of the surface, and minute deviations in robot construction such as a minor difference between wheel sizes. Nevertheless, odometry is crucial for robot navigation and there are methods to reduce the odometry error. These techniques principally increase the confidence level in computations of the robot's whereabouts.

2.3.6 CASE STUDY OF ODOMETRY FOR A DIFFERENTIAL DRIVE ROBOT

For this case study, we will consider the robot presented in Figure 2.8. The robot uses a differential drive system with two caster wheels. Each motor wheel has an encoder that generates a fixed number of pulses when the wheel-driving motor makes a complete rotation. Since the dimension of the wheel is known, by counting these pulses, we can easily calculate the distance robot traveled. In addition, we can calculate the speed of the robot by counting the number pulses within a fixed period of time. The working principles of encoders are presented in the following chapter on sensors.

A simple block diagram of the motion control system for the differential drive mobile robot in Figure 2.8 is presented in Figure 2.19. In the diagram, the path planner gives a new desired position for the robot $[x_d, y_d, \theta_d]$; this position is translated to left and right velocities by the velocity controller, and these velocity commands are passed to the left and right controllers. The velocity controller commands $[v_{ld}, v_{rd}]$ should not be considered as the actual velocities of the system. This is because the left and right controllers may not be able to achieve desired velocity immediately. Instead, the left and right controllers will try to achieve these velocity requirements by comparing feedback from the encoders $[v_{lf}, v_{rf}]$ with the desired velocities $[v_{ld}, v_{rd}]$ and adjust the controllers' output. Normally, the comparison and adjustment of the velocity is the task of a proportional integral derivative (PID) or other type of control system, which we will discuss in the following chapters. Notice that the encoder information is also used for estimating the current position of the robot $[x_f, y_f, \theta_f]$. The current position of the robot is compared against the desired position by the velocity controller to adjust the velocity according to the distance from the desired position, that is, when the robot is far from the desired position it will navigate faster, and when it gets closer, it will navigate slowly. Since the path planner and the controller will be discussed further in the following chapters, here we will emphasize on the odometry function of the system.

The odometry will provide the current position of the robot as coordinates of the tuple $[x, y, \theta]$, which describe the position in a two-dimensional plane and the robot

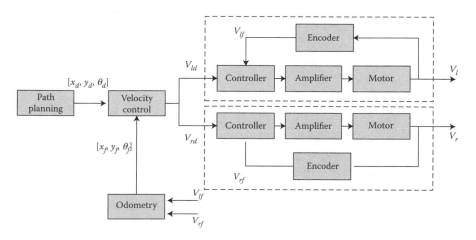

FIGURE 2.19 The differential drive parameters of the robot presented in Figure 2.8 and a block diagram of the odometry system in the motion and navigation system of the robot.

heading. The kinematic equations 2.37 through 2.40 are applicable for this robot; however, to use them in odometry equations 2.52 through 2.54, we need to obtain the differential of the equations. The following equations are derived from the kinematic equations, and they are in terms of displacement of the wheels instead of the velocities.

$$\Delta\theta = \frac{D_l - D_r}{d} \Delta t \tag{2.48}$$

$$\Delta S = \frac{D_l + D_r}{2} \Delta t \tag{2.49}$$

$$\Delta x = \Delta S \cos\theta \, \Delta t \tag{2.50}$$

$$\Delta y = \Delta S \sin\theta \, \Delta t \tag{2.51}$$

where D_l and D_r are the displacements of the left and right wheel readings from the encoder. ΔS is the distance that the vehicle has traveled in Δt period of time. Parameters x and y are the coordinates in a global frame and θ is the robot heading. The odometry equations for this robot are given as

$$x(k) = x(k-1) + \Delta S \cos \theta(k-1)\Delta t \tag{2.52}$$

$$y(k) = y(k-1) + \Delta S \sin \theta(k-1)\Delta t \tag{2.53}$$

$$\theta(k) = \theta(k-1) + \Delta\theta(k-1)\Delta t \tag{2.54}$$

It is necessary to clarify a few constant values before we implement this function. The first value is d, which is the distance between the two wheels. In our robot, this distance was 145 mm as shown in Figure 2.20a. Next, we need to understand the encoder values to calculate the displacement of the left and right wheels D_l and D_r. The encoders used in our robot are incremental with two channels, and each channel has two signals. Each encoder provides 256 counts per turn, with two channels and two signals per channel results in $256 \times 4 = 1024$ counts to complete a turn of the motor. To translate these values into millimeters, we need to calculate the circumference of the wheel. Radius of the robot wheel is 30 mm (see Figure 2.20b), which translates to 188.496 mm of circumference. By dividing the circumference to total counts per turn, we obtain the displacement made by each count in millimeters. This value for the robot shown in Figure 2.20b is 0.1840 mm per count of the encoder. Now that we clarified the encoder values and calculated the millimeters per count, the displacement of the left and right wheels D_l and D_r is calculated by multiplying the difference between the current encoder readings and the previous encoder readings, by 0.1840 mm.

Figures 2.21 and 2.22 present the MATLAB® code for the odometry calculations of the robot. The code could easily be translated into C++, or by writing the appropriate functions like the constructor, the ReadSensors(), and SendMotorCmd(), this code in fact can control an actual robot.

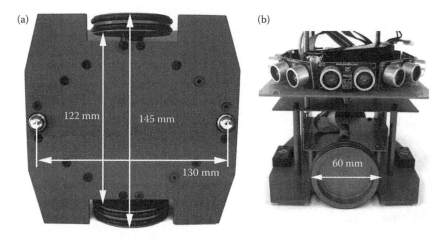

FIGURE 2.20 The physical dimensions of the robot presented in Figure 2.8.

The structure DiffRobot, defined in Figure 2.21, contains all the variables to represent the encoders, motor velocities, and odometry of the robot. The function DifferentialWheels(), shown in Figure 2.22, is the main loop for the robot control. We assumed that the constructor of the DiffRobot class can also initialize any physical communication required to access the physical hardware. The function ReadSensors() updates the encoder values from the hardware. The Odometry() function calculates the odometry values based on the encoder values obtained from the ReadSensors() function. Once the odometry has been calculated, the rest of the program will process the position and heading orientation to calculate the new velocities for the left and right wheels. The code also simulates some of the encoder values to observe odometry calculations.

Table 2.1 shows the actual encoder values obtained by the equivalent of ReadSensors() function running on the robot itself. The position and heading values are calculated by the Odometry() function. When we plot all the (x,y) positions calculated by the odometry code, we can clearly see the path that the robot moved (see Figure 2.23). Figure 2.24 illustrates the calculation steps taking place in the Odometry() function by using the first set of encoder values [100,100] from Table 2.1.

2.4 ROBOTIC ARMS

A significant part of robotics studies the manipulation of objects. Therefore, it is not surprising to find that many robot competitions involve some degree of object manipulation. In this section, we will discuss the robot arm configuration and how to obtain the control equations. There are many ways of controlling a robot manipulator. Here, we will focus on the kinematic calculations for the control of the robot, which deal with the relationship between the joint angles and the Cartesian coordinate positions.

A robotic arm consists of links, joints, and other structural elements. Links are the physical structures that connect the joints. There are different types of joints, such as prismatic, revolute, and spherical. Most robot manipulators have either prismatic

```
% The DifferentialWheels function uses an object of class DiffRobot

% the members of the class robot are declared as:

%

% robot.encoders(2)           - 1 - Left, 2 - Right

%                             - The current encoder values

% robot.oldEncoder(2)         - 1 - Left, 2 - Right

%                             - The previous encoder values

% robot.motor_speed(2);       - 1 - Left, 2 - Right

% robot.X                     - The robot position in X

% robot.Y                     - The robot position in Y

% robot.Theta                 - The robot heading theta

%

classdef DiffRobot < handle

    properties

        encoder = [ 0.0 0.0 ];

        oldEncoder = [ 0.0 0.0 ];

        X = 0.0;

        Y = 0.0;

        Theta = 0.0;

    end

    methods

        function obj = DiffRobot() % constructor

        end

    end

end
```

FIGURE 2.21   The structure used for the DifferentialWheels() function in Figure 2.22.

joints or revolute joints (see Figure 2.25). Prismatic joints are linear; there is no rotation involved in their motion. They are either hydraulic or pneumatic cylinders, or they are linear electrical actuators. Revolute joints are rotary and most rotary joints are electrically driven by motors.

The configuration of revolute and prismatic joints together with the physical properties of the links defines the arm. By using the Denavit–Hartenberg (D–H) algorithm, it is possible to extract four parameters that represent the relation between joints in a robot arm, and thus we can calculate its control equations (Fu et al. 1987; Craig 1989).

```
% This is the main function of the Differential Drive mobile robot. It will
% loop reading the sensors, Processing, and sending the motor command.
%
function DifferentialWheels()


% Initialized robot
robot = DiffRobot();


%while 1      % For Real Robot
for (i=1:5) % for Simulation
    % Read the New Values of the Sensors
    %ReadSensors(robot);      % For Real Robot
    ReadSensors(robot, i);   % for Simulation


    % Follow the Line
    Odometry(robot);


    data(i,:) = [ robot.X robot.Y robot.Theta];


    % Process the robot position to calculate the new speeds...


    % Write to Motors
    SetMotorCmd(robot);
end


% Plot the robot trajectory
plot(data(:, 1), data(:, 2));
end
```

FIGURE 2.22 MATLAB code for the odometry calculation of the robot presented in Figure 2.8.

```
%function ReadSensors(robot)        % For Real Robot

function ReadSensors(robot, i)      % for Simulation

% Obtain the new readings of the encoders

enc = [ 100 100; 400 400; 1200 1000; 2600 2000; 3000 2300]; % for

Simulation

robot.encoder = enc(i,:);    % for Simulation

end



function Odometry(robot)



% Convert the encoder value in mm

robot.encoder = robot.encoder * 0.1840;



% Calculate distances for both wheels

dSl = robot.encoder(1) - robot.oldEncoder(1);

dSr = robot.encoder(2) - robot.oldEncoder(2);



% Integrate robot position

robot.X = robot.X + (dSr + dSl) * cos(robot.Theta) / 2.0;

robot.Y = robot.Y + (dSr + dSl) * sin(robot.Theta) / 2.0;

robot.Theta = robot.Theta + (dSr - dSl) / 145.0;



robot.Theta = mod(robot.Theta, 2*pi);

if (robot.Theta > pi)

    robot.Theta = robot.Theta - 2*pi;

elseif (robot.Theta < -pi)

    robot.Theta = robot.Theta + 2*pi;

end
```

FIGURE 2.22 (continued) MATLAB code for the odometry calculation of the robot presented in Figure 2.8.

```
% Keep track of previous wheel positions

robot.oldEncoder = robot.encoder;

end


function SetMotorCmd(robot)

% Send the motor_speed to the robot

end
```

FIGURE 2.22 (continued) MATLAB code for the odometry calculation of the robot presented in Figure 2.8.

TABLE 2.1

Encoder Values and Their Respective Odometry Values

Left Encoder	Right Encoder	Robot.X	Robot.Y	Robot.Theta
100	100	18.4000	0.0	0.0
400	400	73.6000	0.0	0.0
1200	1000	202.4000	0.0	0.2538
2600	1600	380.5059	−46.1982	−1.2690
3000	2300	410.5895	−142.8234	−0.8883

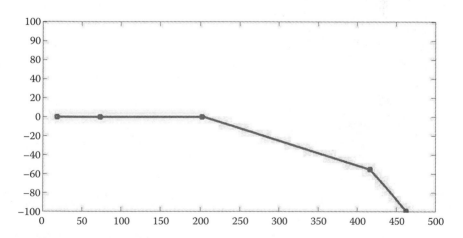

FIGURE 2.23 The plotted path of the odometry values calculated from the encoder values presented in Table 2.1. The units are in millimeters for both axes.

```
function Odometry(robot)                                    robot.encoder =
                                                            [100 100]


% Convert the encoder value in mm
robot.encoder = robot.encoder * 0.1840;                     robot.encoder =
                                                            [18.40 18.40]


% Calculate distances for both wheels
dSl = robot.encoder(1) - robot.oldEncoder(1);               dSl = 18.40
dSr = robot.encoder(2) - robot.oldEncoder(2);               dSr = 18.40


% Integrate robot position
robot.X = robot.X + (dSr + dSl) * cos(robot.Theta) /        robot.X = 18.40
2.0;                                                        robot.Y = 0.0
robot.Y = robot.Y + (dSr + dSl) * sin(robot.Theta) /        robot.Theta = 0.0
2.0;
robot.Theta = robot.Theta + (dSr - dSl) / 145.0;


robot.Theta = mod(robot.Theta, 2*pi);
if (robot.Theta > pi)
    robot.Theta = robot.Theta - 2*pi;
elseif (robot.Theta < -pi)
    robot.Theta = robot.Theta + 2*pi;
end
                                                            robot.oldencoder
                                                            = [18.40 18.40]
% Keep track of previous wheel positions
robot.oldEncoder = robot.encoder;
end
```

FIGURE 2.24 A step-by-step illustration of the calculations using the first set of values from Table 2.1.

D–H algorithm provides a systematic method, based on homogeneous transformations, to describe the position and orientation of each link with respect to its neighboring link (Craig 1989).

The algorithm consists of the following steps:

1. Assign coordinate frames to all links and the end effector of a robot manipulator.

(a)   (b)

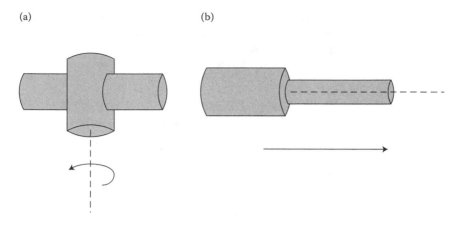

FIGURE 2.25   (a) Revolute joint defined a DOF by the angle. (b) Prismatic joint defined a DOF by the length of displacement.

2. Derive a homogeneous transformation matrix including both rotation and translation to describe the position and orientation of each link relative to its neighboring link.
3. Compute the forward kinematics of the robot manipulator using the post-multiplication rule.
4. Determine the position and orientation of the robot hand with respect to the base frame from the forward kinematic equation.

The basic rules for the assignment of the frame (x_i, y_i, z_i) to the link i are as follows (Craig 1989; Niku 2000):

1. z_i axis is aligned with the motion axis of the rotary joint $i + 1$.
2. x_i axis is normal to both z_{i-1} and z_i axes.
3. y_i axis is chosen from a right-handed frame (x_i, y_i, z_i).

There are two parameters of the link i; they are the link length a_i and the twist angle α_i. The link length a_i defines the common normal between the z_{i-1} and z_i axes. The twist angle α_i defines the rotational angle of the z_{i-1} axis about the x_i axis. There are also two joint parameters: the joint angle θ_i and the joint distance d_i. The joint angle θ_i is the rotational angle of the x_{i-1} axis about the z_{i-1} axis. The joint distance d_i is defined as the translation distance of the frame $(x_{i-1}, y_{i-1}, z_{i-1})$ along the z_{i-1} axis. Both the link parameters and the joint parameters are called arm parameters or D–H parameters.

2.4.1   FORWARD KINEMATIC SOLUTIONS

The forward kinematic equations determine the position and orientation of the robot hand (or end effector) in terms of the joint variables of the arm. Once the frame has been assigned and the arm parameters have been obtained for each link, we calculate a homogeneous transformation matrix H_{i-1}^i (see Equation 2.16 for the format of the homogeneous matrix) from the frame $(x_{i-1}, y_{i-1}, z_{i-1})$ to the frame (x_i, y_i, z_i). This transformation can be obtained by the following sequence of rotations and translations (Niku 2000):

1. Rotate the frame $(x_{i-1}, y_{i-1}, z_{i-1})$ about the z_{i-1} axis by θ_i angle.
2. Translate the frame $(x_{i-1}, y_{i-1}, z_{i-1})$ along the z_{i-1} axis by d_i units
3. Translate the frame $(x_{i-1}, y_{i-1}, z_{i-1})$ along the x_i axis by a_i units.
4. Rotate the frame $(x_{i-1}, y_{i-1}, z_{i-1})$ about the x_i axis by α_i angle.

Since these transformations are consecutive motions about the corresponding mobile axes, the D–H transformation matrix from the frame $(x_{i-1}, y_{i-1}, z_{i-1})$ to the frame (x_i, y_i, z_i) is defined as follows:

$$H_{i-1}^i = H_{ROT}(\theta_i)H_{TR}(d_i)H_{TR}(a_i)H_{ROT}(\alpha_i) \tag{2.55}$$

$$H_{i-1}^i = \begin{bmatrix} \cos(\theta_i) & -\cos(\alpha_i)\sin(\theta_i) & \sin(\alpha_i)\sin(\theta_i) & a_i\cos(\theta_i) \\ \sin(\theta_i) & \cos(\alpha_i)\cos(\theta_i) & -\sin(\alpha_i)\cos(\theta_i) & a_i\sin(\theta_i) \\ 0 & \sin(\alpha_i) & \cos(\alpha_i) & d_i \\ 0 & 0 & 0 & 1 \end{bmatrix} \tag{2.56}$$

Transformation matrix H_{i-1}^i describes the position and orientation of the link i with respect to the link $i - 1$, hence the matrix H_0^n describes the position and orientation of the robot end effector frame with respect to the robot base frame

$$H_0^n = H_0^1 H_1^2 \cdots H_{n-1}^n \tag{2.57}$$

The transformation matrix H_0^n is called arm matrix or solution of forward kinematics of an n-link robot manipulator.

2.4.2 INVERSE KINEMATICS

In the previous section, we discussed how to determine the robot hand (end effector) position and orientation in terms of the joint variables using the forward kinematics. In this section, we are concerned with the opposite problem, that of finding joint variables in terms of the robot hand position and orientation. This is solved by using inverse kinematics. The general problem of inverse kinematics can be stated as follows (Niku 2000; Man 2005):

1. Given a homogeneous transformation matrix H_0^n that represents the forward kinematics expressed as

$$H_0^n = H_0^1 H_1^2 \cdots H_{n-1}^n = \begin{bmatrix} h_{11} & h_{12} & h_{13} & h_{14} \\ h_{21} & h_{22} & h_{23} & h_{24} \\ h_{31} & h_{32} & h_{33} & h_{34} \\ 0 & 0 & 0 & 1 \end{bmatrix} \qquad (2.58)$$

where each h_{ij} is an equation in terms of the joint variables, these variables might be θ_i for revolute joints or d_i for prismatic joints.

2. Specify the desired position and orientation of the robot hand relative to the robot base frame, in terms of the homogeneous matrix.

$$H_d = \begin{bmatrix} d_{11} & d_{12} & d_{13} & d_{14} \\ d_{21} & d_{22} & d_{23} & d_{24} \\ d_{31} & d_{32} & d_{33} & d_{34} \\ 0 & 0 & 0 & 1 \end{bmatrix} \qquad (2.59)$$

3. Solve the following equations:

$$\begin{bmatrix} h_{11} & h_{12} & h_{13} & h_{14} \\ h_{21} & h_{22} & h_{23} & h_{24} \\ h_{31} & h_{32} & h_{33} & h_{34} \\ 0 & 0 & 0 & 1 \end{bmatrix} = \begin{bmatrix} d_{11} & d_{12} & d_{13} & d_{14} \\ d_{21} & d_{22} & d_{23} & d_{24} \\ d_{31} & d_{32} & d_{33} & d_{34} \\ 0 & 0 & 0 & 1 \end{bmatrix} \qquad (2.60)$$

The joint variables, in order to reach to the desired position H_d, will be determined by solving the 12 equations presented in Equation 2.60. However, as we have seen in the previous section, the elements of each h_{ij} of the forward kinematics matrix are often nonlinear functions of the joint variables, and thus it is difficult to solve these equations to find a solution. The solution is even harder when we have to consider constraints of robot motions, singularities, and multiple possible solutions caused by the redundancy of the joints.

There are different approaches to obtain the inverse kinematic equations. An algebraic approach tries to solve the 12 equations presented in Equation 2.60. Another approach uses geometrical decomposition of the spatial geometry of the arm into several plane geometry problems (Man 2005). Both of these approaches have their limitations especially when the robot arm consists of several joints. In those cases, the equation becomes even more complicated and the geometrical analysis becomes too tedious. In those cases, a numerical method can be used to solve the equations presented in Equation 2.60.

2.4.3 CASE STUDY: THREE-LINK ARTICULATED ROBOT ARM

In the RoboCup@Home competition, there are several challenges that require the robot to manipulate objects or at least be able to move them by pushing them. The position of the object is determined by sensors such as a camera, ultrasonic sensors, or a laser range finder. Figure 2.26 shows a three-link articulated robot arm with three rotary joints that was used in RoboCup@Home. It is possible to put a simple gripper actuated by a servo-motor on the robot end to manipulate the objects.

First of all, we need to assign a frame to each robot joint as shown in Figure 2.26b. The z_0, z_1, and z_2 axes are assigned along the motion axes of the three rotary joints. The origin of the frame $x_0 y_0 z_0$ is chosen at the center of the robot base along with the z_0 axis and x_0 and y_0 axes are chosen to form a right-handed frame. For the frame $x_1 y_1 z_1$, the x_1 axis is chosen to be perpendicular to both z_0 and z_1 axes and the y_1 axis is chosen to form a right-handed frame. The frame $x_2 y_2 z_2$ and $x_3 y_3 z_3$ are parallel to $x_1 y_1 z_1$. The origin of the frame $x_3 y_3 z_3$ is located at the end of the third link.

After assigning the coordinate frames for each joint in the robot, we need to determine the arm parameters. Table 2.2 shows the arm parameters where only link

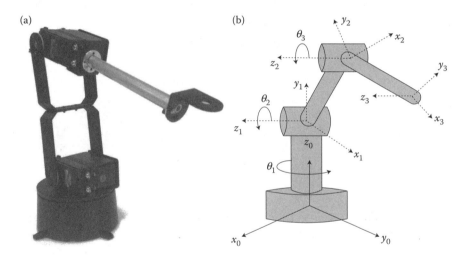

FIGURE 2.26 (a) A picture of the robot arm used. (b) Coordinate frames for three rotary joints.

TABLE 2.2

Arm Parameters for a Three Articulated Robot Presented in Figure 2.26

Link	θ_i	d_i	a_i	α_i
1	θ_1	d_1	0	90°
2	θ_2	0	a_2	0°
3	θ_3	0	a_3	0°

1 has a twist angle of 90°. Notice the change in the position of the axis from z_0 to z_1 in Figure 2.26b.

Using the D–H matrix parameters and the arm parameters given in Table 2.2, we obtain the following transformation matrices for each link:

$$
H_0^1 = \begin{bmatrix} \cos(\theta_1) & 0 & \sin(\theta_1) & 0 \\ \sin(\theta_1) & 0 & -\cos(\theta_1) & 0 \\ 0 & 1 & 0 & d_1 \\ 0 & 0 & 0 & 1 \end{bmatrix}
$$

$$
H_1^2 = \begin{bmatrix} \cos(\theta_2) & -\sin(\theta_2) & 0 & a_2\cos(\theta_2) \\ \sin(\theta_2) & \cos(\theta_2) & 0 & a_2\sin(\theta_2) \\ 0 & 0 & 1 & 0 \\ 0 & 0 & 0 & 1 \end{bmatrix} \tag{2.61}
$$

$$
H_2^3 = \begin{bmatrix} \cos(\theta_3) & -\sin(\theta_3) & 0 & a_3\cos(\theta_3) \\ \sin(\theta_3) & \cos(\theta_3) & 0 & a_3\sin(\theta_3) \\ 0 & 0 & 1 & 0 \\ 0 & 0 & 0 & 1 \end{bmatrix}
$$

Therefore, the forward kinematic solution for this robot arm is given by

$$
H_0^3 = H_0^1 H_1^2 H_2^3
$$

$$
= \begin{bmatrix} \cos\theta_1\cos(\theta_2+\theta_3) & -\cos\theta_1\sin(\theta_2+\theta_3) & \sin\theta_1 & \cos\theta_1\left(a_2\cos\theta_2+a_3\cos\theta_3\right) \\ \sin\theta_1\cos(\theta_2+\theta_3) & -\sin\theta_1\sin(\theta_2+\theta_3) & -\cos\theta_1 & \sin\theta_1\left(a_2\cos\theta_2+a_3\cos\theta_3\right) \\ \sin(\theta_2+\theta_3) & \cos(\theta_2+\theta_3) & 0 & a_2\sin\theta_2+a_3\sin(\theta_2+\theta_3)+d_1 \\ 0 & 0 & 0 & 1 \end{bmatrix}
$$

$$\tag{2.62}$$

Now, let us calculate the inverse kinematics for the arm. We will use the geometric approach since it is a simple configuration with only three joints. Figure 2.27a shows the projection of the arm and the joints in different planes. By considering each of these planes independently, it is possible to calculate the values of the angles. Thus, the plane of (x,y) will be used to obtain θ_1, and the plane (x,y,z) will be used to calculate the values of θ_2 and θ_3.

Figure 2.27b shows the plane (x,y). Observing the triangle formed by (p_x, p_y), the angle θ_1 can be derived by using arctangent. Since the arctangent will produce solutions in the first and the fourth quadrants, and the signs of p_x and p_y are determined by the quadrant, the arc tangent 2 function would be a better choice

$$
\theta_1 = a\tan 2(p_x, p_y) \tag{2.63}
$$

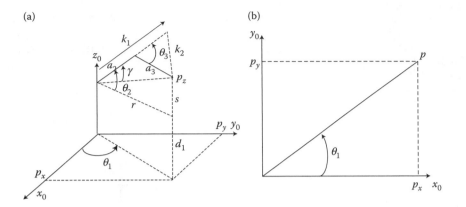

FIGURE 2.27 (a) A 3D view of the robot arm and geometrical projection into the planes (b) projection on (x,y) plane.

In the plane (x,y,z) presented in Figure 2.28a, a triangle is formed from joint 2 to the p_z as shown in Figure 2.28b. Let us calculate the hypotenuse, and by doing this we will try to extract a common factor that can help us to obtain the other two angles.

$$r = \sqrt{p_x^2 + p_y^2} = a_2 \cos\theta_2 + a_3 \cos(\theta_2 + \theta_3) \tag{2.64}$$

$$s = p_z - d_1 = a_2 \sin\theta_2 - a_3 \sin(\theta_2 + \theta_3) \tag{2.65}$$

$$p_x^2 + p_y^2 + (p_z - d_1)^2 = r^2 - s^2$$

where $\cos(\theta_2 + \theta_3) = \cos\theta_2 \cos\theta_3 - \sin\theta_2 \sin\theta_3$ and $\sin(\theta_2 + \theta_3) = \sin\theta_2 \cos\theta_3 + \cos\theta_2 \sin\theta_3$. The solution of θ_3 can be found by expanding and solving the aforementioned equation. Let us start by expanding the right side of the equation. To simplify the representation of solution equations, we will use the following acronyms:

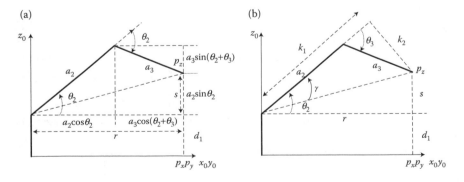

FIGURE 2.28 Geometrical projection of the joint 2 and 3 into the plane (a) xy and (b) z.

$$S_{ij} = \sin(\theta_i + \theta_j)$$
$$C_{ij} = \cos(\theta_i + \theta_j)$$
$$S_i = \sin(\theta_i)$$
$$C_i = \cos(\theta_i)$$

hence,

$$= a_2^2 C_2^2 + a_3^2 C_{23}^2 + 2a_2 a_3 C_2 C_{23} + a_2^2 S_2^2 + a_3^2 S_{23}^2 + 2a_2 a_3 S_2 S_{23}$$

$$= a_2^2 (C_2^2 + S_2^2) + a_3^2 (C_{23}^3 + S_{23}^2) + 2a_2 a_3 (C_2 C_{23} + S_2 S_{23}) \qquad (2.66)$$

Since $\sin \theta^2 + \cos \theta^2 = 1$, the aforementioned equation can be simplified further

$$= a_2^2 + a_3^2 + 2a_2 a_3 (C_2[C_2 C_3 - S_2 S_3] + S_2[S_2 C_3 + C_2 S_3])$$

$$= a_2^2 + a_3^2 + 2a_2 a_3 (C_2^2 C_3 - C_2 S_2 S_3 + S_2^2 C_3 + C_2 S_2 S_3)$$

$$= a_2^2 + a_3^2 + 2a_2 a_3 (C_3[C_2^2 + S_2^2])$$

$$= a_2^2 + a_3^2 + 2a_2 a_3 C_3 \qquad (2.67)$$

and now we can solve all the equations for C_3

$$C_3 = \frac{p_x^2 + p_y^2 + (p_z - d_1)^2 - a_2^2 - a_3^2}{2a_2 a_3} \qquad (2.68)$$

Once again using trigonometric identity $\sin \theta^2 + \cos \theta^2 = 1$, we can also obtain S_3

$$S_3 = \pm\sqrt{1 - C_3^2} \qquad (2.69)$$

Because the sine has a period of 2π, the value of the angle could be positive or negative. This means that there are two solutions for the sine. The positive solution will represent a "joint up" configuration, and the negative solution will be a "joint down" configuration. In either case, θ_3 is obtained with

$$\theta_3 = a\tan 2(S_3, C_3) \qquad (2.70)$$

For the last angle θ_2, let us see the triangle form by k_1 and k_2 with an angle γ (see Figure 2.28b)

$$k_1 = a_2 + a_3 C_3$$

$$k_2 = a_3 S_3$$

$$\gamma = a \tan 2(k_2, k_1) \qquad (2.71)$$

Once the angle γ is calculated, it is evident that θ_2 consists of the angle γ and the angle formed by the component r and s.

$$\theta_2 = a \tan 2(h, r) - \gamma$$

$$\theta_2 = a \tan 2(p_z - d_1, \sqrt{p_x^2 + p_y^2}) - a \tan 2(a_3 S_3, a_2 + a_3 C_3) \qquad (2.72)$$

From the equation, we can see that θ_2 will also have two possible values for the inclusion of S_3, thus the final inverse kinematics equations for the robotic arm can be described as

$$\theta_1 = a \tan 2(p_x, p_y)$$

$$C_3 = \frac{p_x^2 + p_y^2 + (p_z - d_1)^2 - a_2^2 - a_3^2}{2 a_2 a_3}$$

$$S_3 = \pm\sqrt{1 - C_3^2}$$

$$\theta_3 = a \tan 2(S_3, C_3)$$

$$\theta_2 = a \tan 2(p_z - d_1, \sqrt{p_x^2 + p_y^2}) - a \tan 2(a_3 S_3, a_2 + a_3 C_3) \qquad (2.73)$$

Finally, it is possible to calculate the values of θ_1, θ_2, and θ_3 from a target position $[p_x, p_y, p_z]$. As mentioned earlier, in practice, the target position is obtained by the sensors of the robot. The inverse kinematic equations calculate the values of the joint angles to reach that position with the robotic arm. The forward kinematics can be used to know exactly where the arm is in motion.

REFERENCES

Craig, J.J. 1989. *Introduction to Robotics: Mechanics and Control*. Reading, MA: Addison-Wesley.

Fred, G.M. 2001. *Robotic Explorations: An Introduction to Engineering through Design*. Upper Saddle River, NJ: Prentice-Hall.

Fu, K.S., Gonzales, R.C., and Lee, C.S.G. 1987. *Robotics, Control, Sensing, Vision and Intelligence*. New York: McGraw-Hill.

Man, Z. 2005. *Robotics*. Singapore: Prentice–Hall, Pearson Education Asia Pte Ltd.

Mataric, M.J. 2007. *The Robotics Primer*. Cambridge, MA: The MIT Press.

Niku, S.B. 2000. *Introduction to Robotics: Analysis, Systems, Applications*. Upper Saddle River, NJ: Prentice-Hall.

Siegwart, R. and Nourbakhsh, I.R. 2004. *Introduction to Autonomous Mobile Robots*. Cambridge, MA: The MIT Press.

3 Sensors

A fundamental component of the robot is its sensors, which are used for acquiring information about the robot itself and its environment. There are a vast number of sensors available for the robotics and automation industry. Depending on the application, we employ a combination of them in robot design. A detailed description of these sensors and their properties can be found in Soloman (2009) and Fraden (1996). In this chapter, we will discuss sensors commonly used in game robotics and competitions.

Sensors give measures about physical properties of the environment such as illumination, temperature, distance, size, and so on. In other words, a sensor is a measurement tool that converts physical quantities from one domain to another. In robotics, we are interested in sensors that convert physical phenomenon, electrical or nonelectrical in nature, to an electrical signal so that it can be processed by a microprocessor. A number of parameters determine the characteristics of sensors, which indicate their capabilities and limitations. It is important to know these parameters for the proper selection of a sensor for a robot design. Some of these parameters are discussed below:

Range: Maximum and minimum values that can be measured by the sensor. For instance, the range for the Microchip MCP9501 temperature sensor is from −40° to +125°. The term dynamic range refers to the overall range of the sensor from its minimum to maximum reading.

Resolution or discrimination: The resolution of a sensor is the smallest distinct change in the measured value it can reliably detect. For example, the Honeywell HR3000 digital compass has a resolution of 0.1°. Resolution does not indicate accuracy. A sensor can have high resolution, but may not be accurate. A sensor's ability to detect minute changes is mainly limited by the electrical noise.

Error: Difference between the measured and actual values.

Accuracy/inaccuracy/uncertainty: Accuracy indicates the maximum difference between the actual value of the measured parameter and its measured value by the sensor. In other words, it is an indicator of the maximum expected error.

Linearity: Maximum deviation from a linear response. The term linearity of a sensor implies the extent that measured curve of a sensor departs from an ideal straight-line curve. Sensors with linear response simplify robot design and programming.

Sensitivity: It is a measure of change at the output of the sensor for a change in the amount being measured.

Precision/repeatability: Precision of a sensor, also called repeatability, shows the difference in measurements on the same thing and under the same conditions. A sensor can have high precision/repeatability yet poor accuracy if there is a systematic error in the measuring system. On the other hand, a highly accurate sensor will not have poor repeatability since repeatability is a requirement for accuracy.

Response time: Sensor output does not change immediately when there is a change in the input parameter. Its response will change over a period of time, which is known as response time. Fast response time is desirable for robotics application.

Output: The type of output from the sensors determines the peripheral circuitry needed to be built for the robot. As listed in Table 3.1, many sensors deliver analog as well as digital output. The latest sensors include built-in circuitry not only for producing digital output, but also to do part of the signal processing, such as filtering, on the sensor unit. Many of these sensors also come with data bus compatible output, which makes their interfacing to processors very straightforward.

Frequency of measurement: The other parameter of concern in robot design is the sensor frequency, which indicates the number of readings that can be done per second. This is a major bottleneck preventing robots from doing fast motions in their environment.

When a robot is negotiating its environment, information obtained from its sensors determines a great deal of its actions and behavior. Therefore, reliability of its sensors is very crucial. We look at accuracy and repeatability parameter of a sensor as a measure of its reliability. Ideally, a good sensor is sensitive to the measured property only and it is not affected by other environmental parameters. However, in practice, data obtained from sensors are noisy and prone to errors.

3.1 SENSORS USED IN GAME ROBOTICS

The type of sensors used in robotics varies depending on the application. However, robots designed for robotic games are mostly low-cost mobile robots, and they need a small set of these sensors. In this section, we will first look into sensors that give a reading of the robot itself.

The sensors used in game robotics are mainly of two types. A set of sensors is used to get information about the robot itself such as its speed and orientation (proprioception). Another set of sensors is used to obtain information about its surrounding, such as distance from an obstacle (exteroception). Table 3.1 shows a classification of these sensors and their potential use in robot design.

3.1.1 MEASURING ROBOT SPEED

Sensors are placed in the motors or wheels of a robot to obtain feedback on its motions. In mobile robotics, encoders are assembled to its driving motors, and a direct reading of the angular speed is obtained. This enables control of the position, direction, and the speed of the motor-controlled wheels. Encoders are also used to gather position information from rotary parts of the robot such as the pole-balancing robot that will be discussed in Chapter 10, or joint angles of a robotic arm. Optical and magnetic encoders are the most commonly used encoders in robotics.

TABLE 3.1
Classification of Sensors Used in Game Robotics

Sensor	Application	Type of Sensor	Output Type
Contact switches, strain gauge, infrared sensors	Detection of physical contact and closeness	Tactile sensors	Binary on/off analog
Optical and magnetic encoders, potentiometers	Rotation, motor speed, and position	Wheel/motor sensors	Digital
Compass, gyroscopes, accelerometers, inclinometers	Detecting inclination, acceleration, and orientation of the robot	Heading sensors	Analog/digital
Ultrasonic sensors, laser rangefinders, infrared sensors, optical triangulation sensors	Detecting proximity to objects, map generation, obstacle detection	Range sensors	Analog/digital
Camera, color sensors, linear sensor array	Object recognition and manipulation, analysis of robot environment	Vision-based sensors	Analog/digital
Altimeter, depth gauge, GPS	Flying robots, underwater robots, land robots	Position and navigation sensors	Analog/digital

Optical encoders: An optical encoder is made of a rotor disk with engraved optical grids, a light source, and a photo sensor as shown in Figure 3.1a. The rotor moves with the angular motion of the motor and causes a change in the amount of light received by the optical sensor. The resulting sine wave-like signal is converted to square wave by thresholding. High and low cycles of square wave indicate the black and white areas passing through the optical sensors. The resolution of encoders is described as the number of pulses per revolution (PPR). The encoder illustrated in Figure 3.1a is of incremental type. By placing another optical unit 90° apart from the original, quadrature-type encoders are obtained. This arrangement produces two square waves with 90° phase shift. By utilizing the phase shift between them,

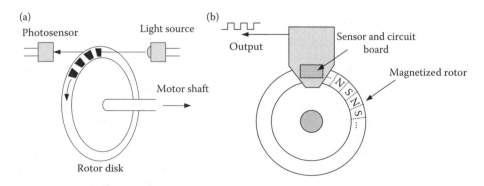

FIGURE 3.1 (a) Optical and (b) magnetic encoder.

the resolution (PPR) is increased by four times. Furthermore, by determining the sequence of square pulses, the direction of rotation is also found easily.

Magnetic encoders: These encoders convert mechanical motion into a digital output by means of magnetism. Typically, a magnetized disk with a flux pattern is attached to the rotor; by detecting the change of flux, a magnetic encoder generates a signal at the output as shown in Figure 3.1b. Hall effect sensors or magnetoresistive sensors are widely used for detecting the flux change. A magnetized rotor is attached to the shaft, and there is a thin air gap between the rotor and the sensor. Sensors produce a sine wave signal when the rotor turns and flies over the sensor. The sine wave signal is then converted to a square wave and delivered as output from the encoder system. Magnetic encoders have good reliability and durability. They are not easily affected by environmental factors such as dirt, dust, and oil. However, they are subject to magnetic interference. Apparently, encoders used in robotics must be fast enough to be able to count the shaft speed. Many encoders available in the market have no limitations to use in robotics, and they are reliable.

3.1.2 Measuring Robot Heading and Inclination

Heading sensors are mainly used for determining orientation as well as inclination of a robot.

By knowing the orientation and the speed of a mobile robot, we can estimate its whereabouts. Similarly, knowing the inclination of a humanoid robot, we can correct its motions so that it does not topple.

Compass: The principle of a digital compass is based on measuring the direction of Earth's magnetic field. Many cost-effective digital compasses are built with Hall effect sensors, which are based on the principle that electric potential changes in a semiconductor when it is exposed to a magnetic field. An example of this type of sensor is the Allegro A132X family Hall effect sensors, where the presence of a south pole magnetic field perpendicular to the IC package face increases (decreases in the case of north pole) the output voltage from its neutral value, proportional to the magnetic field applied depending on the sensitivity of the device. A single Hall effect sensor measures flux in one dimension. To measure the two axes of magnetic fields, two of these sensors are placed at 90° angles. The resolution obtained with Hall effect sensors is low and prone to errors, particularly due to interfering magnetic fields.

Another technology used in digital compasses exploits magnetoresistivity, which is the property of change in resistivity of a current-carrying magnetic material in the presence of a magnetic field. Assume that the current is passing through the ferromagnetic material as shown in Figure 3.2. When the material is exposed to an external magnetic field, the internal magnetization vector changes its position. The strip resistance depends on the angle θ between magnetization and the direction of the current flow. This resistance will have the largest value if the current flow and magnetization vector are parallel. Conversely, it will be the smallest when the angle between them is 90°.

To measure magnetic field, four sensors are connected in a bridge configuration with each resistor oriented to maximize the sensitivity and minimize the temperature effects. The values of the resistors will change when they are exposed to a magnetic field and the bridge will be imbalanced, thus generating an output voltage proportional to the

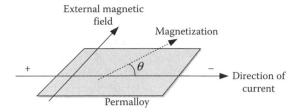

FIGURE 3.2   The magnetoresistive effect.

magnetic field strength. Digital compasses developed with this technology are reliable, and they have good resolution and fast response. Nevertheless, they are also sensitive to interfering magnetic fields. Therefore, using them in manmade environments requires caution. Examples of these sensors are Devantech's CMPS03 magnetic compass, which uses the Philips KMZ51 magnetic field sensor and the Honeywell HMR3000 digital compass module that provides heading, pitch, and roll outputs for navigation.

Gyroscope: It can measure the angular motion of a robot relative to an inertial frame of reference; hence, the gyroscope is also a device for measuring orientation. There are various types of gyroscopes available; however, digital gyroscope using MEMS (small microelectromechanical systems) technology is the most popular and cost-effective sensor used in many electronic devices as well as in robotics. MEMS gyroscopes detect rotational rate about the X, Y, and Z (or roll, pitch, and yaw) axes. When the gyroscope is rotated about any of these axes, the Coriolis effect causes a deflection, which is detected, demodulated, and filtered to produce a voltage that is proportional to the angular rate. Analog Device's ADIS16485 is an example of a MEMS-based gyroscope, which provides three axes gyroscope readings in digital form via serial parallel interface (SPI) bus.

Accelerometer: An external force acting upon a system, such as gravity, causes a change in the velocity. This sensor measures acceleration caused by such external forces. They are mainly used for sensing robot motions. Dynamic balancing of a walking robot is a good example of accelerometer use in robotics. An accelerometer can be considered as a damped mass on a spring. When the sensor faces acceleration, the mass will be displaced toward a point that the spring permits. By measuring this displacement, the acceleration is found. By arranging three of them orthogonally, it is possible to detect acceleration in all three axes. There are various types of commercial devices using piezoelectric, piezoresistive, and capacitive components. However, the latest accelerometers are often MEMS devices. Analog Device's ADXL202 is an example of such accelerometers. The sensor has ±2 g sensing range with a pulse width modulated (PWM) or analog signal output.

3.1.3   Measuring Range

The sensors we have discussed so far provide feedback about the robot itself. In the following, we will discuss sensors that gather feedback about the robot's environment. One of the key components used in mobile robots is the range sensor. These sensors measure the distance of the objects from the sensor. In robotics, they are mainly used for detecting objects, generating a map of the environment, and avoiding collisions.

Ultrasonic sensors: Ultrasonic sensors transmit an ultrasonic wave package and receive the reflected signal. The time taken for a signal to travel and return gives an indication of the distance. The ultrasound a frequency range is between 40 and 180 kHz. It permits more concentrated direction of the sound since at higher frequencies sound dissipates less in the environment. We can consider the ultrasonic sensor as a pair of speaker and microphone, one produces the sound and another receives the echo (see Figure 3.3a and b). A short ultrasonic signal is generated as shown in Figure 3.3c, and the timer is triggered. The receiver captures the echoing

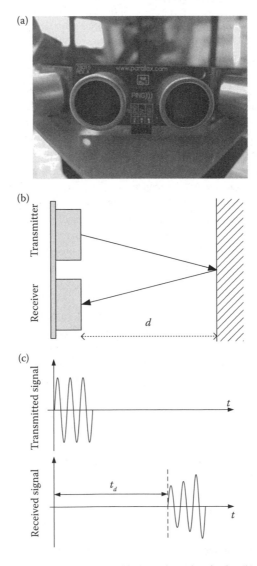

FIGURE 3.3 (a) Ultrasonic sensor assembled on the robot body. (b) Ultrasound signal is reflected from an object at distance *d*. (c) Sent and received signals and the time difference between them.

sound and stops the timer. This period of time, known as time of flight (ToF), is given as t_d. Hence, the distance d is calculated as

$$d = \frac{c_{air} \times t_d}{2} \tag{3.1}$$

Here, c_{air} indicates the speed of sound. It is about 330 m/s in air and 1500 m/s in water.

There are a number of issues that have to be understood when working with ultrasonic sensors.

i. Maximum range: One of them is the maximum distance that can be sensed by the sensors and it is related to the frequency used in the sensor design. Depending on the application, the appropriate one should be selected. For example, the sensor shown in Figure 3.3a is "Ping" from Parallax and it has a maximum range of 3 m, while Maxbotics MB1260 has a range of 10 m. Ultrasonic sensors cannot accurately measure the distance to an object, which is further than the sensing range.

ii. Blind zone: An inherent issue with ultrasonic sensors is the blind zone, which is the close range in front of them. The readings in this range are not reliable. The blind zone varies from sensor to sensor. For example, the length of blind zone is 2 cm for Parallax's Ping and 20 cm for Maxbotics MB1260.

iii. Reflection: The basic operation principle of an ultrasonic sensor is based on detecting reflected sound. There are many situations when this reflection may not take place. For example, the objects with a soft or irregular surface may absorb sound. Objects may be too small to reflect enough sound back to the sensor. The object surface may be at a shallow angle; hence, not enough sound reflection occurs. The shape of the beam is like a cone, so if the sensor is mounted very low on the robot there may be wrong readings from the reflections off the floor.

iv. Temperature: Temperature affects the speed of sound in air. If the temperature change in the environment is large, then the errors can be significant.

Laser rangefinder: The operation principle of a laser rangefinder is similar to ultrasound sensors except that these sensors use a laser beam, typically a near infrared light, instead of sound. In the case of a laser beam, time of flight is very short. Measuring such short time of flight requires very fast circuitry, which operates at the picosecond range and it makes the sensor expensive. The low-cost laser rangefinders utilize the phase shift between transmitted and received signals. Commercially available laser rangefinders have a wide choice of maximum range reaching up to hundreds of meters. Since the operation principle of these sensors is based on sending a light beam and detecting its reflection, color and texture of objects may affect their accuracy consequently. Shiny, bright colored objects reflect light better; on the other hand, dark objects absorb the light and reflect a lot less to the sensor, causing a reduction of the sensing range. Similarly, rough or smooth surfaces also affect the specular reflection. Nevertheless, laser rangefinders are a lot better in accuracy compared to sonar sensors discussed earlier. The laser rangefinder shown in Figure 3.4b is the Hokuyo URG-04LX-UG01, which uses infrared laser of wavelength 785 nm. The direction of light is altered by a rotating mirror. The reflected light is captured by the photo diode. The phases of the emitted and received light are compared, and the distance between the sensor and the object is calculated. A rotating mirror sweeps the laser beam horizontally over a range of 240°, with an angular resolution of 0.36°, which corresponds to 683 measured points in its scanned region. The sensor has a scan area of 240° semicircle with a maximum distance of 4000 mm radius. Figure 3.4c shows a scan of the robot at the end of a corridor. Owing to the high resolution of the sensor, even the corners of the corridor can be seen clearly. Each scan takes 100 ms, which gives a 10 Hz scan rate. A thorough analysis of this sensor can be found in Okubo et al. (2009). A sensor of this range is satisfactory for many indoor robotics applications such as the service robot shown in Figure 3.4a, which uses the SICK LMS 110 laser scanner.

Optical sensors: Infrared transmitters and receivers are simple and inexpensive sensors that are used for detecting objects. They are normally used for detecting the existence of an object rather than measuring the distance. For instance, an infrared transmitter and receiver pair is used for line following (reflective tapes placed on the competition platform) in pole-balancing robot (see Figure 1.1) or the

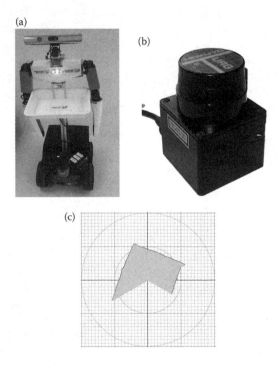

FIGURE 3.4   (a) Service robot using multiple sensors such as laser rangefinder, camera, and ultrasonic sensors. (b) Laser rangefinder from Hokuyo. (c) 240° scan of an indoor environment where the robot is at the end of a corridor.

presence of a pellet in the gripper of the robot designed for colony competition (see Figure 3.7), or detecting the existence of walls in micromouse competition. The principle of operation is based on detecting reflected infrared light emitted by the transmitter diode from the surface of objects. The reflected light amount depends on the object color and surface as well as the distance. In robotic games, competition platforms are well defined and uniform; therefore, simple infrared sensor pairs can be used to measure the distance, to sense objects, and even the hue of the object color. Figure 3.5 shows a typical arrangement of infrared emitter (D_1) and photodiode (D_2) receiver circuit used for detecting lines drawn on the competition platform. The transistor T_1 is used for switching on IR emitter D_1 when necessary. The resistor R_4 provides means to compare light intensity by the receiving diode. If the light intensity is above a threshold value, the output of the operational amplifier will be high. By tuning this resistor, minor reflections can be omitted; the robot only responds to strong reflection. The competition platforms are usually painted in contrasting colors, such as a black field with white reflective tapes or a white field with black tapes, and are used for indicating boundaries. For instance, the platform for robot colony competition is black and boundaries are marked with reflective tapes. As black color returns no light, the output from the circuit will be logic zero indicating that robot is in the black zone. The robot shown in Figure 3.7a has three pairs of transmitter and receiver for reliable sensing of reflected light from boundaries placed on the floor.

The circuit shown in Figure 3.5 produces binary output by comparing received signal strength with a threshold value. The same IR sensor pair can also be used for measuring the distance of objects at close proximity by measuring the strength of the reflected light. The circuit for such an application is shown in Figure 3.6. The current passing through D_2 is proportional to the reflected light; hence, the voltage is induced on resistor R_3. By using an analog-to-digital convertor (ADC) port of the microcontroller, this voltage can be measured and calibrated to measure the distance. However, there are also compact infrared sensors such as the Sharp GD2D02 series, which is very popular in game robotics due to its low cost.

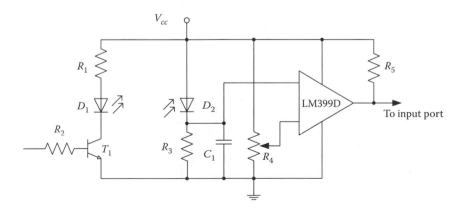

FIGURE 3.5 Basic circuit used for an IR sensor pair.

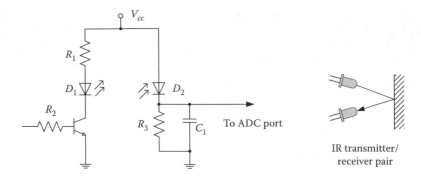

FIGURE 3.6 A simple circuit that can be used for measuring the distance to objects at close distance.

It works on the same principle of emitting infrared light and detecting its reflection from objects. The emitted infrared light is high in intensity and collimated. When the light is reflected from the surface of the object at the receiver end, a lens setup directs it to a photo sensor array strip. The position on this array in which the light falls is used to calculate the distance from the transmitter. The sensor is highly accurate, although it has limitations. For instance, the reliable operating range of the sensor is 8–80 cm. The measurements are also affected by the color of objects. The output from the sensor is an analog signal proportional to the distance. However, it is not a linear output, therefore a look-up table is necessary to derive the distance measured. The output of the digital version of this sensor, GP2D15, is a pulse and its width is proportional to the distance. Figure 3.7b shows the robot using Sharp sensors for distance measurement.

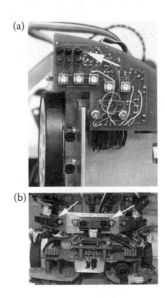

FIGURE 3.7 (a) IR sensor pairs for line following. (b) Robot using Sharp GD2D02 sensors.

3.1.4 DETECTING COLOR

In many robotic competitions, robots are expected to detect and handle objects. The color is a prominent feature that can be used to identify them. For example, in the robot colony game, robots collect pellets and deposit them into bins based on their color. In an intelligent robot game, the robot collects colored objects and deposits them in allocated bins based on their shape and color. A quick solution to color detection is to use infrared sensors as mentioned earlier. The reflected light intensity depends on the object color. This property can be exploited to identify color since the objects used in these competitions are standard. The color of the objects can also be detected using camera and some basic image-processing algorithms. This topic will be discussed in the following chapter. Here, we will consider another low-cost color sensor, which we can consider as a one pixel camera. These sensors are mainly developed for the automation industry, and they are also convenient to use in robotic games. MRGBiCT and MTCSiCT from Mazet, S9706 and the S10942 series sensors from Hamamatsu are some examples of color sensors. The sensor produces analog signals corresponding to red, green, and blue components of a color. Apparently, a wide range of colors can be detected with these three components. However, sensing distance is rather short, about 2–6 cm.

Colors to be detected in robotic games are well specified. In the robot colony game, robots need to identify pellets in blue and green color; therefore, a comparator circuit was enough to complete color recognition in hardware. For example, the color sensor used in robot colony design is the Mazet MTCSiTC (Mazet 2012). The sensor is made of SI-PIN diodes, and it is covered with RGB filters, a microlens array, and an imaging microlens. The four terminals are the RGB signal outputs and a common cathode (5 V). MT104Bx is a complementing device for amplifying the signal. It is necessary to illuminate objects during detection phase for a better performance of the sensor. The signals from the amplifier MTI04B are fed to a comparator circuit to produce a binary value indicating the detection of a particular color as shown in Figure 3.8.

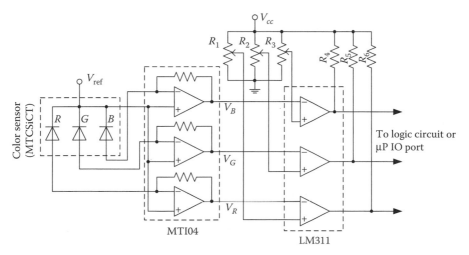

FIGURE 3.8 Color sensor (Mazet MTCSiCT) and circuit to detect pellet color in hardware for robots designed for robot colony game.

Alternatively, if the microcontroller used in the robot design has an ADC port, then RGB signals can be connected to ADC inputs. The color detection is completed in software with look-up tables. This is a much more flexible method, and a wide spectrum of colors can be identified. Refer to videos that show robot competitions that rely on color sensing (Intelligent Robot 2010; Robot Colony 2010).

REFERENCES

Fraden, J. 1996. *Handbook of Modern Sensors*. New York: Springer-Verlag.
Intelligent Robot. 2010. http://www.youtube.com/watch?v=A34vIthvFSY.
Mazet Electronic Engineering and Manufacturing. 2012. http://www.mazet.de.
Okubo, Y., Ye, C., and Borenstein J., 2009. Characterization of the Hokuyo URG-04LX laser rangefinder for mobile robot obstacle negotiation. In *Proceedings of SPIE Conference on Unmanned, Robotic, and Layered Systems*. Orlando, FL, Vol. 7332.
Robot Colony. 2010. http://www.youtube.com/watch?v=_R7iyWzOOzA
Soloman, S. 2009. *Sensors Handbook*. New York; Singapore: McGraw-Hill.

4 Robot Vision

4.1 INTRODUCTION

There is a vast amount of research in the field of computer vision, which has great potential in various fields of engineering, in particular robotics and automation. Computer vision units are effectively used in industrial robotics for inspection and handling of manufactured parts. Most of these systems operate in a controlled environment with good illumination and minimum interference. However, mobile robots mostly operate in unpredictable environments. Typically, a vision unit is employed for steering a robot, avoiding obstacles, detecting landmarks, handling objects of interest, tracking objects, or navigation (Chen and Tsai 1997; DeSouza and Kak 2002; Kosaka and Kak 1992; Kumagai and Emura 2000). In early mobile robots, visual feedback was commonly used for robot navigation and obstacle avoidance. Recently, particularly in service robotics, vision is used for interacting with people, recognizing the environment and handling objects. Needless to say, vision is one of the most desirable features that we would like to have in a robot.

Vision capability can be incorporated in a robot by using a camera and a processor setup. A camera can be treated as another sensory unit for the robot. However, a camera alone will not be enough. To extract useful information from this sensor, a series of sophisticated computations has to be performed on the image data, typically using an on-board processor. There is an immense amount of material available on this topic, which could be overwhelming when a robot designer wants to incorporate vision into robot design. This chapter is solely dedicated to this topic, and our objective is to provide basic information on robot vision for a quick start. We present algorithms and techniques that could be useful for incorporating vision as a sensory input for game-playing robots such as in robot soccer where a vision unit is used for identifying relevant objects in the game such as the ball, opponents, and teammates based on their features.

In the following, we will first look into the available hardware, particularly inexpensive systems, suitable for use in game robotics. We will also discuss how an image is formed and the parameters that affect the quality of the image acquired. We will also present some of the fundamental algorithms and their use in robot vision. The image-processing algorithms that we will consider in this chapter can be classified into two categories. The first category of algorithms deals with pixels only. That is, the input and the output of the algorithms are all pixels. We will refer to them as basic image-processing algorithms or low-level image processing as commonly known in the literature. The second type of algorithms takes pixels as input, but delivers symbolic representations as output. For instance, an algorithm that detects circular objects in the image takes pixels as input, but delivers centroid and radius of circles at the output. The output in this case is not pixels, but a symbolic representation of the object with its parameters. We will consider this type of algorithms as symbolic feature-extracting algorithms or intermediate-level algorithms as commonly known in the computer vision community. Typically, a combination of these algorithms is used to enable a robot to detect certain objects of interest in its vicinity and take action accordingly.

4.2   CAMERA SYSTEMS FOR ROBOTICS

In the early days, hardware for robot vision used to be bulky and costly. This was a major disadvantage in applying artificial vision in robots designed for education or entertainment purposes. With the advances made in very large scale integration (VLSI) technology, it became viable to incorporate vision systems in the design of robots. There are many inexpensive camera systems in the market with an embedded preprocessing unit and with very low power consumption. Figure 4.1 shows examples of such systems. The vision unit of the robot shown in Figure 4.1a is the

(a)

(b)

(c)

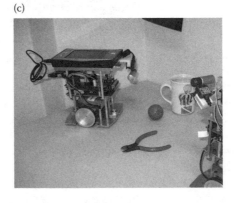

FIGURE 4.1   (a) Humanoid robot with CMU cam. (b) Mobile robot with SRV-1 Blackfin. (c) Autonomous robot with USB camera.

CMUcam, the latest of which is CMUcam4 (CMUCAM 2012), which employs a propeller P8X32A processor with eight built-in parallel microprocessors. This compact vision unit has a resolution of 640 × 480 color pixels, and it can be programmed through a serial port. Another embedded vision system, SRV-1 Blackfin (Surveyor 2012), is shown in Figure 4.1b, and it employs a powerful digital signal processor for vision algorithms. It has a resolution of 640 × 480 with plenty of I/O options. The advantage of these systems is their low cost, small size, and programmability. Their software is open source, that is, the source code and many fundamental image-processing algorithms are readily available. If a robot is controlled by an embedded PC such as PC104 or a notebook computer, a simple webcam can be utilized as a vision unit (Figure 4.1c). In all these systems, a video signal is delivered to the host system through various means such as USB, RS-232, or wireless connection.

4.3 IMAGE FORMATION

Image formation comprises three major components as illustrated in Figure 4.2. The first component is the illumination which makes the environment visible. The second component is the optical system which transmits the illuminated scene onto a sensor unit. The third component is the sensor unit which is made up of a matrix of photo-sensors that responds to the light and transforms the image into an electrical signal.

Illumination: Illumination is the first step, and it directly affects the image quality, which consequently determines the performance of the vision process. In industrial applications, illumination is customized for the application. Back lighting, front lighting, and structured lighting are some of the illumination techniques that are used to improve image quality, enhance the object features, and maximize the signal-to-noise ratio. These illumination techniques improve the reliability of the vision algorithms and reduce computation time by eliminating the need for many preprocessing algorithms to enhance the image quality. On the other hand, mobile robots rarely use such illumination techniques as they operate under available ambient light. It means that

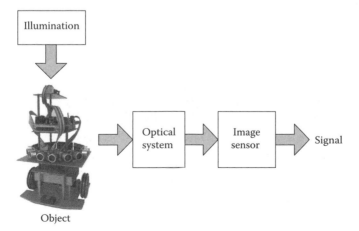

FIGURE 4.2 Three components of image formation.

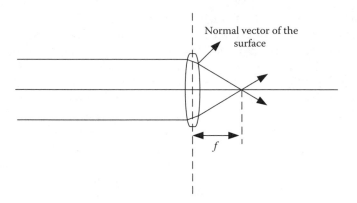

FIGURE 4.3 Path of light rays through a lens.

they are likely to work with images of poor illumination and noise. Therefore, some preprocessing is required to improve the image quality in software.

Optical system: A lens is the common optical tool for focusing an image on the sensor array. An important parameter associated with this optical system is its focal length. The focusing property of a lens is the result of light waves having a higher velocity of propagation in air than in glass or other optical material. The lens converges the light rays on a point known as focal point (see Figure 4.3). The distance from the lens to the focal point f is also known as the focal length. In other words, the focal length is an indicator of the convergence power of a lens. A smaller f value indicates more severe convergence; it also means a wider angle of view. For instance, a 28 mm lens has a wider angle of view then a 50 mm lens. If an imaging sensor or photographic film is placed at the focal point as shown in Figure 4.3, then we can get a sharp image of the distant object. However, if an object is at near distance to the lens as shown in Figure 4.4, then a relation to the focal length is defined in terms of distance of the lens to the object, D_o, and distance of the lens to the image plane, D_i, using the Gaussian lens formula

$$\frac{1}{f} = \frac{1}{D_i} + \frac{1}{D_o} \tag{4.1}$$

Assume that a camera system is using a 35 mm lens. If the object is placed at a distance very far from the lens, that is, $D_o = \infty$, then the image distance will be at the focal point $f = D_i$. However, if the object is at 1 m distance, then we get

$$\frac{1}{35\,\text{mm}} = \frac{1}{D_i} + \frac{1}{1000\,\text{mm}} \quad \text{or} \quad D_i = 36.26\,\text{mm}$$

This result implies that the lens has to be shifted 1.26 mm away from the sensor so that a sharp image of it appears on the sensor surface. The cameras that we intend to use in our robotic applications are simple, inexpensive cameras; naturally, they do not have a built-in autofocus mechanism. In other words, the lens has to be focused

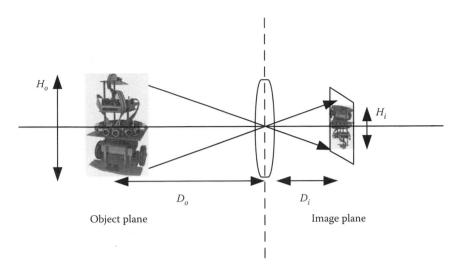

FIGURE 4.4 Magnification factor of lens.

manually up to a certain range once, and it will remain the same during the entire operation of the robot whether it is facing an object at a 10 or 100 cm distance.

In Figure 4.4, we can see that the size of the object image formed on the sensor is determined by the focal length of the lens. A proper focal length depends on the sensor size. For instance, typical sensor sizes used in charge-coupled device (CCD) or complementary metal–oxide–semiconductor (CMOS) cameras are 4.9×3.7 and 8.8×6.6 mm. Now, let us compare these sensors with a full frame sensor, which is the size of photographic film of 36×24 mm, used in the latest DSLR cameras. Apparently, a lens used in a CCD or CMOS camera has to converge a lot more to cover the same area of a full frame DSLR lens. For instance, the focal length of a lens, equivalent to 18 mm with 90° view angle, used in a photographic camera should be a lot smaller in CCD or CMOS cameras used in our application. In this case, a proper lens can be chosen using the magnification ratio, which is defined as

$$M = \frac{H_i}{H_o} \quad \text{or} \quad M = \frac{D_i}{D_o} \tag{4.2}$$

Here, H_o indicates the size of the real object and H_i indicates the size of its image. By substituting the magnification ratio in Equation 4.1, we get

$$f = \frac{D_o M}{1 + M} \tag{4.3}$$

If M is considerably <1, then we can approximate

$$f \approx \frac{D_o H_i}{H_o} \tag{4.4}$$

The above formula can be used to calculate a suitable lens. For example, the sensor size of SRV-1 Blackfin (Surveyor 2012) used in the robot shown in Figure 4.1b is 4.14×3.29 mm. If a 30 cm tall object is to be imaged on this sensor from a distance of 50 cm, the magnification ratio will be

$$M = \frac{3.29}{300} = 0.011$$

and the focal length will be

$$f = \frac{500 \times 0.011}{1 + 0.011} = 5.49 \text{ mm}$$

A suitable lens should be <5.5 mm. The default lens on SRV-1 is 3.6 mm; hence, the object will fit into the image frame.

The other parameters that play an important role in image formation are relative aperture, f-number, and depth of field. The light-gathering capabilities of a lens are determined by the f-number, which is the ratio of the focal point and the lens aperture (largest usable diameter). Lenses manufactured for photography come with a pupil mechanism assembled in front of them, and the amount of light that will pass through the lens depends on the diameter of the pupil mechanism (for a fixed focal point). For example, a 50 mm lens with an aperture setting of 4 will have a pupil diameter of 12.5 mm as calculated below:

$$f_{num} = \frac{f}{D}, \quad 4 = \frac{50}{D} \quad \text{hence } D = 12.5$$

The light-gathering capability of a circular lens is proportional to the square of its diameter. Therefore, to receive twice the amount of light for a fixed focal point, the aperture has to be $\sqrt{2}$ or 1.141 times larger (Hecht 2002). So, typical f-numbers are a sequence of the powers of $\sqrt{2}$, that is, $f / \sqrt{2}^0, f / \sqrt{2}^1, f / \sqrt{2}^2, f / \sqrt{2}^3, \ldots$. For convenience, these numbers are approximated as 1, 1.4, 2, 2.8, 4, 5.6, $\ldots$. The lower f-number means a larger aperture; hence, more light will pass through the optical system. In contrast, higher f-numbers will limit the amount of light passing through the lens. Another effect of changing the aperture is the depth of the field, or the difference between the farthest and nearest points in focus. Simply, the depth of field is larger for a shorter focal length and for a higher f-number. By varying the f-number, we are able to manipulate the depth of the focused area in the image as well as the contrast between the object and the background. However, the simple camera systems utilized in game robotics usually come with a fixed f-number allowing a suitable depth of field.

Image sensing: The image-sensing unit is a solid-state sensor array that transforms the image formed by the optical system into an electrical signal. The sensor array is made of photosensitive sensors, namely pixels. The number of pixels that are fit into a sensor array naturally affects the resolution and the quality of the image obtained. The pixel size of the sensors used in the latest digital cameras reached 7360×4912 pixels (approximately 36 megapixels). A large number of pixels naturally means more computation time and large memory requirement. The typical camera resolution used in robotics application is about 640×480 pixels.

CCD image sensors have long been the dominant imaging sensor in most of the state-of-the-art vision units. A CCD image sensor is simply a matrix of photosensors, with adequate circuitry, which brings out the signal from each column of pixels in serial form. Recently, CMOS image sensors have been developed drastically. CMOS sensors are now widely used in consumer electronics, such as digital cameras and mobile phones. Consequently, they are also used in robot vision. The operation of these sensors basically has two phases. The first stage is the charging of individual sensors proportional to the intensity of light focused on them, and the second stage is to transfer the charge information to the camera output as an electrical signal. A major advantage of CMOS sensors is that they do not need a clock synchronization and relevant circuitry to deliver image data since each pixel data is transmitted in parallel. They also consume significantly less power, which is favorable in robotics. It is obvious that future developments in image-sensing technology will only benefit robot vision applications more. For instance, color image processing was not common in early systems because of the high cost. Recent advances in imaging technology brought color image processing into robotics. Since color is one of the distinguishing features of objects, in the following, we will also discuss color detection and tracking techniques, which are very useful in robotic games.

4.4   DIGITAL IMAGE-PROCESSING BASICS

A visual scene is a continuous function of reflectance, which is an analog quantity. Such representation of image cannot be processed in a computer since computers are intrinsically discrete. A digital image on the other hand represents the scene in a sampled and quantized form. Sampling is naturally done at the image sensor during the sensing, and the scene is divided into a matrix of pixels. The sampling density, that is, the number of sampling points per unit measure, is the spatial resolution and is usually measured in terms of the number of pixels in both horizontal and vertical directions such as 640×480 pixels as mentioned earlier. Quantization is done while converting the electrical charge of photosensors into integer values using analog-to-digital converters. The resulting digital image is a matrix of numbers representing the scene in terms of gray levels, RGB, YUV, or HSV color formats. In summary, a digital image is a two-dimensional array of values representing the reflectance function of the actual scene. In the following sections, digital images will be expressed in a discrete function form of $f(x,y)$.

4.4.1   COLOR AND COLOR MODELS

Color is an important feature that makes the identification of objects and shapes easier in robotic games. It also provides plenty of information that can be used to analyze an image. The color of objects is primarily determined by their reflectance properties. Our sensors (either human eye or CMOS sensors) are able to respond to reflected light rays.

A color model is a standard way of describing colors, and they are typically defined in a three-dimensional coordinate system. Three-dimensional space also shows all the possible colors that can be constructed by mixing primary colors from three axes. Color models are regulated toward specific hardware or software applications.

RGB: The RGB model is one of the most common color model employed in many digital cameras, color monitors, and most video cameras. In this model, an image is made of three independent image planes. These three image planes represent the primary colors: red, green, and blue. All the colors in the image are defined by combining these three primary colors. The Cartesian coordinate system shown in Figure 4.5 represents the geometry of the RGB color model for specifying colors. A color is made up of different amounts from these three axes. For example, grayscale lies on the line that connects white and black points. As it can be seen in the figure, all the color components have the same magnitude along this line.

Grayscale: A digital RGB image is composed of three matrices to represent a scene where each matrix represents one primary color channel. Naturally, a color image requires large memory space, and processing them requires longer computation times. Therefore, many robot vision systems still use monochrome images. The light intensity is represented with different shades of gray. The number of gray levels in grayscale is called the gray-level resolution of the system, and it is bounded by two gray levels, black and white. Black and white correspond to the minimum and the maximum measurable intensity level, respectively. Typically, 256 discrete gray levels are used, each of which can be easily represented by a single byte; hence, a monochrome image needs a single matrix of grayscale values. However, if the captured image is in color, then it can be converted from the RGB image to grayscale image by simply averaging three color components as follows:

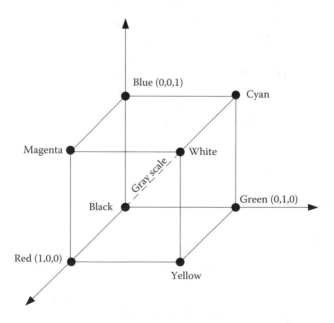

FIGURE 4.5   The RGB color cube where each axis represents one primary color.

$$I_{\text{gray}}(x,y) = \frac{R(x,y) + G(x,y) + B(x,y)}{3} \qquad (4.5a)$$

Another method of grayscale conversion gives more weight to the green component (since the human eye is more sensitive to green) then others when averaging:

$$I_{\text{gray}}(x,y) = 0.299R(x,y) + 0.587G(x,y) + 0.114B(x,y) \qquad (4.5b)$$

Equation 4.5b simply gives a measurement of brightness and would be suitable if brightness is a feature to be exploited.

YUV: When the vision application is based on analyzing colors, it needs to be robust against alterations in illumination. In that case, using YUV color space is more practical since the color components and the illumination are represented separately. The U and V channel represent color, and the Y channel represents brightness. The conversion between the RGB and the YUV color space is also defined with a linear transformation:

$$Y(x,y) = 0.299R(x,y) + 0.587G(x,y) + 0.114B(x,y)$$
$$U(x,y) = -0.147R(x,y) - 0.289G(x,y) + 0.436B(x,y) \qquad (4.6)$$
$$V(x,y) = 0.615R(x,y) - 0.514G(x,y) - 0.101B(x,y)$$

HSV: Recently, the HSV color model has been more widely used in robotics, in particular when identifying objects based on their colors. It is effective in filtering out unreliable color information in low illumination or low saturation areas of the image (Cheng and Sun 2000). The color space transformation from RGB to HSV can be done with an algorithm as described in Russ (2002).

The conversion between color spaces can be done simply using MATLAB® functions. For instance, "rgb2hsv" converts RGB to HSV. Similarly, the function "rgb2gray" converts RGB color values to grayscale, and the function "rgb2ycbcr" converts to YCbCr color space (or YUV color space).

4.5 BASIC IMAGE-PROCESSING OPERATIONS

The very first step of the vision process deals directly with pixels. The objective of such operations would be to improve the quality of image (such as increasing the contrast, reducing the noise level, etc.) or finding pixels that contain some object features such as corners, colors, edges, and so on. As mentioned, a distinct feature of the basic image-processing algorithms is that the input of the algorithms and the outputs are both image pixels.

The basic image-processing techniques broadly manipulate image in the spatial or frequency domain. The spatial domain algorithms deal with image itself and operate on pixels directly. The frequency domain operations require the image to be transformed to the frequency domain by Fourier transform. In the following, we will present spatial domain operators as they are commonly used in robot vision. However, sometimes the frequency domain methods are preferred to speed up computation time, especially for large images.

The spatial operators can be point type or neighborhood type. The point-type operators deal with individual pixels such as thresholding and contrast stretching. The neighborhood operations, also known as group operations, make use of immediate neighbors of a processed pixel. This group of operators frequently employs a convolution mask (other common names for convolution mask are templates, windows, and filters). In the following sections, we will present some of the common spatial domain operators and algorithms used in image processing. However, an important mathematical tool used in these operations is two-dimensional convolution, and we will describe it first.

4.5.1 CONVOLUTION

The convolution operation is defined by the following function:

$$g(x, y) = \sum_{m=-(M-1)/2}^{+(M-1)/2} \sum_{n=-(M-1)/2}^{+(M-1)/2} f(x + m, y + n) \times h(m, n) \tag{4.7}$$

Here, f is the image matrix, h is the convolution mask (or template), and g is the resulting image and they are all two-dimensional arrays. By observing the above equation, we see that matrix h scans through the entire image while performing a series of multiply and add operations. Figure 4.6 illustrates the convolution operation graphically. The template used in this illustration is a 3×3 matrix. It is superimposed upon the image, and each image pixel is multiplied with the corresponding weights in the mask. The resulting nine values from these multiplications are summed to produce the new pixel value for the output image. Referring to Figure 4.6, the new value for the pixel at position $x = 1$ and $y = 1$ is $11 + 7 + 10 - 3 - 6 - 7 = 12$. The operation continues by shifting the mask to the next pixel and by scanning the entire image sequentially.

From the figure, we can see that if we align the center of the template with top left most corner of the image, part of the template will fall outside the image boundaries. This problem will arise when calculating convolution at image borders. The typical

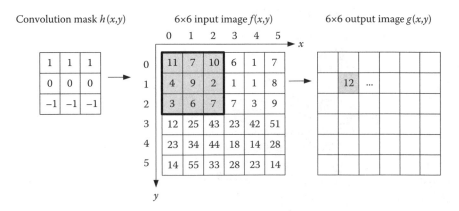

FIGURE 4.6   Two-dimensional convolution of image with a mask.

solution for this is to leave the borders blank. Another approach is to assume that the image is periodic and it gets the missing pixel values from a cyclic shift from the opposite border. For the example image in Figure 4.6, convolution starts from $x = 1$ and $y = 1$ and finishes at $x = 4$ and $y = 4$. For the larger templates, we may need to leave out border pixels accordingly. For instance, if the template is 5×5, then two rows and columns of border pixels are left blank.

The program shown in Figure 4.7 is a rough illustration of convolution operation in MATLAB. It can be easily translated to C++ or any other programming language. In MATLAB, two-dimensional convolution can also be done by simply calling the "conv2" or "imfilter" functions.

```matlab
%2D convolution example
% 'IMG_IN' is input image
% 'h' is template matrix


function IMG_OUT=convolute2D (IMG_IN,h)
[row,col]=size (IMG_IN)                    %Get image size
IMG_OUT=zeros(row,col)                     %Create array for result
[ydim,xdim]=size (h)                       %Get template size
border=floor(ydim/2)
offset=ydim-border;
sum=0.0;
for x=1:xdim                               %sum convolution template
    for y=1:ydim
    sum=sum+h(y,x);
    end
end
norm=sum;
IMG_IN=double(IMG_IN);                     %Calculations need double
for i=(1+border):(row-border)
    for j=(1+border):(col-border)
        sum=0.0;
        for x=1:xdim
            for y=1:ydim
            sum=sum+h(y,x)*IMG_IN(i+(y-offset),j+(x-offset));
            end
        end
        IMG_OUT(i,j)=sum;
    end
end
if (norm>0)                                %if template sum is > 0
    IMG_OUT=IMG_OUT/norm;                  %normalize the result
end
IMG_OUT=int8(IMG_OUT);
end
```

FIGURE 4.7   Convolution of image with a mask.

4.5.2 SMOOTHING FILTERS

Smoothing eliminates noise and other fine changes and variations in the image due to quantization, environmental effects, or poor data-acquiring conditions. It is necessary to eliminate fine details in the image, particularly when we are searching large shapes and forms in the image.

Mean filter: Mean filter, also known as moving average filter, is the simplest approach in image smoothing. In this operation, a pixel is replaced with the average of the pixels in its $m \times m$ neighborhood. A mean filter can be realized using convolution operator using a template with equal weights. For example, a 3×3 template for mean filter is defined as

$$h(x,y) = \begin{bmatrix} 1/9 & 1/9 & 1/9 \\ 1/9 & 1/9 & 1/9 \\ 1/9 & 1/9 & 1/9 \end{bmatrix} \tag{4.8}$$

Convoluting the image with this mask blurs the image and suppresses the noise. Consequently, it also erodes fine details in the image; since it is as a low-pass filter. For more severe blurring effects, the dimension of mask size, m, is increased.

Gaussian filter: The Gaussian filter employs Gaussian function in calculating weights in filter kernel. The mask obtained in this method has higher weights for pixels at the center pixel and lower weights at the edges. The Gaussian function is defined as

$$G(x,y) = \frac{1}{\sqrt{2\pi}\sigma} e^{-\frac{x^2+y^2}{2\sigma^2}} \tag{4.9}$$

where x,y are the mask coordinates and σ is a standard deviation of the Gaussian distribution. Once the filter mask is determined using Equation 4.9, it can be implemented on the image using convolution operator. The template for a 5×5 Gaussian filter is given as

$$h(x,y) = \frac{1}{53} \begin{bmatrix} 0 & 1 & 2 & 1 & 0 \\ 1 & 3 & 4 & 3 & 1 \\ 2 & 4 & 9 & 4 & 2 \\ 1 & 3 & 4 & 3 & 1 \\ 0 & 1 & 2 & 1 & 0 \end{bmatrix} \tag{4.10}$$

In robotic application, edges of objects, obstacles, or a robot path can be detected more reliably after a Gaussian smoothing. More explicitly, when the application requires only the global edges to be detected, then standard deviation is increased. For example, Figure 4.8 shows the resulting effect of smoothing filters for a test image. An edge-detection filter, which will be discussed in the following sections, can detect the edges of objects. Figure 4.8b shows the result of an edge detection algorithm where white pixels indicate edge points over the image and in this example edge detection produced many minor details, and it is not needed. However, after

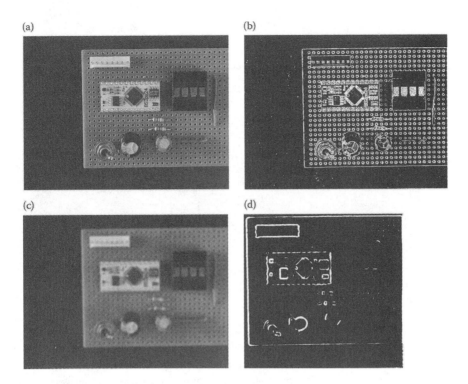

FIGURE 4.8   An application of Gaussian smoothing. (a) Original image. (b) The result of edge detection. (c) Smoothed image with 13×13 Gaussian filter with $\sigma = 4.5$. (d) Edge detection after Gaussian smoothing where only major object boundaries are detected.

smoothing the image with a Gaussian filter and then applying the edge detector, we get the major boundaries of the object.

Median filter: In some applications, we would like to eliminate noise while keeping the sharp edges and fine details of the objects in the image. The mean filter and Gaussian filters blur the image, while eroding the sharp edges and other details. To overcome this problem, a median filter is employed. The median filter is a nonlinear approach, and it reduces the noise with minimal effect on the edge pixels. When we implement a median filter of $m \times m$ neighborhood, we sort all the pixels in ascending or descending order and take the central one as the new value for the resulting image. Figure 4.9 depicts an example to 3×3 median filtering. Sorting pixels in the shaded area will result in {11, 10, 9, 8, 7, 6, 4, 3, 2}, then the pixel value replacing the center pixel will be 7. This operation will continue by scanning the entire image from left to right and top to bottom. Figure 4.10 shows a MATLAB code for comparing all the image smoothing filters described above.

4.6   ALGORITHMS FOR FEATURE EXTRACTION

To identify objects in a scene, we need to identify certain features of them. The main objective of this section is to describe some of the essential algorithms that can

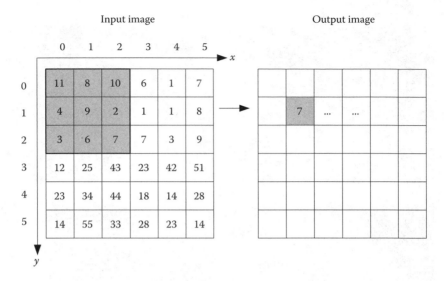

FIGURE 4.9 Example of 3 × 3 median filter.

deliver object features in an image. There are plenty of algorithms that fall into this category. Choosing a proper set of algorithms is the role of the application developer. Basic algorithms like thresholding can be used for segmenting an image into object and background; on the other hand, more complex algorithms can deliver specific object features such as corners, lines, circles, and so on.

4.6.1 THRESHOLDING

Thresholding is one of the most basic image-processing operations. It is also an elementary tool used in image segmentation. It segregates the image into uniform regions, mainly to mark objects or features and background pixels, based on some threshold value. The algorithm scans the image. If a pixel is greater than the threshold value, T, it is marked as "1" to indicate the object or it is marked as "0" to indicate the background or vice versa. The output of a threshold operator is a binary picture containing two levels of intensity. Briefly, we create a threshold image by defining

$$g(x,y) = \begin{cases} 1 & f(x,y) \geq T \\ 0 & f(x,y) < T \end{cases} \tag{4.11}$$

The above definition can be quickly altered to separate a range of gray values between T_1 and T_2 in an image. This is known as multilevel thresholding and is described as follows:

$$g(x,y) = \begin{cases} 0 & 0 \geq f(x,y) > T_1 \\ 1 & T_1 \geq f(x,y) \geq T_2 \\ 0 & f(x,y) > T_2 \end{cases} \tag{4.12}$$

```
IMG = imread( 'testIMG.jpg' );              %open a test image
mask_ave=[1 1 1 1 1;
          1 1 1 1 1;
          1 1 1 1 1;
          1 1 1 1 1;
          1 1 1 1 1];
mask_ave=mask_ave/25;
mask_gauss=[0 1 2 1 0;
            1 3 4 3 1;
            2 4 9 4 2;
            1 3 4 3 1;
            0 1 2 1 0];
result_ave=imfilter(IMG,mask_ave);
result_gau=imfilter(IMG,mask_gauss);
result_med=medfilt2(IMG, [5 5]);
% Display the original image and filtered images.
subplot(2,2,1);
imshow(IMG);
title('Original Image');
subplot(2,2,2);
imshow(result_ave);
title('average filter');
subplot(2, 2,3);
imshow(result_med);
title('gaussian filter');
subplot(2,2,4);
imshow(result_med);
title('median Image');
```

FIGURE 4.10 MATLAB code for smoothing filters.

An important issue in thresholding is to choose a proper threshold value to optimally segregate an object from the background. Often, a single global threshold value is applied to the whole image. However, objects in the image may not have uniform intensity due to various reasons such as poor illumination. In this case, a global threshold value may not produce a good result. To overcome this problem, there are various techniques developed, one of them being adaptive thresholding (Haralick and Shapiro 1992). In adaptive thresholding, a threshold value for each pixel in the image is calculated. There are a number of ways to calculate this threshold value, although many of them are based on using smaller overlapping image regions. It is more likely that smaller image regions will have nearly uniform illumination. One of the adaptive algorithms is known as the Chow and Kaneko method (Chow and Kaneko 1972). After dividing the image into subimages, histograms of these subimages are analyzed, and an optimum threshold is obtained for each region. Since the subimages overlap, a threshold value for each pixel is then obtained by interpolating the thresholds found for the subimages. Another approach, which is a lot less computationally intensive, is by a statistical analysis of the local neighborhood of each pixel. The calculation of threshold value based on this statistical analysis can be as simple as the mean, median, or the average of the maximum and minimum values of the local intensity distributions.

Figure 4.11 shows the implementation of the threshold operator to segregate object from background. The available MATLAB command for this purpose is "im2bw." Figure 4.12 shows a test image and the output image after thresholding.

```
% Function: Threshold operator

% 'IMGgray' is input image

% 'IM_BW' is resulting binary image

% 'T' threshold value


function IM_BW=threshold(IMGgray, T)


[row,col]=size (IMGgray);                    %Get image size

IM_BW=zeros(row,col);                        %Create blank image

for i=1:row

    for j=1:col

        if (IMGgray(i,j) > T) IM_BW(i,j)=255;

        else IM_BW(i,j)=0;

        end

    end

end

end
```

FIGURE 4.11   MATLAB code for threshold operation.

(a)   (b)

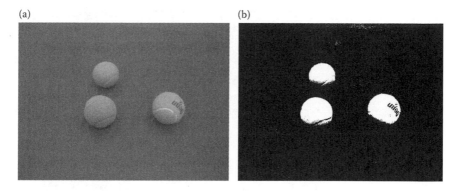

FIGURE 4.12   (a) Original image. (b) Thresholded with a global value at $T = 128$.

4.6.2 EDGE DETECTION

Edge detection is one of the fundamental operators in image processing and computer vision. Typically, in edge detection, the gradient of image, which is change in intensity, is measured as an indicator of edges using a "gradient operator." The basic idea employed in most edge-detection techniques is the computation of a local derivative operator. Ideally, an edge can be modeled as a step function although this is unlikely for natural images. Therefore, an edge is usually modeled as a ramp. The first-order derivative of a region with uniform intensity will be zero. If there is a change in intensity, then the result of the first-order derivative will be nonzero. If we calculate the second-order derivative, the result will be nonzero at the beginning and at the end of an intensity transition. Referring to the magnitude of the first-order derivative, we can detect the presence of an edge. Utilizing the second-order derivative, we can determine the direction of change as well. The gradient of image is defined as

$$G(f(x,y)) = \begin{bmatrix} G_x \\ G_y \end{bmatrix} = \begin{bmatrix} \dfrac{\delta f}{\delta x} \\ \dfrac{\delta f}{\delta y} \end{bmatrix} \tag{4.13}$$

To identify edges, calculating the magnitude of this vector G is satisfactory.

$$G(f(x,y)) = \sqrt{\left|G_x^2 + G_y^2\right|} \tag{4.14}$$

Gradients G_x and G_y can be obtained by convoluting the image with two masks known as Sobel operators, and they are illustrated in Figure 4.13 (see Fu et al. (1987) and Gonzales and Woods (1992) for further reading on this topic).

The edge-detection algorithm based on the second-order derivative of the image is known as the Laplacian operator. The second-order derivative of the image $f(x,y)$

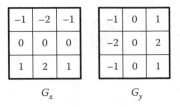

FIGURE 4.13 Sobel operators.

with respect to x and y directions is defined as

$$\nabla^2 f(x,y) = \frac{\delta^2 f(x,y)}{\delta x^2} + \frac{\delta^2 f(x,y)}{\delta y^2} \tag{4.15}$$

By definition, the first-order derivatives in the x and y directions, considering that the smallest increment in x and y direction is 1 unit, will be as follows:

$$\frac{\delta f(x,y)}{\delta x} = \frac{f(x+1,y) - f(x,y)}{1}$$

$$\frac{\delta f(x,y)}{\delta y} = \frac{f(x,y+1) - f(x,y)}{1} \tag{4.16}$$

The second-order derivative of image $f(x,y)$ in the x direction will be

$$\frac{\delta^2 f(x,y)}{\delta x^2} = \frac{f'(x+1,y) - f'(x,y)}{1} = \frac{[f(x+2,y) - f(x+1,y)] - [f(x+1,y) - f(x,y)]}{1}$$

$$\frac{\delta^2 f(x,y)}{\delta x^2} = \frac{[f(x+2,y) - 2f(x+1,y) + f(x,y)]}{1} \tag{4.17}$$

Similarly, the second-order derivative for the y direction will be

$$\frac{\delta^2 f(x,y)}{\delta y^2} = \frac{[f(x,y+2) - 2f(x,y+1) + f(x,y)]}{1} \tag{4.18}$$

Using the shifting property, we replace $x \to x-1$ $x+1 \to x$ $x+2 \to x+1$ and $y \to y-1$ $y+1 \to y$ $y+2 \to y+1$. We now obtain

$$\frac{\delta^2 f(x,y)}{\delta y^2} = f(x+1,y) - 4f(x,y) + f(x-1,y) + f(x,y+1) + f(x,y-1) \tag{4.19}$$

Computation of the second-order derivative of a pixel at position x and y is defined with Equation 4.18. This equation can be transformed in a convolution mask as

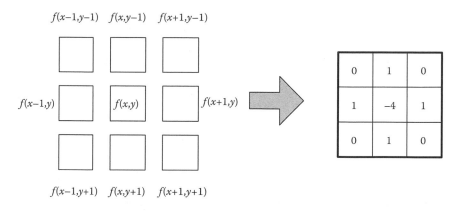

FIGURE 4.14   Pixel coordinates and corresponding mask for the Laplacian operator.

illustrated in Figure 4.14. The Laplace edge detector requires only one mask and finds edge pixels in a single convolution.

The algorithms described above give a starting point for edge-detection techniques. Other common edge detectors are Prewitt's edge detector, Robert's edge detector, and Canny's edge detector (Gonzales and Woods 1992). The MATLAB code in Figure 4.15 illustrates edge detection using Sobel operators. The output of the algorithm for a test image is also shown in Figure 4.16. The same result can be achieved using the "edge" function from MATLAB image-processing library.

4.6.3 COLOR DETECTION

Color is one of the important features of objects, and it is highly utilized in game robotics. For example, when a soccer robot is chasing a ball or a humanoid robot is performing a penalty kick, exploiting the color feature of the objects is very convenient. A straightforward method of color detection is done by using HSV color space. Normally, an image is captured in RGB color space first then converted to HSV color space. Hue specifies the intrinsic color, saturation defines the purity of the color, and value gives a measure of the brightness of the color. Subsequently, the detection of a particular color is performed by referring to the hue and value components of the object. Pixels within a certain range of object hue and value are marked as white pixels, and the remaining pixels are marked as black. Figure 4.17 shows an example of MATLAB code for color detection. The HSV values of the tennis ball are measured as $H = 0.33$, $S = 0.38$, and $V = 0.96$. Threshold values are chosen as ±30% of hue and value components. Figure 4.18 shows the detection result.

Commonly, inexpensive and less powerful processors are used for image processing in game robotics. The processor speed being a constraint, a fast method for color segmentation is necessary. A good example of such an algorithm is given in Leclercq and Bräunl (2001) where the authors present a color segmentation and an object localization method using a look-up table to speed up pixel classification in color classes.

```
% Function: Sobel edge detector

% 'IMGgray' is input image

% 'IM_BW' is resulting binary image


function IM_BW=edgeSobel(IMGgray)


sobel1=[-1 -2 -1; 0 0 0; 1 2 1];

sobel2=[-1 0 1; -2 0 2; -1 0 1];


s1=conv2(IMGgray,sobel1);              %Apply Sobel operators

s2=conv2(IMGgray,sobel2);

s3=s1.*s1+s2.*s2;                      %calculate gradient

s3=sqrt(s3);

s3=uint8(s3);


[row,col]=size (s3);                   %Create blank image

IM_BW=zeros(row,col);

for i=1:row

    for j=1:col

        if (s3(i,j) > 80) IM_BW(i,j)=255;  %identify strong edge points

        else IM_BW(i,j)=0;

        end

    end

end

end
```

FIGURE 4.15   MATLAB code for implementing Sobel edge detector.

(a)   (b)

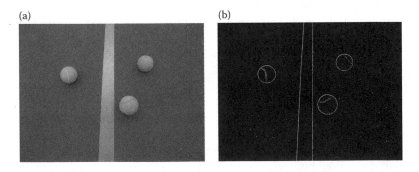

FIGURE 4.16   An example of edge detection with Sobel operators. (a) Original picture.
(b) Result after edge detection.

```
IMG = imread( 'Ball_13.jpg' );            %Open test image

IMGhsv = rgb2hsv( IMG );                   %Convert image to HSV


IMGhue = IMGhsv( : , : , 1 );              %Separate hue

IMGsat = IMGhsv( : , : , 2 );              %Separate saturation

IMGval = IMGhsv( : , : , 3 );              %Separate value


%Thresholds values for Hue and Value for the yellow colour

hueThresholdLow = 0.1;

hueThresholdHigh = 0.3;

valueThresholdLow = 0.6;

valueThresholdHigh = 0.9;


hueMask = (IMGhue >= hueThresholdLow) & (IMGhue <= hueThresholdHigh);

valueMask = (IMGval >= valueThresholdLow) & (IMGval <= valueThresholdHigh);

yellowObjectMask = uint8(hueMask & valueMask);


maskedImageR = yellowObjectMask .* IMG(:,:,1);

maskedImageG = yellowObjectMask .* IMG(:,:,2);

maskedImageB = yellowObjectMask .* IMG(:,:,3);

maskedRGBImage = cat(3, maskedImageR, maskedImageG, maskedImageB);


% Display all the images.

subplot(3,2,1);

imshow(IMG);

title('Original Image');

subplot(3,2, 2);

imshow(IMGhue);

title('Hue Image');

subplot(3, 2,3);

imshow(IMGsat);

title('Saturation Image');
```

FIGURE 4.17   MATLAB code for detecting colors using HSV color space.

```
subplot(3,2,4);

imshow(IMGval);

title('Value Image');

subplot(3, 2,5);

imshow(yellowObjectMask, []);

title('Yellow pixels detected');

subplot(3,2,6);

imshow(maskedRGBImage);

title('Show only yellow object in the image');
```

FIGURE 4.17   (continued) MATLAB code for detecting colors using HSV color space.

(a)   (b)

FIGURE 4.18   (a) Test image with yellow- and orange-colored tennis balls. (b) Detected yellow object in the image.

4.7 SYMBOLIC FEATURE EXTRACTION METHODS

A common property of the algorithms discussed so far is that they take an image as input and produce another image as output. The pixels in the output image indicate certain object features detected in the input image. For example, nonblack pixels in Figure 4.18b imply that a yellow-colored object is detected in that region. Similarly, white pixels in Figure 4.16 indicate the edges of objects found in the image. However, these results do not say much about intrinsic object features. For example, on the basis of these results, a robot will not know whether the objects are circles or straight lines, their location, their size, and any other useful information. There are more elegant algorithms that can deliver such features. Consequently, the output of these algorithms is not pixels, but object parameters in symbolic form. In the following, we will describe the Hough transform algorithm, which is a popular method used for detecting useful object features.

4.7.1 HOUGH TRANSFORM

The Hough transform identifies linear line segments (Hough 1962). It has been extended to detect circular shapes in an image and any other arbitrary shape that can be represented by a set of parameters (Ballard 1981). In general, the computation of Hough transform has two phases. The first phase is a voting process for collecting evidence. The result of voting is accumulated in a parameter space. In the second phase, parameter space is elaborated and strong candidates are selected as objects in the image.

We first look into a well-known line detection algorithm. The voting phase involves calculating the prospective line candidates, which are represented in terms of parameters. The typical Hough transform method employs a polar form of lines. Figure 4.19a illustrates a point $p1$, which is along a straight line in a given image. This point can be defined in polar form by

$$r = x \cos \theta + y \sin \theta \tag{4.20}$$

Here, θ is the angle of the line normal to the line segment in the image, and r is the distance to origin. It is obvious that the range for θ is up to 180° and r is limited to $r = \sqrt{n^2 + m^2}$ for $n \times m$ image. During the voting phase, a point along the line will map to a curve in parameter space using Equation 4.20. Two points along the same line will intersect at the parameters space. This intersection point also describes the line that is connecting them. As shown in Figure 4.19b, the intersection point is now having two votes for points $p1$ and $p2$. When we continue these calculations for all the points along the line segment in the image, we will observe a large vote count in the parameter space at the intersection point, which clearly implies a linear line segment and its parameters.

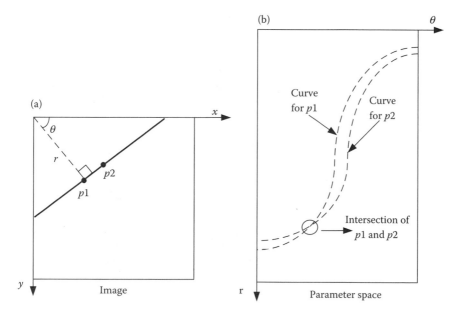

FIGURE 4.19   (a) A line in the image and polar form representation in Hough transform. (b) Curves build up in parameter space due to the voting process for each pixel along the line.

```
% Function: Hough Transform for lines
% 'image' is original grey image
% 'edge_image' is binary image after an edge detection


function image=houghTF_lines (image,edge_image)
[rows,columns]=size(edge_image);
rho=ceil(sqrt(rows^2+columns^2));        %calculate maximum r value
H_space=zeros(rho,180);                  %create accumulator space


for k=1:rows
    for l=1:columns
        if (edge_image(k,l)==1)          %check if it's an edge pixel
            for i=1:180
                th= i*180.0/(180-1)-90.0; %displace so that range +/-90
                th=th*180/pi;
                r=k*cos(th)+l*sin(th);    %find r value
                r=(((r+rho)/(rho*2.0))*(rho-1))+0.5;
%normalize/displace
                temp=floor(r);         % and find position in Hough space
                j=temp;
                H_space(j,i)=H_space(j,i)+1;
            end
        end
    end
end


figure, surfl(H_space);                  %Display Hough space
shading interp
colormap(gray);
```

FIGURE 4.20 MATLAB code for Hough transform.

```
max_value=max(H_space);

peak=max(max_value);


for k=1:rho

    for angle=1:180

        y=0;

        if (H_space(k,angle)>0.60*peak)    %select strong candidates

            for i=1:rows                   %draw corresponding line over

            r=(k*2.0*rho/(rho-1)-rho);     %original image

            th=angle*180/(180-1)-90;

            th=th*180/pi;

            if (sin(th)==0)

                y=y+1;

            else y=(r-(i*cos(th)))/sin(th);

            end

                y=y+0.5;

                j=floor(y);

            if (j>0 && j<columns)

                image(i,j)=255;

            end

                end

        end

    end

end

end
```

FIGURE 4.20 (continued) MATLAB code for Hough transform.

The MATLAB code for the Hough transform is given in Figure 4.20, which is adapted from C language code given in Pitas (2000). The calculation of parameters and mapping into an accumulator array is done by Hough function call. Once the accumulator array is filled, the line segment extraction is done by simply thresholding the accumulator array and selecting strong peaks as linear line candidates. To illustrate better, they are superimposed to the image in the second part of the program. An example image and the processing result are also shown in Figure 4.21.

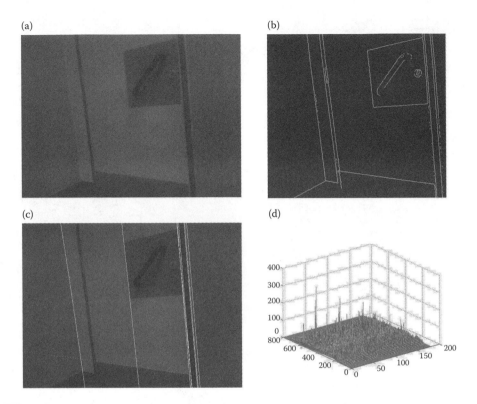

(a)

(b)

(c)

(d)

FIGURE 4.21 (a) Original picture. (b) Edge picture. (c) Strong candidates for linear line segments detected with Hough transform and superimposed on the image. (d) Hough space.

The Hough transform algorithm presented above is also extended for detecting circles in an image. The equation for a circle is given by

$$r^2 = (x - x_0)^2 + (y - y_0)^2 \tag{4.21}$$

The above equation defines the locus of points with a radius of r from a given center point (x_0, y_0). The first phase of the Hough transform is the same as before, but this time voting is done for the above equation, which is defined in a parametric form as

$$x_0 = x - r\cos(\theta)$$
$$\tag{4.22}$$
$$y_0 = x - r\sin(\theta)$$

The MATLAB code shown in Figure 4.22 illustrates the two phases of Hough transform calculations for detecting an example circle of 38 pixels radius. In the accumulator space, the votes counted for such a circle will appear as the highest peak as shown in

```
% Function: Hough Transform calculation for circles

% 'image' is original grey image

% 'edge_image' is binary image after an edge detection

% 'r' is the radius of circle being searched


function image=houghTF_circles (image, edge_image, r)

[rows,columns]=size(edge_image);

accumulator=zeros(rows,columns);        %create accumulator space


for x=1:columns

    for y=1:rows

    if (edge_image(y,x)==1)                  %check if it's an edge pixel

        for angle=0:360

        theta=(angle*pi)/180;                %convert angle to radian

        x0=round(x-r*cos(theta));

        y0=round(y-r*sin(theta));            %calculate r

        if ((x0<columns && x0>0) && (y0<rows && y0>0)) %is it valid?

        accumulator(y0,x0)=accumulator(y0,x0)+1; %increment the vote at bin

        end                                    % corresponding r and theta value

        end

        end

    end

 end

figure, surfl(accumulator);

shading interp

colormap(gray);

max_value=max(accumulator);

peak=max(max_value);

x0=0;

y0=0;
```

FIGURE 4.22 Hough transform for detecting circles.

```
for x=1:columns

    for y=1:rows

        if (accumulator(y,x)>0.9*peak)   %check if it's a strong candidate?

            image (y,x)=255;             %mark centre of the circle

            x0=x;

            y0=y;

                for angle=0:360          %draw detected circle in the image

                theta=(angle*pi)/180;

                x_coord=round(r*cos(theta)+x0);

                y_coord=round(r*sin(theta)+y0);

                    if ((x<columns && x>0) && (y<rows && y>0))

                    image(y_coord,x_coord)=255;

                    end

                end

            end

        end

end

end
```

FIGURE 4.22   (continued) Hough transform for detecting circles.

Figure 4.23b for the test image shown in Figure 4.23a. The position of these peaks in the accumulator space also shows the center coordinates of the circles. In this example, the nearest ball to the camera is detected. When we repeat the Hough transform calculations for a radius of 36 pixels, we are able to detect the ball on the left-hand side of the image as shown in Figure 4.23d. Its radius is smaller since it is slightly further away than the ball in front. In robotic games, the size of the circular objects in the field, pellets, balls, and so on are prefixed; hence, the possible radius of a circular object to look for is usually known to the robot developer. For example, the ball shown in Figure 4.23 is used in robot soccer games. By manipulating the code in Figure 4.22 and scanning for a range of radius and ranking corresponding peak votes in the accumulator space, we can enhance this algorithm to detect any circular object in the image.

4.7.2 CONNECTED COMPONENT LABELING

Connected component labeling is a fundamental and frequently used technique for categorizing chunks of pixels (or blobs) identified as object features. For instance, in Figure 4.12, after a threshold operation, three blobs of white pixels are obtained. It would be difficult to analyze these shapes and make a decision unless we provide a

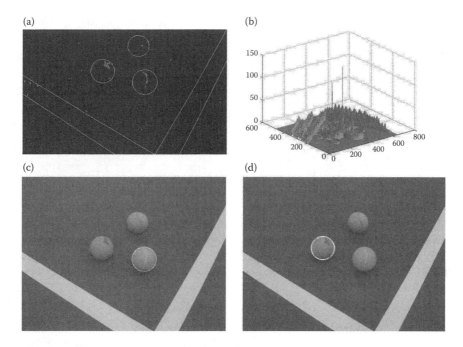

FIGURE 4.23 (a) The edge picture. (b) Hough space. (c) Strong candidate for a circle of radius 38 is highlighted on the actual image. (d) Repeated calculations for a circle of radius 36 are highlighted on the actual image.

means to segregate one blob from another. The connected component labeling algorithm identifies each blob and gives it a unique number. A typical definition of connectivity is given as for every pair of vertices i and j, in an undirected graph P, if there is a path from i to j, then the graph P is defined as connected (Haralick and Shapiro 1992). The objective of the connected component labeling algorithm is to find such connected pixels and assign them a label so that all the connected pixels share the same label. Hence, a collection of four or eight adjacent pixels with the same intensity value will be grouped into a connected region. In four-connected pixels, neighbors of a pixel will be touching horizontally and vertically. A pixel at (x,y) is four-connected to pixels at $(x+1,y)$, $(x-1,y)$, $(x,y+1)$, and $(x,y-1)$. Similarly, in the case of eight-connected pixels, the neighbors of a pixel will be connected horizontally, vertically, and diagonally. Hence, a pixel at (x,y) is eight-connected to pixels at $(x-1,y-1)$, $(x,y-1)$, $(x+1,y-1)$, $(x-1,y)$, $(x+1,y)$, $(x-1,y+1)$, $(x,y+1)$, and $(x+1,y+1)$. In the example binary image (i.e., black and white image) shown in Figure 4.24a, a pixel value of 0 indicates a black pixel and a pixel value of 1 indicates a white pixel. Typically, the objects of interest are marked as white pixels. The connected component-labeling algorithm will assign a unique number to each blob of white pixels that are connected. We can conclude that this operation performs a transformation from pixels to regions, which make the analysis of these regions a lot easier.

There are many connected component algorithms published in the literature. A collection of classical methods is given in Haralick and Shapiro (1992), more

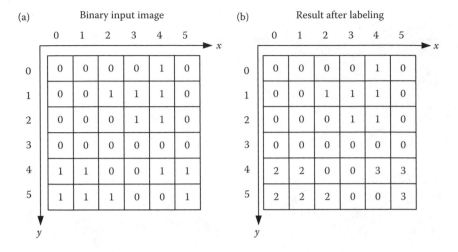

FIGURE 4.24 (a) Binary image with three blobs. (b) The result after component labeling algorithm.

recent techniques are described in Suzuki et al. (2003) and Hea et al. (2009). Figure 4.25 shows the MATLAB function for an iterative connected component labeling algorithm described by Haralick (1981). The algorithm has three phases; at the initialization phase, a top down left to right scan is performed over the image and each pixel with a value of 1 is given a unique label in an incremental fashion. The next two phases are the iterative part of the algorithm, where a top down pass, followed by a bottom up pass, are performed. At the first top down pass, the image is scanned from top to bottom and left to right by replacing each pixel value by the minimum value of its nonzero neighbors. Similarly, the bottom up phase image is scanned from bottom to top and right to left by replacing each pixel value by the minimum value of its nonzero neighbors. Iteration stops when no pixel value change is performed during these passes. The MATLAB code shown in Figure 4.25 is an implementation of the iterative connected component labeling algorithm for the illustration purpose. Nonetheless, in the following sections, we will employ built-in MATLAB functions for connected component labeling such as "bwlabel" and "bwconncomp" for the same purpose. Figure 4.26 shows an example image and connected component labeling result where the pixels of each blob are replaced with the label given to that blob.

As we have seen above, the connected component algorithm produced regions that are individually labeled. After this step, a number of properties of those regions such as area, centroid, and boundaries can be obtained. Furthermore, many statistical properties such as mean and variance can be studied by referring to pixels in that region in the original image. Let us take a look at the labeled image given in Figure 4.27 and study the region labeled as 1. By simply counting the pixels labeled as 1, we obtain the area of that blob or region. In the example image, the area of an object, labeled as 1, is 6 pixels. Summing up all the x coordinates of the pixels in this blob

```matlab
% Function: Connected component labelling

% 'IMG' is binary image

% 'result' output of labelling operation


function result=ConnectedComp(IMG);

[rows,columns]=size(IMG);

result=zeros(rows,columns);

CHANGE=1;

number=1;

for y=1:rows

    for x=1:columns

        if (IMG(y,x)>0)

            result(y,x)=number;

            number=number+1;

        end

    end

end


while (CHANGE ~=0)

CHANGE=0;

for i=2:rows-1        % Top down pass

for j=2:columns-1

    if (result(i,j)>0)

    minimum=result(i,j);

    for ii=-1:1:1

    for jj=-1:1:1

        if ((result((i+ii),(j+jj))>0) && (result((i+ii),(j+jj))<minimum))

        minimum=result((i+ii),(j+jj)); % find minimum valued neighbour

        end

    end

    end

    end

    end
```

FIGURE 4.25 MATLAB code for two pass connected component labeling algorithm.

```
     if ( minimum < result(i,j))

     result(i,j)=minimum;              % replace the pixel with minimum value

     CHANGE=1;                         % flag the change

     end

     end

end

end

for i=rows-1:-1:2    % Bottom up pass

for j=columns-1:-1:2

     if (result(i,j)>0)

     minimum=result(i,j);

     for ii=-1:1:1

     for jj=-1:1:1

         if ((result((i+ii),(j+jj))>0) && (result((i+ii),(j+jj))<minimum))

         minimum=result((i+ii),(j+jj)); % find minimum valued neighbour

         end

     end

     end

     if ( minimum < result(i,j))

     result(i,j)=minimum;              % replace the pixel with minimum value

     CHANGE=1;                         % flag the change

     end

     end

end

end

end

end
```

FIGURE 4.25 (continued) MATLAB code for two pass connected component labeling algorithm.

and dividing it to the area, we obtain the x coordinates of its centroid and similarly repeating it for y coordinates, we obtain the y coordinates of its centroid. For the example image, the centroid of the blob (x_0, y_0) will be

$$x_0 = \frac{4 + 2 + 3 + 4 + 3 + 4}{6} = 3.33 \approx 3$$

(a)

(b)

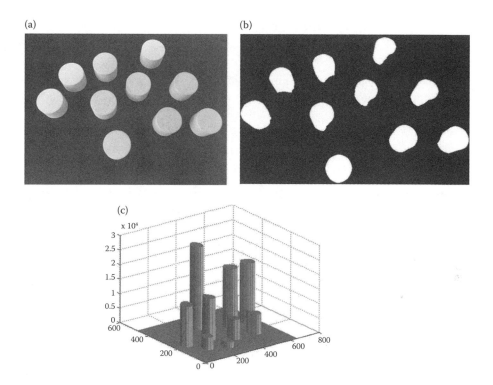

(c)

FIGURE 4.26 (a) Original picture. (b) Binary image after threshold. (c) Connected component output in 3D where each blob is labeled with a different number.

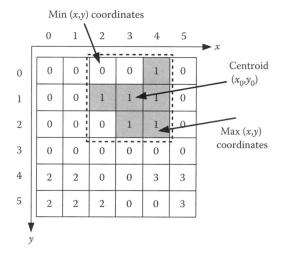

FIGURE 4.27 An example of labeled regions in an image and their properties.

$$y_0 = \frac{0+1+1+1+2+2}{6} = 1.16 \approx 1$$

Furthermore, by finding the minimum x and y coordinates of the blob, we determine the top left most corner of the rectangle that encloses this region likewise by finding the maximum of x and y coordinates, we obtain bottom right most corner of it.

4.8 CASE STUDY TRACKING A COLORED BALL

In this section, we will put what has been discussed above into practical use. A soccer-playing humanoid robot is expected to detect the ball in the field, approach it, and perform a kick to score a goal. Detecting and locating the ball is done by the vision unit of the robot. The flow of operations for this case study is demonstrated in Figure 4.28 and the corresponding MATLAB code is given in Figure 4.29.

We first convert the captured image to HSV color space and separate the yellow objects based on their hue and value range, which is determined experimentally. The result is a binary image where white pixels imply yellow-colored objects. However, before we perform connected component labeling, we try to eliminate single isolated white pixels that are due to noise and reflections by using the "imerode" function, which is also known as the erosion algorithm (Haralick and Shapiro 1992; Russ 2002). In the following, we perform "imfill" to fill out empty pixels in blobs, in a process also known as the dilation algorithm (Haralick and Shapiro 1992; Russ 2002). These two steps make the blobs in the binary image more compact and tidy. This way we can also speed up the labeling process by avoiding isolated insignificant pixels. The final stage is the analysis of blobs, which is done with the dedicated

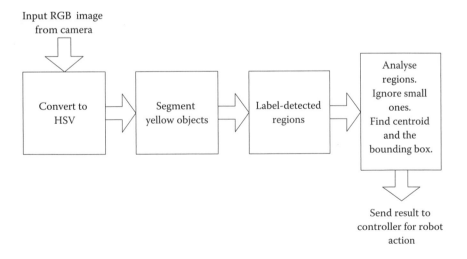

FIGURE 4.28 An example set of operations for ball tracking based on its color.

```
%Thresholds for the yellow coloured tennis ball

hueThresholdLow = 0.19;

hueThresholdHigh = 0.3;

valueThresholdLow = 0.19;

valueThresholdHigh = 0.3;

se = strel('disk',2);

flag=1000;


vidobj = videoinput('winvideo', 1, 'RGB24_320X240');

while (flag ~=0)

IMG = getsnapshot(vidobj);


IMGhsv = rgb2hsv( IMG );                    %Convert image to HSV

IMGhue = IMGhsv( : , : , 1 );               %Seperate hue

IMGsat = IMGhsv( : , : , 2 );               %Seperate saturation

IMGval = IMGhsv( : , : , 3 );               %Seperate value


hueMask = (IMGhue >= hueThresholdLow) & (IMGhue <= hueThresholdHigh);

valueMask = (IMGval >= valueThresholdLow) & (IMGval <= valueThresholdHigh);

yellowObject = uint8(hueMask & valueMask);% resulting binary image

                                     % white pixels imply yellow objects

IM0 = imerode(yellowObject,se); % eliminate isolated pixels


IMG0 = imfill(IM0,'holes');       % fill empty pixels inside blobs


[L N]=bwlabel(double(IMG0));        % connected  component labelling

prop=regionprops(L,'Area','Centroid','BoundingBox');

                                  % obtain properties of blobs

imshow(IMG)

hold on

numObj = numel(prop);
```

FIGURE 4.29 MATLAB code for tracking an object based on its color.

```
for k = 1 : numObj

    if (prop(k).Area>100)          % select worthy blobs

    plot(prop(k).Centroid(1), prop(k).Centroid(2), 'bo');

    rectangle('Position', prop(k).BoundingBox, ...

        'EdgeColor','y'); %superimpose bounding box and centroid on image

    end

end

hold off

flag=flag-1;

end

delete(vidobj)

clear vidobj
```

FIGURE 4.29   (continued) MATLAB code for tracking an object based on its color.

(a)   (b)

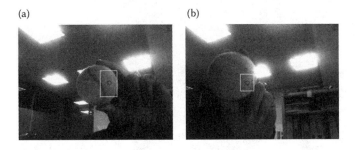

FIGURE 4.30   Screen shots from object tracking application.

MATLAB function "regionprops." It delivers the desired features of the labeled regions such as area, centroid, and so on. The program given in Figure 4.29 captures the image from a webcam attached to a PC and performs the color detection on the captured image frame. To track an object continuously, the sequence of operations shown in Figure 4.28 is put in a loop and repeated for each captured frame. Figure 4.30 shows the screen captures from the experiments with a tracking program. The program is very simple using the image-processing library functions of MATLAB. However, as we describe the details of the algorithms used in these functions, rewriting this code in C or other relevant programming language to implement on a stand-alone system is rather straightforward.

4.9   SUMMARY

In this chapter, we have discussed the available hardware and software tools for robot vision. By employing a low-cost camera and some fundamental image-processing techniques, we can provide a higher level of sensing to the robot. As camera and image-processing tools are becoming more widely available and cheaper, their use in robotic games is also becoming popular. Nevertheless, image-processing litera- ture encompasses a huge number of algorithms and techniques. In this chapter, we presented the fundamental techniques that can provide a quick start in realizing a vision unit for the robot.

REFERENCES

Ballard, D.H. 1981. Generalising the Hough transform to detect arbitrary shapes. *Pattern Recognition* 13:111–122.

Chen, K.H. and Tsai, W.H. 1997. Vision-based autonomous land vehicle guidance in out- door road environments using combined line and road following techniques. *Journal of Robotic Systems* 14:711–728.

Cheng, H.D. and Sun, Y. 2000. A hierarchical approach to colour image segmentation using homogeneity. *IEEE Transactions in Image Processing* 9:2071–2082.

Chow, C.K. and Kaneko, T. 1972. Automatic boundary detection of the left ventricle from cineangiograms. *Computers and Biomedical Research* 5:388–410.

CMUCAM 2012. http://www.cmucam.org/projects/cmucam4/

DeSouza, G.N. and Kak, A.C. 2002. Vision for mobile robot navigation: A survey. *IEEE Transactions on Pattern Analysis and Machine Intelligence* 24:237–267.

Fu, K.S., Gonzales, R.C., and Lee, C.S.G. 1987. *Robotics Control, Sensing, Vision and Intelligence*. New York: McGraw-Hill.

Gonzales, R.C. and Woods, R.E. 1992. *Digital Image Processing*. Reading, MA: Addison-Wesley.

Haralick, R.M. 1981. Some neighbourhood operations. In: *Real Time/Parallel Computing Image Analysis*, eds. M. Onoe, K. Preston and A. Roselfeld. New York: Plenum Press.

Haralick, R.M. and Shapiro, L.G. 1992. *Computer and Robot Vision*. Reading, MA: Addison-Wesley.

Hea, L., Chao, Y., Suzuki, K. and Wu, K. 2009. Fast connected-component labelling. *Pattern Recognition* 42:1977–1987.

Hecht, E. 2002. *Optics*. 4th edition. San Francisco: Addison-Wesley.

Hough, P.V.C. 1962. Methods and means for recognizing complex patterns. U.S. Patent 3,069,654.

Kosaka, A. and Kak, A.C. 1992. Fast vision-guided mobile robot navigation using model based reasoning and prediction of uncertainties. *Computer Vision and Image Processing: Image Understanding* 56:271–329.

Kumagai, M. and Emura, T. 2000. Vision based walking of human type biped robot on undulat- ing ground. *Proceedings of International Conference on Intelligent Robots and Systems*, Takamatsu, Japan, Vol. 2, 1352–1357.

Leclercq, P. and Bräunl, T. 2001. A color segmentation algorithm for real-time object localiza- tion on small embedded systems. *Lecture Notes in Computer Science* 1998:69–76.

Pitas, I. 2000. *Digital Image Processing Algorithms and Applications*. New York: John Wiley and Sons.

Russ, J.C. 2002. *The Image Processing Handbook*. Boca Raton, FL: CRC Press.

Surveyor 2012. http://www.surveyor.com/blackfin/

Suzuki, K., Horiba, I. and Sugie, N. 2003. Linear-time connected-component labelling based on sequential local operations. *Computer Vision and Image Understanding* 89:1–23.

5 Basic Theory of Electrical Machines and Drive Systems

This chapter presents a brief description of electrical machines used in robots. The term "electrical machine" defines devices that convert mechanical energy to electrical energy or vice versa. In this chapter, we will cover the basic concepts and describe the relevant issues required to choose a suitable system for powering robots. Since the requirements of motion differ from robot to robot, it is imperative to know the particular requirements of a robot before choosing a drive system. For example, a robot designed to work in a car-manufacturing factory may need the same speed requirements in all the joints, but may need different requirements of torque depending on the joint. (Torque is defined as the rotational force that rotates a shaft and it is usually measured in newton-meters. In comparison, force pushes along a straight line and is measured in newtons.) The actuators at the robot end effector may need only relatively less torque when compared to the base motion joints. On the other hand, a humanoid robot may need high torque in its "roll" joints (joints that sway the robot sideways), however, rotating with less speed. The knee joints may have exactly opposite requirements.

The main objective of this chapter is to provide a starting point for the robot designer, without spending too much time on machine theory, which is a broad subject. In the following, we will first describe the principle of operation of common actuators used in game robotics and later discuss issues concerning their control. Clayton (1969), Say (1984), Cotton (1970), McKenzie-Smith and Hughes (1995), Fitzgerald et al. (1990), and Langsdorf (2001) are valuable resources for further reading on electrical machines.

5.1 ACTUATORS FOR ROBOTS

Robots need some source of torque and power to accomplish the desired motion; in other words, every robot needs some form of actuation. The devices that provide that actuation are in general called "actuators." When a robot moves on a terrain on wheels, the motion needs to be generated by a drive system with one or more prime movers. When a robot moves its arm, in most cases, there is a power source inside every joint, which is actually moving the arm. In comparison, when a human being is moving each joint in an arm, the motion is generated by some pushing and pulling caused by muscles. This type of mechanism governs the motion of many living organisms. Accomplishing the same mechanism is very complex for man-made systems. In most cases, a joint itself is self-powered and provides the motion desired. There are various types of actuators that can be used in robotics. They can be largely classified as electrical, pneumatic, and hydraulic. In mobile robots, pneumatic actuators are used occasionally if there is a need for a very high torque. A main disadvantage is that the system requires an external air supply to operate. Hydraulic actuators are common in earthmoving machines. We may come across hydraulic actuators on mobile robots as well, although these are exceptions. In this book, we will focus only on electrical devices as most game robots work with electrical actuators.

5.2 ELECTRICAL ACTUATORS

Electrical actuators are electrical motors, which are mainly classified as AC (alternating current) and DC (direct current) motors. AC drives are seldom used in mobile robots or game-playing robots. Usually, powerful industrial robots operating in manufacturing lines use AC drives. Similarly, underwater robots and ROVs (remotely operated vehicles) with power cables may also use AC drives.

5.2.1 FUNDAMENTAL CONCEPTS OF GENERATING AND MOTORING

Before we explore further, we need to go through some basic concepts of generating and motoring. One surprising fact is that in principle there is no major difference between a motor and a generator. Both are energy conversion devices converting electrical energy to mechanical energy and vice versa. There is only a handful of rules that we need to understand their operation even though the construction of such machines is quite complex and still labor intensive even in this age of automation. Let us recall some of the basic laws needed to understand the operation of motors. Even though we are only interested in motors, these laws apply to both motors and generators.

Faraday's law: The law states that whenever there is a change of flux linkages associated with a coil, an electromotive force (EMF) is induced in the coil, which is proportional to the rate of change of flux linkages.

Lenz's law: This law states that the direction of that EMF induced (above) acts in such a way as to oppose the cause (whatever is causing the EMF to be induced). The above laws can be combined into one equation:

$$e(t) = -N \frac{d\phi}{dt} \tag{5.1}$$

where $e(t)$ represents the instantaneous value of the voltage induced, N is the number of turns in the coil in question, and φ is the flux linking through the coil. The negative sign is due to Lenz's law.

The right-hand rule (RHR) and induced voltage: The rule is applied for generation. Figure 5.1 shows a conductor in a magnetic field, which is acting perpendicularly toward the plane of the figure. The flow direction of magnetic field is illustrated with an arrow. In this case, since it is pointing toward the plane of the figure, it is indicated by a circle with a cross in the center. The conductor segment, $a–a'$ is moving to the right with a velocity of V. Imagine that a voltmeter is connected to the

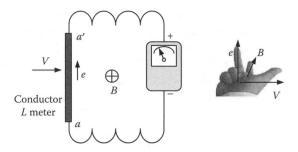

FIGURE 5.1   Right-hand rule for induced EMF.

circuit as shown in Figure 5.1. Then, the induced EMF that will be measured with the voltmeter is defined as

$$e(t) = BLV \qquad (5.2)$$

where B is the flux density in webers/m² or tesla, and L is the length of the conductor. The direction of EMF induced is found by the RHR. Stretching the three fingers of the right hand with thumb pointing toward the direction of motion of the conductor and first finger (index) pointing toward the direction of the flux, we find that the second finger shows the direction of the EMF. The example in Figure 5.1 shows the application of the RHR to find the direction of the induced EMF.

The left-hand rule (LHR) and the force: Whenever a current-carrying conductor is placed in a magnetic field, it experiences a force. Figure 5.2 shows such a situation. The magnitude of this force (in newtons) is given by

$$f = BLI \qquad (5.3)$$

where B is the flux density, I is the current, and L is the length of the conductor. The direction of this force is found using the LHR. By stretching three fingers of the left hand with the first finger pointing toward the direction of the flux and the second finger pointing toward the direction of the current flow, the direction of the thumb indicates the direction of the force. Hence, following the LHR, the direction of the force will be toward the right-hand side for the setup shown in Figure 5.2.

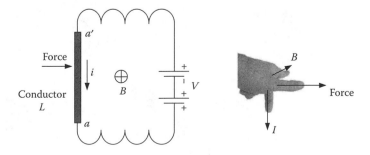

FIGURE 5.2   Left-hand rule for force on a conductor.

Motoring and generating happens simultaneously: We can learn something more by looking at Figure 5.2, assuming that a voltage *V* is driving the current and that creates the force to the right. The RHR implies that because the conductor is moving to the right (as shown in Figure 5.1), there must be an EMF of *E* induced in the conductor, which is actually acting upward, thus obeying Lenz's law that the induced voltage "will oppose the cause." As the conductor moves, only the difference between the applied voltage *V* and the induced EMF (often called "back EMF") *E* is available to drive the current. Hence, it is obvious that both RHR and LHR will simultaneously come into play. In conclusion, in an electrical machine, the generating and motoring actions are inseparable. When the applied voltage is higher than the back EMF, the machine is motoring and when back EMF is higher than the applied voltage, the machine is generating.

In the above discussion, when the conductor moves or is moved toward the edge of the magnetic field, the action will stop. There will be no more generating or motoring actions. To get a working machine, we need a proper structure and it will be described in the following sections.

5.2.2   DC Machines

DC machines are still the mainstay in robotic drives. Hence, it is important for us to become familiar with their theory and operation. Let us start the process of understanding DC machines from a simple case. We have already mentioned that when a DC motor is working, both generating and motoring actions happen simultaneously. To understand the operation of a DC motor, we need to understand its operation as a generator. Hence, in the following section, we first look at it as a generator.

Primitive DC machine as a generator: In this section, we examine the back EMF induced in a primitive motor. Figure 5.3 shows a very basic structure of a DC machine. In this figure, we assume that there is a magnetic field of *B* acting from top to the bottom of the north pole to the south pole. Assume that at the starting instant, the coil is perpendicular to the magnetic field as shown. Assume that the coil has *N*

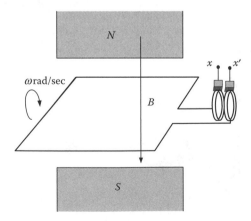

FIGURE 5.3   Basic slip ring generator.

turns, its area is A, and it is rotating with an angular velocity ω, then the maximum flux linkage is given as

$$\lambda = NAB \tag{5.4}$$

At the instant shown in Figure 5.3, the flux linkage as a function of time can be written as

$$\lambda(t) = NAB \cos(\omega t) \tag{5.5}$$

Then, defining that the maximum flux, $\varphi = AB$, we can write

$$\lambda(t) = N\varphi \cos(\omega t) \tag{5.6}$$

Then, invoking Faraday's law and Lenz's law, we derive the induced EMF as

$$e(t) = -\frac{d\lambda}{dt} = \omega N\phi \sin(\omega t)$$

$$e(t) = V_{max}\sin(\omega t) \tag{5.7}$$

The voltage pattern that can be tapped from the brushes touching the slip rings is a sine wave with a maximum magnitude of value V_{max} and angular frequency of ω, with which we are all familiar. Since the waveform is sinusoidal, the device is a primitive AC generator. Furthermore, the waveform is periodic and the period of the waveform is $T = (2\pi/\omega)$. We realize that we started our discussion to understand how DC motors are made, but we ended up on a device that produces AC waveform as back EMF. Such a device cannot work on a DC power supply. In fact, by simply replacing the slip rings shown in Figure 5.3 with split rings shown in Figure 5.4, we can obtain a primitive DC motor, which will produce a waveform of back EMF

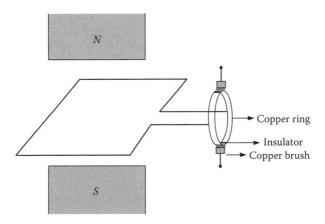

FIGURE 5.4 Split ring DC machine that can act as a DC motor.

with DC average value. From the machine shown in Figure 5.3, we get a sinusoidal alternating voltage wave form since brushes are always in touch with the same terminals of the winding. If we can interchange connection at each zero crossings of the waveform by swapping the brushes from one part of the ring to the other, the brush polarity will remain the same.

We know that the split ring is rotating continuously. The split positions of the ring are placed appropriately on the motor/generator shaft so that the brushes on the frame of the machine touch the rings. Thus, when the metal ring, connected to the terminal with positive voltage, is about to turn negative, the positive brush slips to the other half of the ring. When the metal half ring, connected to the terminal with negative voltage, is about to turn positive, the negative brush slips to the other half of the ring. This is achieved by mechanically placing the split part in the ring and brushes appropriately. The above process is called "commutation." The split ring shown in Figure 5.4 is a primitive commutator. The resulting voltage will look like a rectified AC wave form. Nevertheless, such a DC voltage will create many problems in usage. Over the years, DC machine design has gone through many milestones, and now it has been stabilized. DC voltage induced is made smooth, constant, and steady by distributing the winding over the armature surface and by increasing the commutator segments to much more than just two.

We will now try to describe the basic operations of a modern DC motor. As we mentioned earlier, there should be a magnetic field established. This is usually realized by field winding and pole configuration as shown in Figure 5.5. This pole configuration is called a stator consisting of a drum-like yoke fitted with salient poles. Figure 5.5 shows a machine with two poles. Obviously, the number of poles must be even. The magnetic field is established by field windings. The field winding can be a separate circuit, or as in some small DC machines, such a field is established by using permanent magnets. The yoke completes the magnetic path.

The rotating structure is called the armature, and it is made of a laminated magnetic material. This assembly provides axial slots on the surface for windings to be

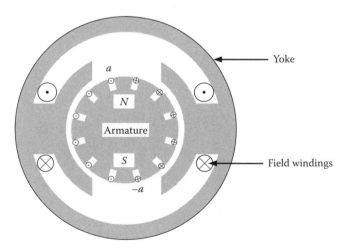

FIGURE 5.5   The complete DC machine that can act as a motor as well as a generator.

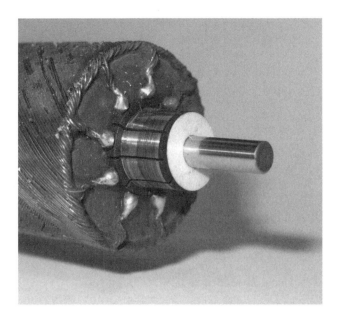

FIGURE 5.6   A typical commutator and armature setup.

placed. The copper conductors are placed in those slots and form the armature winding. The ends of these windings are terminated on commutator segments. The setup is much more sophisticated than what we have shown in Figure 5.5. A picture of a typical commutator is shown in Figure 5.6.

Back EMF induced in a DC motor: Figure 5.7 shows a skeleton winding on the armature of four-pole DC machine. Let us assume that one coil consisting of conductors a and a' is in position as shown in Figure 5.7. The conductor pair $a-a'$ forms one

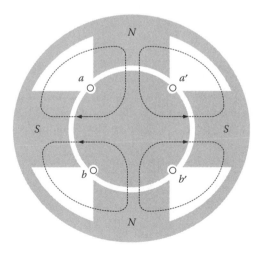

FIGURE 5.7   Primitive four-pole winding.

turn. We assume that the flux per pole is φ, emanating from the north-pole on the top. This flux enters the coil and divides into two parts and enters the S-poles on two sides. The same thing happens to the flux originating from the north-pole below, but linking the coil formed by b–b'. The mechanical commutator connects the turns a–a' and b–b' in series so that whatever EMF induced in them add to each other.

Let us focus on the turn a–a'. The flux linkage for that coil is given by $\lambda = \varphi$ since there is only one turn. Let us assume that the armature rotates through 90° clockwise so that the turn a–a' comes directly under an S-pole taking a time of T seconds. The flux linked by the same turn would have reversed from $+\varphi$ to $-\varphi$. We can write that the average rate of change of flux linkage per turn in coil formed by a–a' is $2\varphi/T$, which is the same as the average EMF induced in one coil, e_{av}, according to Faraday's law. However, a typical DC machine armature has a large number of conductors, say, Z of them, thus producing $Z/2$ turns. The commutator setup groups them into "A" number of parallel paths. Then, the number of turns in any parallel path is $Z/2A$. There are two categories of windings. Depending on the winding type, A can be 2 or P, where P is the number of poles in the machine. Hence, the EMF induced in one of the parallel elements of the winding will be

$$E_{av} = \frac{Z}{2A} \times \frac{2\phi}{T} = \frac{Z\phi}{AT} \tag{5.8}$$

The time taken by a coil to move from one pole center to the next pole center is given by

$$T = \frac{1}{\text{rps}} \frac{1}{P} = \frac{60}{N} \frac{1}{P} \tag{5.9}$$

where P is the number of poles, N is the speed (in revolutions per minute, rpm), and rps is revolutions per second. Hence, by substituting this in Equation 5.8, we get the EMF equation as

$$E_{av} = \frac{P\phi ZN}{60A} \tag{5.10}$$

An alternative form of EMF equation: We may encounter different forms of EMF formulas in motor data sheets, especially for permanent magnet motors commonly used in robotics where φ is fixed. For instance, the angular velocity can be written as

$$\omega = 2\pi \frac{N}{60} \tag{5.11}$$

By substituting Equation 5.11 in Equation 5.10 for EMF, we can write

$$E = \frac{P\phi Z[60]\,\omega}{[60]A\,2\pi} = \frac{P\phi Z\omega}{2\pi A} = K_b\omega \tag{5.12}$$

where K_b is defined as the EMF constant in terms of volt/radian per second such that

$$K_b = \frac{P\phi Z}{2\pi A} \quad \text{V/rad/s} \tag{5.13}$$

This information is often provided in the data sheets as volts/unit angular speed or in a different scale as volts/1000 rpm.

Torque equation: The average value of the flux density crossing the air gap between the armature surface and the pole surface can be written as a ratio of total flux divided by the total armature surface area. Hence, we obtain

$$B_{av} = \frac{P\phi}{L2\pi r} \tag{5.14}$$

where r is the radius of the armature and L is the length of the armature. Using the LHR, we can obtain the average force on one conductor as

$$f = \frac{P\phi}{2\pi r} i \tag{5.15}$$

where i is the current passing through a conductor. Since there are A parallel paths, i can be written in terms of total input current I as I/A.

Hence, the expression for force is given as

$$f = \frac{P\phi I}{2\pi A r}$$

Since there are Z number of such conductors, the total peripheral force can be written as

$$F = \frac{P\phi I Z}{2\pi A r} \tag{5.16}$$

Since torque is the product of force and armature radius, then the torque equation becomes

$$T = \frac{P\phi I Z}{2\pi A} \tag{5.17}$$

Alternate torque equation: Once again considering the case of permanent magnet motors, the above torque equation can also be written as

$$T = \frac{P\phi I Z}{2\pi A} = K_t I \tag{5.18}$$

where K_t is defined as the torque constant such that

$$K_t = \frac{P \phi Z}{2 \pi A} \text{ Nm/A} \qquad (5.19)$$

Referring to Equation 5.13, we can see that K_t and K_b are numerically the same.

DC motor types: In nonpermanent magnet DC motors, there are two electrical windings, one for the field excitation and the other for armature. Moreover, the field excitation can be from a high-voltage source such as its own armature, or it can be excited by the armature current using only a few turns. The former is called shunt winding and the latter is called series winding. There are many ways these three windings (armature, shunt-field, and series field) can be interconnected. For a given DC motor, there is no need for both kinds of field windings to be available. DC motors are classified by the way the windings are connected with each other. They are permanent magnet motors, separately excited DC motors, shunt DC motors, series DC motors, and cumulatively compounded DC motors, where series winding field supports the shunt winding field, and differentially compounded DC motors, where the series winding field opposes the shunt winding. Each of these motors will have different speed versus torque characteristics. However, our main interest lies in separately excited motors and permanent magnet motors. The speed variation of permanent magnet motors and separately excited motors will be somewhat flat which implies that the change of speed with respect to load is minimal. They both are suitable for robotic applications. Permanent magnet motors are the most preferred since there is no need to supply magnetization current.

Nowadays, to reduce the rotational inertia, air-core armatures are used, instead of rotating heavy magnetic core armatures. These advanced techniques are all due to the research in machines and materials technology.

5.2.3 AC Motor Drives

As mentioned earlier, AC motors are mainly used in large stationary robots in industry. We rarely encounter them in robotic games. These motors are meant for fixed load, and they run at a given fixed speed. For the sake of completeness, we will briefly describe their operating principle here. AC motors can be of synchronous or induction type. Rotating magnetic field is the most important concept used in understanding synchronous motors. It is possible to show that when three-phase currents are passed through the three windings spaced 120° apart on the stator (outer part) of synchronous motor, a magnetic field that rotates at a particular speed is created. Similarly, when two-phase currents are passed through the windings spaced 90° apart on the stator, again a rotating magnetic field is created. This magnetic field rotates with a speed given by $N_S = (120/P)f$, where P is the number of poles and f is the frequency of the polyphase power supply. The speed of rotation of the magnetic field is called the synchronous speed. This concept has been well discussed in the literature (Cotton 1970; McKenzie-Smith and Hughes 1995; Fitzgerald et al. 1990; Langsdorf 2001).

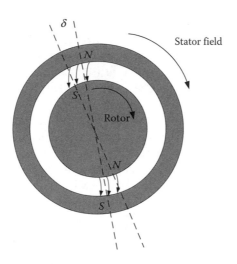

FIGURE 5.8   Operation of a synchronous motor and the concept of load angle.

Operation of a synchronous motor: A three-phase, two-pole synchronous motor with pole pattern is shown in Figure 5.8. The windings are not shown in order to simplify the picture. Assume that a rotor has a field winding that establishes a pair of *N–S* poles. This field winding in the rotor is fed with a DC supply through slip rings. Then, the stator poles and rotor poles will lock with each other and hence, the rotor rotates along with the rotating magnetic poles at synchronous speed. If there is no load, the stator and rotor poles will be fully aligned. As the load increases, the rotor poles will start to lag behind through an angle called load angle, δ, as shown in Figure 5.8. As the load further increases, the rotor may no longer be able to follow the stator, and it will simply stop.

Induction motors: An induction motor also has a stator winding similar to that of a synchronous motor and produces a rotating magnetic field. However, the rotor does not have magnetic poles. Instead, the rotor has a core and short-circuited windings on it. Torque is produced due to the induced currents in the rotor windings. Unlike synchronous motors, induction motors always run at less than the synchronous speed. They are widely used in industry.

5.3   SPECIFIC NEEDS OF ROBOTICS DRIVES

In the previous sections, we discussed operations of motors, but we did not consider their manipulation by computers. They are all designed to operate under planned, that is, known loading conditions and once they are switched on they will operate without any intervention.

When these motors are used for actuating motions of robots, they have to be controlled using computers and appropriate electronics. One of the important characteristics of these motors is their speed. The speed of AC motors is bounded by the supply frequency. For instance, with a 60 Hz supply, a two-pole AC machine can run up to a speed of 3600 rpm. On the other hand, DC machines run at designated speeds, which can vary from very slow to very high speeds. At this point, it is necessary to point out that when we use electronics to switch winding currents the distinction between AC and DC motors gets somewhat blurred. For example, the power supply to a stepper motor may be just DC, but the current passing through the winding is in fact alternating. The same is the case for brushless DC (BLDC) motors.

In many robotics applications, we need a very close control of speed and position. A major issue in robotics practice is that we need smaller motors to build robots of reasonable size. The power needs of such motors may vary from milliwatts to several tens of watts. The work done per revolution by a motor depends on the size of the motor, the quality of the material, insulation type, and the waste power it can dissipate. However, the power rating of the motor is the product of the work done per revolution and the speed of the motor. The conflict is that they have to be small with weights not more than few hundred grams. A small motor can only produce small power at low speed. However, if the same motor runs safely at a higher speed, then it can deliver higher power. The logical conclusion is that to satisfy the size and power requirements, motors used in robotics have to run at very high speeds.

The above argument easily explains why many DC motors for robotics drives offered by manufacturers have ratings of low torque of around 75 milli-newton-meters, but speeds of around 10,000 rpm or more. The motion in robotics does not require such high speeds but requires high torque. These motors are always used with speed reduction gears. Many DC motor manufacturers offer motors with built-in gears and boast elaborate catalogs of many power, speed, torque, and gear ratio combinations (see, e.g., Faulhaber 2011).

5.3.1 DC Permanent Magnet Motors

Among the DC motors, permanent magnet motors are commonly used in game robotics. They are used along with driver circuits and their actions are continuously monitored using encoders or potentiometers attached to their shafts. They are seldom directly connected to a supply since in robotics we need to control position and speed precisely. More details on their control will be discussed later in this chapter.

5.3.2 Servo Motors

Servo motors are very popular in game robotics. They are nothing but DC motors with built-in control electronics. Unlike DC motors, they are not meant for continuous rotation, but used for a fixed angular rotation. Their built-in electronics provides the means of controlling angular position by using a potentiometer, and they are all encased as a part of the motor body. In applications such as a humanoid limb or a grabber where joints need to make fixed angular motions, a servo motor is a good choice to power it. Such motors eliminate the trouble of position control, since the position control circuit is built in and the user only has to provide a specific signal to achieve the angular rotation. Figure 5.9 shows the assembling of a servo motor to a robot joint. Note that a flange is fitted to the rotor shaft, and the load is fastened to this flange.

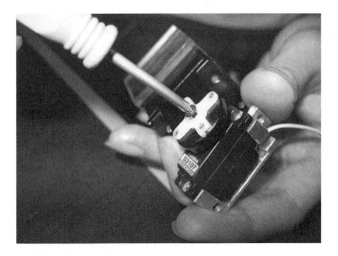

FIGURE 5.9 Servo motor and its assembly to a robot joint.

There are three wires connected to the motor, usually color coded as red for motor power supply line V_s, black for ground line G, and white for control signal line C. Figure 5.10a shows the control signal for a Hitec 422 servo motor (Hitec 2012). The control signal is a series of square pulses with a frequency of 50–100 Hz, or a period of 10–20 ms. When the "ON" period is 1.5 ms, the load flange is in the neutral

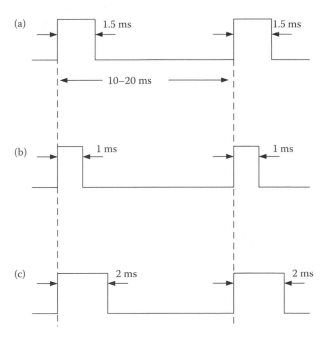

FIGURE 5.10 (a) Pulse train to achieve neutral position. (b) Pulse train to achieve −90° position. (c) Pulse train to achieve +90° position.

position. When the pulse width is reduced to a lower value of 1.0 ms, as shown in Figure 5.10b, the wheel rotates through 90° in one direction and stops. If the pulse width is increased to 2.0 ms, as shown in Figure 5.10c, the flange wheel rotates in the opposite direction through 90° and stops. These actions are repeatable, and the range of the angular rotation is from −90° to +90° which is 180°. This range may change according to the specifications supplied by the manufacturer. Here, we described only a typical example.

Apparently, controlling of a servo motor with computers is very straightforward. If the train of pulse stops, the angular position may drift due to the load. Hence, for achieving and holding a certain angular position, the computer has to continue providing the required pulse width periodically. Typical specifications provided for servo motors are as follows: Holding torque (in units of kg-cm), speed of response (in units of s/60°), input voltage range, gearing type, and overall size and weight.

5.3.3 STEPPER MOTORS

Stepper motors provide easy and precise control of motions. As the name indicates, the motions are in steps activated by pulse trains (Kenjo and Sugawara 1994). The number of pulses decides the number of angular steps the motor rotates. The step size of these motors can have a wide range. As long as the pulses are fed, the motor will keep rotating. When the pulses stop, the motor will also stop. It is in major contrast to DC motors, which will continue to rotate as long as there is power connected. Hence, stepper motors are used in many applications where we need exact position control. The ratings of such motors are rather limited to small power values. There are various types of stepper motors. A widely acceptable classification of them is as follows:

a. Variable reluctance (VR) motors (further classified as single stack and multistack)
b. Permanent magnet stepper motors
c. Hybrid stepper motors

This is not a complete classification, although it is sufficient for our understanding of their application in robotics.

Single-stack VR stepper motors: A diagram of a single-stack VR stepper motor (Kuo 1979; Edwards 1991), illustrating its features, is given in Figure 5.11. The single-stack VR stepper motors are compact. They usually have two or three phases. The step size obtainable is quite large, with a practical number of teeth on stators and rotors. There are more stator teeth than the teeth on the rotor or vice versa. Obviously, this difference is important for incremental motion. The achievable step size ranges from 0.9° to 30°. The example motor to be used here for discussions has 12 stator teeth, and winding is provided for three phases. Hence, there are four teeth per phase. Thus, three phases are covering all the 12 teeth. The motor shown in Figure 5.11 has 16 teeth on the rotor. We can observe that five rotor teeth cover four stator teeth. Phase *A* winding is covering four teeth. Similarly, phase *B* and phase *C* also cover four teeth; hence, all the 12 teeth are covered. Let us assume that phase *A*

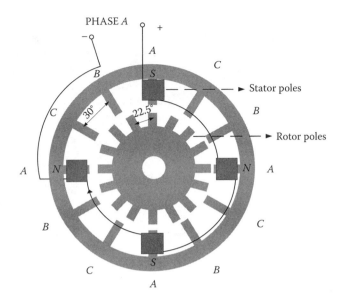

FIGURE 5.11   Single-stack three-phase VR stepper motor.

is energized. Magnetic polarity for phase A winding is marked on Figure 5.11. The tooth pitch of the stator is $360/12 = 30°$, and the tooth pitch of rotor is $360/16 = 22.5°$. Figure 5.11 shows that only four rotor teeth are aligned with the stator teeth and the rest of them are all misaligned. Now, assume that phase B is energized at the same time when phase A is switched off. This will result in rotor to rotate 7.5° counter-clockwise so that four rotor teeth will again align with teeth wound with phase B. Consequently, others will be misaligned. This motion goes on as phases are cyclically switched on and off. Apparently, in this example, the step size is 7.5°. By changing the switching sequence ABC to ACB, the direction of rotation can be changed.

To draw some conclusions, let us go through some requirements for proper operation. Let us define some terms as below:

P_s = the number of teeth on the stator
P_r = the number of teeth on the rotor
n = the number of phases wound on the stator

The number of stator teeth per phase will be

$$p = \frac{P_s}{n} \tag{5.20}$$

Or

$$P_s = np \tag{5.21}$$

It is obvious that p must be an even integer in order to have a viable distribution of phases on stator teeth. When any one of the phases is energized, there should be

p stator teeth that get exactly locked with the p rotor teeth. The stator teeth found *per* adjacent pairs of locked teeth are the same as the number of phases which is just n and those teeth would be wound with one phase each. However, if the number of teeth on the rotor *per* adjacent pairs of locked teeth are also the same as n, no motion will be possible when windings are switched. There should be a misalignment to facilitate motion, and hence the number of teeth on the rotor per those two pairs of locked teeth should be one less $(n - 1)$ or one more $(n + 1)$. This is also clear from Figure 5.11.

Case I (when $P_r > P_s$):

$$P_r = (n + 1)p \tag{5.22}$$

Then step size is given by

$$\theta = \frac{360}{P_s} - \frac{360}{P_r} \tag{5.23}$$

$$\theta = \frac{360}{n\,p} - \frac{360}{(n+1)\,p} \tag{5.24}$$

Case II (when $P_r < P_s$):

$$P_r = (n - 1)p \tag{5.25}$$

Then step size is given by

$$\theta = \frac{360}{P_r} - \frac{360}{P_s} \tag{5.26}$$

$$\theta = \frac{360}{(n-1)\,p} - \frac{360}{n\,p} \tag{5.27}$$

EXAMPLE 5.1

For the motor shown in Figure 5.11, we see that $P_r > P_s$ and $n = 3$ and the number of teeth per phase, $p = 4$. Hence, the step size can be calculated by applying the formula

$$\theta = \frac{360}{np} - \frac{360}{(n+1)p} = \frac{360}{3 \times 4} - \frac{360}{4 \times 4} = 30 - 22.5 = 7.5°$$

The number of steps per revolution will be

$$S = \frac{360}{7.5} = 48$$

Design issues: Given the step size, we should be able to decide the number of teeth on stator, rotor teeth, and the number of phases. For this, we simply rewrite Equations 5.24 and 5.27 to obtain p.

Case I (when $P_r > P_s$):

$$p = \frac{360}{n\theta} - \frac{360}{(n+1)\theta} \tag{5.28}$$

We can start from the value of $n = 3$ onwards for the given value of θ and iterate until we get an even integer for p. Then

$$P_s = np$$
$$P_r = (n+1)p \tag{5.29}$$

Case II (when $P_r < P_s$):

$$p = \frac{360}{(n-1)\theta} - \frac{360}{n\theta} \tag{5.30}$$

After getting an appropriate value for p, we can obtain

$$P_s = np$$
$$P_r = (n-1)p \tag{5.31}$$

Caution: For $n \geq 3$ and for p being an even integer, it is possible to realize a motor design and compute the step size. On the other hand, it is not possible to design a motor for any arbitrary value of step size. Some step sizes may be impossible to realize.

EXAMPLE 5.2

Let us assume that in a robot design, we need a smaller step size, say 0.9°. Can we achieve that? Let us try. Assume that $P_r > P_s$. Using the equations in case I, design calculations will yield to

n	$p = \dfrac{360}{n \times 0.9} - \dfrac{360}{(n+1) \times 0.9}$	
3	33.333	Discard
4	20	Accept

Then the number of teeth required will be

$$P_s = 4 \times 20 = 80$$
$$P_r = 5 \times 20 = 100$$

From the above example, we can conclude that to have smaller and smaller step size, enormous number of teeth are needed. To avoid this problem, multistack VR stepper motors came into existence.

Multistack VR stepper motors: The design concepts used in multistack VR stepper motors are much simpler. On a common rotor axis, n sets of rotors each having their own sets of poles are mounted on a single shaft. In one design, all the teeth of n rotors are aligned. However, each rotor segment is covered by a separate stator phase and its own winding. The stator teeth are misaligned from one another by $1/n$ of the stator tooth pitch, where n is the number of phases, which is the same as the number of rotor segments. This situation is shown with an example in Figure 5.12, where $n = 4$ and the stator tooth pitch is 60°. In this design, the step size obtainable is 15°, which is rather large. Obviously, when the power is switched from phase A to phase B, the rotor moves by 15° to be aligned with the midstator section. This will continue as phases are switched. The operation principle is rather straightforward.

Alternative design: There are other alternative designs that can offer even much smaller step size. In one such design, the stator poles of each stack are provided with teeth, and they are energized by different phases. The stator teeth are all aligned. The rotor has three stacks, and each stack is misaligned from one another. The cut sections of three phases of such design are shown in Figure 5.13. Note that the tooth pitch of rotor teeth and stator teeth of all the phases is the same and in this case it

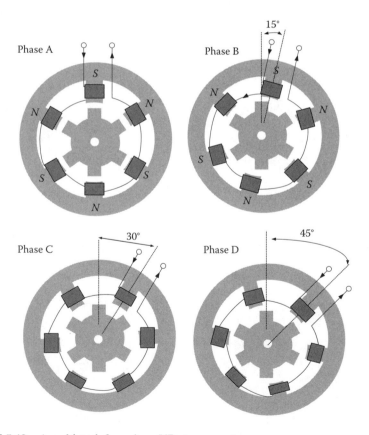

FIGURE 5.12 A multistack four-phase VR stepper motor.

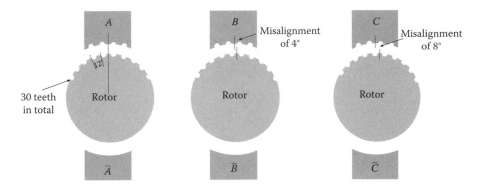

FIGURE 5.13   Stack cut section view showing misalignment.

is 12°. Assume that n is the number of phases, which is the same as the number of stator stacks. Furthermore, phase B rotor teeth are misaligned by $1/n$ of the tooth pitch from phase A stack, similarly the teeth of phase C rotor stack are misaligned by another 4° from phase B teeth. At the start, let us assume that the first stack is energized with a current in phase A. The whole stator will be aligned with stack A as shown in the first diagram of Figure 5.13.

At this instance, owing to the arrangement of stacks, phase B teeth will fall behind by an angle of 4° and phase C teeth will be by a further 4°. Hence, when the phase B is energized, while switching off phase A at the same time, the rotor will move by 4° counterclockwise. Similarly, when phase C is energized, while switching off phase B at the same time, the rotor will move counterclockwise by another 4°. Thus, if we cyclically energize A, B, C, A,..., the rotor continues to move in steps of $1/n$ of the tooth pitch, which is 4° in this example. This type of multistack stepper motors are easy to manufacture, and the step size only depends on the tooth pitch and the number of phases. Some designers arrange rotor teeth aligned, but stator teeth misaligned though the principle of operation is still the same.

Permanent magnet stepper motors: The operational principle of permanent magnet stepper motors is rather straightforward. One such motor is shown in Figure 5.14 (Kenjo and Sugawara 1994). The machine has a cylindrical permanent magnet magnetized radially and mounted on a shaft. The stator of the machine has four poles and two phase windings. Pole 1 and pole 3 have windings A and A' connected in series appropriately so that their fields will be on the same direction. Similarly, pole 2 and 4 have windings B and B' connected in series appropriately so that their fields will support each other. This implies for a particular direction of current in windings A, if pole 1 is south then pole 3 will be north. If the current is reversed, the polarity will be opposite. Let us assume that windings A is excited with pole 1 having south polarity and pole 3 having north polarity. Then, the rotor will be aligned vertically. If windings B are excited such that pole 2 is south and pole 4 is north before switching off winding A, the rotor will move clockwise 45°. If winding A is switched off at this time, the rotor will move further to align with poles 2 and 4. At this time if winding A is excited in reverse direction, the rotor will move further 45°. Now, if winding B is switched off, the rotor will move further 45°. It is important to notice that by this

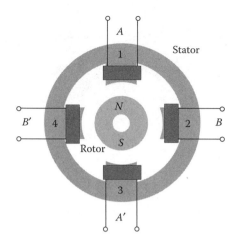

FIGURE 5.14   A simple permanent motor stepper motor.

method two step sizes can be achieved. A step size of 90° can be achieved by switching off *A* before switching on *B*. A step size of 45° can be achieved by keeping *A* and *B* "on" for a brief period and then switching off *A*.

Other stepper motor designs: There are many other designs of stepper motors such as hybrid stepper motors, linear stepper motors, and so on. Some of them are designed to achieve higher holding torque; some of them are designed for precision applications such as printers.

5.3.4   BRUSHLESS DC MOTORS

Another development due to the availability of cheap and reliable electronics and computing power is the advent of BLDC motors (Edwards 1991). They are just DC motors where the mechanical commutator is replaced by an electronic commutator. We know that usually armature windings are placed in rotors. However, since electronics can be easily fixed to a nonmoving part, the armature windings of BLDC motors are placed on the stator. The rotor has poles of permanent magnets. The only difference is that in a classic DC motor, there would be large number of commutator segments. In BLDC motors, it is uneconomical to duplicate this since large number of semiconductor switches will be needed. Hence, BLDC motors have usually three phases supplied by a three-phase inverter. There are two possible ways to connect these three phases: star or delta. The switching of the inverter must be according to the rotor position. This can be achieved by using optical sensors, or Hall effect sensors. Another alternative is to use electronic methods of measuring the back EMF, and hence finding the rotor position.

Star-connected BLDC motor: Figure 5.15 shows a typical simplified cross-sectional diagram of a motor with two poles, along with winding connections. The stator carries a three-phase winding. Typically, winding *A* consists of conductors distributed in segment "*a*" and conductors distributed in segment. "*ā*" placed in a diagonally opposite position. A similar arrangement is shown for windings of phases *B* and *C*. The rotor is mounted with a permanent magnet with its poles facing radially

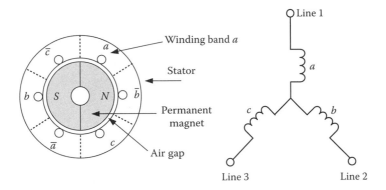

FIGURE 5.15   Star-connected BLDC motor.

outwards. Each pole spans 180°, and the flux density is usually constant through the pole surfaces. Current is supplied to the windings from a controller.

All the diagrams discussed in this section show the cross-sectional view of the BLDC motors. See Figure 5.15 for example. Typically, the conductor segments a and $\bar{a}$ are connected behind to form winding A. What we see up front are the terminals of the winding. Since the conductor segments are axially placed, they are shown as small circles, which indicate their cross-sectional view. If the current in a specific conductor segment flows toward the observer or away from the plane of the figure, it is marked as a dot inside the circle. If the current flows away from the observer or toward the plane of the figure, it is marked as a cross inside the circle. Hence, positive current I_a is denoted with a cross on segment a and a dot on segment $\bar{a}$.

The following discussion is based on Figure 5.16. In the first row the rotor position is shown on the left. Winding currents are shown on the right hand side. The current flowing from terminal toward center is considered positive. When the controller sends current from line 1 to line 2, I_a is positive, $I_b = -I_a$ and $I_c = 0$. The current directions in conductors are shown in Figure 5.16a. N pole is under conductor segments a and $\bar{b}$. Applying the LHR, we can say that the conductors a and $\bar{b}$ suffer forces that tend to move them clockwise, but they are fixed and cannot move. So the reaction moves the magnetic rotor counterclockwise. This motion is further aided by currents in the conductor segments b and $\bar{a}$, since they are under the S pole. Hence all the forces are such that the rotor rotates counterclockwise.

The situation does not change until the rotor turns 60 degrees in the counterclockwise direction. Figure 5.16b shows the position at the end of 60 degrees. For the rotation to continue, the conductor segments c and $\bar{c}$ should have currents in the appropriate direction and winding B can now be disconnected. The appropriate currents are shown in Figure 5.16b. This switching is done based on position sensor information. In this case, after the switching, currents will be as $I_c = -I_a$ and $I_b = 0$. In this manner, the rotation continues. Respective rotor positions and corresponding winding currents are shown in the subsequent parts of the diagram in Figure 5.16. After the sixth row, one cycle is completed; the condition reverts back to Figure 5.16a, and the sequence continues cyclically. In all these cases, at the end of every

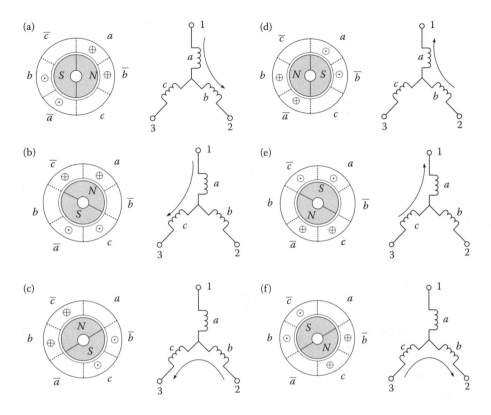

FIGURE 5.16 (a–f) Winding switching sequence for star-connected BLDC motor.

60° of rotation, the current switching is done based on either encoder feedback, back EMF sensing or by Hall effect sensor feedback.

One disadvantage of star connection is that one of the windings is not carrying current at any given time. In addition, one-third of the magnet surface is not utilized.

Delta-connected BLDC motor: Figure 5.17 shows a typical simplified cross-sectional diagram of a delta-connected BLDC motor with two poles, along with winding connections. The rotor magnets span only 120°. As before, the flux density is

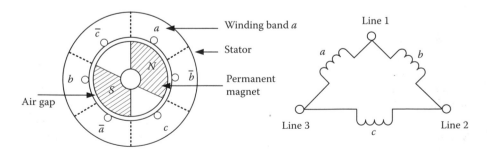

FIGURE 5.17   Delta-connected BLDC motor.

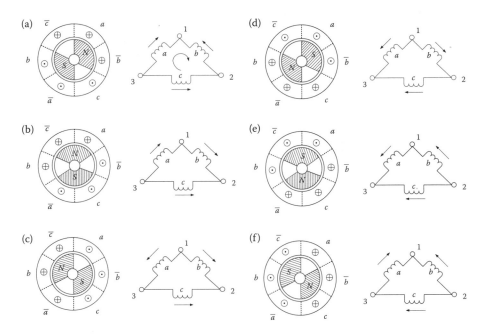

FIGURE 5.18 (a–f) Winding switching sequence for delta-connected BLDC motor.

mostly constant through the pole surfaces. Current is supplied to the winding from a controller. Winding A is formed by the conductor segment a and conductor segment $\bar{a}$. Similar notation applies to windings B and winding C. To establish a convention and a basis for determining the force directions, we assume that clockwise current that flows in the loop formed by the delta is positive. It implies that currents from terminals 3 to 1 in winding A, from terminals 1 to 2 in winding B, and from terminals 2 to 3 in winding C are positive. The sequence of rotor positions and winding and conductor currents are shown in Figure 5.18. Let us consider the rotor position shown in Figure 5.18a and the current flow shown next to it. Here, we use the terms line1, line2, and line3 to indicate supply lines connected to delta points 1, 2, and 3. The current flows from line 3 to line 1, and no current flows through line 2, since it is disconnected. In this case, $I_c = I_b$ since they are in series. According to the conventions we established earlier, they are both negative. But, I_a is positive and has a higher value since phase a is directly across the supply. The current markings in the conductor segments shown in Figure 5.18a are also according to the conventions we have established. Applying the LHR, we see that the force on conductor segments a and $\bar{b}$ is clockwise. As discussed before, since the conductors are fixed, the reactive force moves the magnets in the counterclockwise direction. As the rotor rotates through the next 60°, conductor segment $\bar{c}$ is covered and conductor segment $\bar{b}$ is uncovered by N-pole without any change in torque. Similar things happen to the conductors under S-pole where conductor segment c is getting covered and conductor segment b is getting uncovered without any change in force or torque, since both phases carry the same current.

At the end of this 60° counterclockwise rotation, rotor position is shown in Figure 5.18b. At this point, the current in winding B, which is formed by conductor segments b and $\bar{b}$, alone should be reversed to achieve further motion. Therefore, a new connection is made such that current flows from line 3 to line 2. Now, the motion continues counterclockwise as before. At the end of the 60° rotation, rotor position is now shown in Figure 5.18c. These principles are similar until the sixth row in Figure 5.18, after which the situation reverts back to the conditions shown in the first row. In all these cases, at the end of a 60° rotation, the current switching is done based on either encoder feedback, by back EMF sensing or by Hall effect sensor feedback.

The advantages of a delta-connected machine over star-connected machines are twofold. First, current flows in all three windings resulting in a better use of windings. Second, the volume of permanent magnet material used is only two-thirds the equivalent of star-connected machine. Hence, the majority of BLDC motors are delta-connected machines.

5.4   DRIVE SYSTEMS

In the previous section, we discussed the operation principles of commonly used motors in robotics systems. These motors need control mechanisms to deliver speed, motion, and position requirements of the robot designed. In this section, we describe some of drive schemes used for this purpose.

5.4.1   DC MOTOR CONTROL

In robotics, a very tight control of speed and position of the motors is needed. This is normally achieved by a feedback mechanism using the optical encoders discussed in earlier chapters and a driver circuit. Many commercially available DC motors come invariably with optical encoders and gear heads attached to them. To control DC motors, there are many drive circuits commercially available. Some of them are analog devices based on FETs and power transistors. However, the majority of such devices are bridge drivers, which are very cost effective and energy efficient.

Basic principle: The basic principle behind a bridge driver (commonly known as H bridge) can be explained using the circuit shown in Figure 5.19. When the

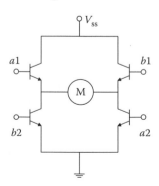

FIGURE 5.19   H-bridge driver principle.

transistors $a1$ and $a2$ are activated, current flows from left to right through the motor. On the other hand, when $b1$ and $b2$ are activated, the current flow through the motor is reversed and thus it becomes a bipolar device. When there is no transistor activated, there will be no current fed to the motor. During the transition, there is a possibility of dead-short of the power supply. Assume that transistors $a1$ and $a2$ are on and $b1$ and $b2$ are off. To reverse the current, $a1$ and $a2$ should be turned off and $b1$ and $b2$ should be turned on. Under fast-switching conditions, there is often a situation where turn-on happens before the other pair has completely turned off. This results in a transient dead-short of the supply, which causes heating. The number of such dead-shorts will increase as switching frequency increases, thus limiting the switching frequency. There are electronics as well as software means to avoid a dead-short. Commercially available driver ICs avoid dead-short problems with built-in circuitry and furthermore they provide many other safeguarding features.

DC motor controllers and their operation: The switching components shown in Figure 5.19 are incorporated in commercial DC motor controller ICs that are supplied by many manufacturers. These integrated units also provide additional conveniences such as protection from overheating, line to ground short circuits, current limiting (chopping) features, and so on.

Figure 5.20 shows the input signals and motor connections of typical DC motor controllers available commercially. The H bridge shown in Figure 5.19 is the core of such DC motor drivers. The protections, current chopping, and so on are built around the H-bridge core and only the signal and output lines, relevant for the user, are provided as pins. The motor terminals are floating. This circuit is designed for pulse width modulation (PWM) control of DC motors. Many commercially available devices are capable of supplying a current of around 2.5 A through the load at the operating voltage of up to 40 V. For control purposes, "phase" and "enable" lines are provided. A high in the "enable" line and a high in "phase" line will send current through from $m0$ to $m1$, through the motor. With the enable line still being high, if the "phase" signal goes low, the applied voltage appearing across the motor will reverse. When "enable" goes low, the output terminals go into high impedance mode.

Type 1 application: In one type of application, the PWM is connected to the "enable" line. The enable signal may be of high active or low active type. In general,

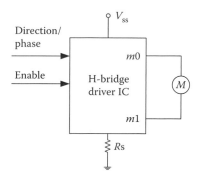

FIGURE 5.20   Block diagram of a commercial DC motor controller.

the "phase" line controls the direction of the motor and "enable" (PWM) controls the speed of the motor.

When a computer calculates the manipulated variable, the magnitude is used to fix the PWM duty cycle and the phase signal is set high or low depending on the sign, high for positive or low for negative. Hence, if a high positive manipulated variable produces high current from, say, $m0$ to $m1$ through the motor, a high negative manipulated variable will produce high current from $m1$ to $m0$ through the motor. This is the most popular method of using the motor controller.

Type 2 application: Some users prefer to use the enable line for on/off control, while supplying the PWM to the phase line. By this method, a 50% duty cycle in the phase line commands the motor to move forward and reverse at the PWM frequency, for which the motor does not respond and stands still. A PWM with duty cycle higher than 50% moves the motor forward and a PWM with duty cycle <50% moves the motor in the opposite direction. This method begets faster response of the motor, but heats up the switching devices, needing heavier heat sinks. A further analysis of this method will be presented in Chapter 10. There are many other DC drive devices, employing similar principles frequently used in robotic applications.

Some examples of commercial DC motor controllers: Two popular examples of DC motor controllers are Allegro A4973 (Allegro 2012) and L6203 (ST-Microelectronics 2012). They both have an H bridge as their core and other circuits are built around it in a single package. The A4973 provides the input signals as explained earlier. The IC provides low active enable line $\bar{E}$ and phase line distinctly. Full schematic and application notes can be found in Allegro (2012). In addition to its basic motor control, this IC also provides internal circuit protection including motor lead short-to-supply/short-to-ground, thermal shutdown with hysteresis, undervoltage monitoring, and crossover-current protection.

Another popular full bridge driver is L6203. The full schematic of this device can be found in ST-Microelectronics (2012). In L6203, high active enable line is provided, but the phase effect is achieved by two input signals IN1 and IN2. That is, instead of a phase input, two other inputs, namely, IN1 and IN2 are provided. Referring to the basic H bridge shown in Figure 5.19, when IN1 input is set to high and IN2 is set to low, the top left and lower right FETs of the bridge circuit will conduct and pass the current through the motor in one direction. If IN1 is low and IN2 is high, then the top right and lower left FETs will conduct and thus reverse the current. Hence, we need to generate IN1 and IN2 from a single external phase signal. L6203 also provides the protections mentioned above.

5.4.2   STEPPER MOTORS DRIVERS

The main functions needed for stepper motor controllers are switching sequence generation, power current driving, and current limiting and regulation. All the sequence signals can be generated using computers, microprocessors, or microcontrollers. This will need programming efforts, and execution will consume processor time. Employing a separate sequence generator will enable the processor to deal with higher-level functions rather than performing low-level power-driving functions. Consequently, this also simplifies the programing task. As discussed previously, there are many types of stepper motors. We will not be able to describe drivers for all types of motors, but we will give some examples.

5.4.2.1 Sequence Generator

Most stepper motor drive systems come in pairs. One of them is a sequence generator that cannot drive the motor on its own. The other one is a driver which uses the signals from the sequence generator and drives the currents through the windings appropriately.

The sequence generator should be able to

- Provide drive signals to the power driver in appropriate sequence for forward or reverse motion.
- Run the stepper motor in half or full steps.
- Provide the current control signals to the driver for safety of motor and driver.

For this purpose, the sequence generator normally takes the following signals: clock for deciding the speed, half/full step, direction, enable, current feedback signals, and current reference signals. One such popular sequence generator is SGS Thompson L297 IC (ST-Microelectronics 2012), which may be used independently using discrete power semiconductor components or along with other bridge-based chopper drivers. We will describe functions, capabilities, and applications of L297. In particular, we will see drivers for four-phase unipolar motors and two-phase bipolar motors. A block diagram of L297 is shown in Figure 5.21 to serve our purpose. The L297 can be used with any other power stage or standard power driver. As we mentioned above, this sequencer takes in basic input from a microprocessor and generates output signals required for the power stage using the built-in internal logic. The outputs are phase signals A, B, C, and D, and inhibit signals $\overline{INH}1$ and $\overline{INH}2$. The input signal "control" decides the mode of current control.

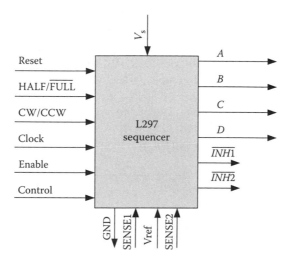

FIGURE 5.21 Block diagram of a sequence generator L297.

In any power driver system, the current pumped into the motor needs to be regulated for the sake of torque control and for the safety of the motors and electronics. This needs current feedback. For the current feedback, three inputs Vref, SENSE1, and SENSE2 are provided. Using these signals, L297 is capable of producing two types of current control either for fast current decay mode or slow current decay mode. The choice is made by means of input line "control."

5.4.2.2 Operating Modes of L297

There are four logic outputs (A, B, C, and D) from L297, which can produce 16 states. But the hardware limits it to only eight states as listed below in Table 5.1. The device transits through the states cyclically according to the choice of user control. The home state is stage 1 where $ABCD = 0101$. Depending on the instruction, the sequencer output combination can move

1. Through all states from state 1 to state 8 sequentially
2. Only through even states, that is, 2,4,6,8,2
3. Only through odd states, that is, 1,3,5,7,1

The two additional output signals $\overline{INH1}$ and $\overline{INH2}$ are generated according to the following logic in all modes:

$$\overline{INH1} = A + B$$
$$\overline{INH2} = C + D$$

$$(5.32)$$

These two signals are used for current termination. However, a robot designer has the choice of using either phase signal or inhibit signal for current control. In following, we will see the modes of operation one by one.

Transition through all states (half-step mode): This is achieved by providing a logic high signal to the *HALF/FULL* line of L297 at any time. When the sequence generators move through all states 1, 2, 3, 4, and so on, the ON states of the output signal change as A, AC, C, CB, B, BD, D, DA, A, AC, and so on. It means that states transit as ABCD = 1000, 1010, 0010, 0110, 0100, 0101, 0001, 1001, 1000, 1010, …. This is also called the half-step mode, since when it is applied, a stepper motor moves in half steps.

Transition through even states (wave mode—full step): Transition through even states is *not* straightforward. This is achieved by providing a logic low signal to the *HALF/FULL* line of L297 at an even state. When the sequencer moves through even

TABLE 5.1
L297 States

Stage	1	2	3	4	5	6	7	8
ABCD	0101	0001	1001	1000	1010	0010	0110	0100
Logic ones	BD	D	AD	A	AC	C	BC	B

states, it results in "one phase ON" states 4, 6, 8, 2, which implies that only one phase is ON and other phases are OFF at any given instant. The sequence of switching is $A,C,B,D,A,\ldots$. This can be used to run unipolar four-phase stepper motors. $INH1$ and $INH2$ are generated according to the same logic given in Equation 5.32. Application of this mode typically produces the full step motion of stepper motors. We will see examples in a later section.

Transition through odd states (full step mode): Once again, the implementation of this mode is not straightforward. A logic low signal is provided to $HALF/\overline{FULL}$ line of L297 at an odd state to achieve this mode. Thus, the states move through only odd states such as $1,3,5,7,1,\ldots$. This results in two phases ON at any given instant. The states move through $DA, AC, CB, BD, DA,\ldots$. This mode is usually applied to motors where winding currents can be reversed. As discussed in the previous section, the application of this mode results typically in the full step motion of stepper motors. It is important to note that for the logic sequence given above that $INH1$ and $INH2$ lines are always high which forces the designer to use only phase lines for current regulation.

5.4.2.3 Applications

We have seen the three modes of operation of the sequence generator in the previous section. Here, we will present how those sequences generated can be used to control a stepper motor. We use a driver using discrete components as shown in Figure 5.22.

Operation of four phase VR stepper motor with unipolar windings: The stepper motor in our example has four windings, each passing through two poles; hence, there are eight poles in the motor. Furthermore, currents cannot be reversed. The motor and its windings are shown in Figure 5.23. From an earlier discussion on stepper motors, we can deduce that the step size of the motor shown in this figure is 9° in full step mode. Since there are four windings, the four-phase signals can be readily used for this motor and L297 can be used to create the sequencing in all the three modes of the L297. The power stage is built using four power-switching devices as shown in Figure 5.22.

In half step, "all" state mode, the switching ON sequence is A, AC, C, CB, B, BD, . . . , which will result in half-step motions of stepper motor. Referring to Figure 5.23, the step size in this case will be 4.5°. In full step wave mode of even state

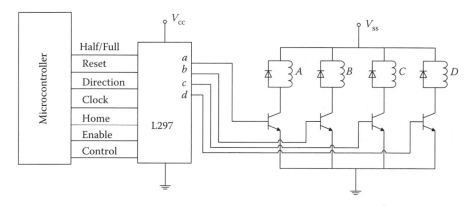

FIGURE 5.22 Typical discrete component-based driver for unipolar four-phase stepper motor.

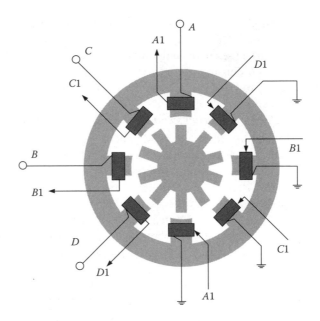

FIGURE 5.23   A unipolar four-phase stepper motor example.

operation, the switching ON sequence is A, C, B, D, A,... , and it will yield full step motions of the motor. In full step odd state operation, the sequence moving through AC, CB, BD, DA, ... , and the rotor step size is 9°. In this mode, at no stage is the rotor pole aligned with any stator pole.

Current control: There are two ways of current control offered by L297. One method uses INH1 and INH2 signals, and the other method uses phase signals along with chopping oscillator and a complex circuitry. While using discrete component implementation, it is rather difficult to use these methods.

Direction control: When input signal CW/CCW is low, the switching ON sequence of states will reverse, and this in turn will reverse the motor direction.

Four-phase permanent magnet stepper motor with unipolar windings: An example of such a motor is shown in Figure 5.24. Consider the upper and lower poles in the figure, with coils wound on them where each coil has two windings. Individual windings on the upper pole coil are connected in series to the windings in the lower pole coil. When the current flows from A to A', a magnetic field is established in one direction. Similarly, when the current flows B to B', the magnetic field will be in the opposite direction. Hence, we can reverse the magnetic field without having to reverse the current in any coil. This type of winding is called bifilar winding. The same configuration of the driver circuit shown in Figure 5.22 can also be used here. However, the full step size is 90° for this motor.

In Figure 5.24, we have assumed that when A is ON, upper pole is S and lower pole is N and when C is ON the right-side pole is S and left-side pole is N. When B is ON, upper pole is N and lower pole is S; when D is ON, right pole is N and the left pole is S. When two ON states occur, for instance, A and C are ON, then the upper pole and the right side pole both become S and left and the bottom poles become N.

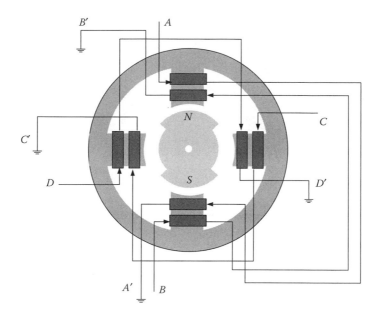

FIGURE 5.24 Permanent magnet motor with unipolar windings (each pole has two unipolar windings) controllable in all three modes by 297 using circuit shown in Figure 5.22.

In half step, all state mode, the switching sequence is *A, AC, C, CB, B, BD,* ... , which will result in half-step operation. In full step wave mode (even states), the sequence is *A, C, B, D*. This will produce a full step operation. In full step mode (odd states), the outputs move through the sequence of *DA, AC, CB, BD, DA,* .... However, the rotor poles do not align with stator poles. They move from one midpoint between the two stator poles to the other midpoint, still maintaining the step sizes of 90°.

Two-phase permanent magnet stepper motor with bipolar windings: Bipolar windings should have reversible currents on each winding; this requires both terminals of each winding to be floating. The circuit shown in Figure 5.22 will not serve the purpose. In such cases, L297 is used in collaboration with a driver L298 or any other dual H-bridge driver, which can reverse the current in a winding (refer to ST-Microelectronics (2012) for more application details). The example step motors we gave here are all have large step sizes, and this is for simplicity of illustration. In real applications, the number of poles is usually quite large, thus resulting in small values of step sizes.

5.4.3 BRUSHLESS DC MOTOR DRIVE

BLDC motors are applied in many fields of engineering apart from robotics. There are two broad types of controllers depending on whether the direction needs to be changed or not. In some applications, such as drones and hobby planes, the motors need not change direction. Surprisingly, owing to enormous demand for such applications, there are many ready-to-use "sensor-less" controllers in the market. The second type of controllers provides bidirectional rotations, and they use feedback

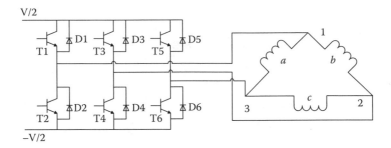

FIGURE 5.25 Bridge circuit to drive a BLDC motor (delta connection).

devices. These feedback devices are encoders, Hall effect sensors, or back EMF sensors, whose outputs provide appropriate locations of the rotor at which the switching of the windings to be effected. A generic bridge drive circuit used for driving BLDC motors is shown in Figure 5.25.

5.4.3.1 Back EMF Sensing-Based Switching

It is easier to describe the operation of EMF-based sensing switching with star-wound machines. At any given time, one winding is free from any electrical excitation, but that winding is placed in a moving magnetic field. Hence, there is EMF induced on that winding. This property can be easily exploited to find proper instants of switching.

We look at the back EMF transitions for winding C. Let us look at the case depicted in the first row of Figure 5.16. The current in winding C is just switched off. Figure 5.26 shows the current flow directions in conductor segments a and b in this case. Since, winding "C" is disconnected, there is no current in that winding. However, EMFs are always induced in all the windings, which apply to unconnected winding C also. At the instant shown in Figure 5.26b, segment $\bar{c}$ is marked with a cross and segment c is marked with a dot to indicate the direction of induced EMFs. The two segments are connected behind to form coil C. The EMFs add to each other in the loop formed by the segment $\bar{c}$ and c. The front end of segment c is positive, and hence the EMF induced in winding C is positive maximum. As the rotor continues to rotate counterclockwise,

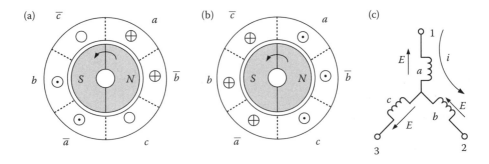

FIGURE 5.26 The situation when c is not connected and the rotor is rotating counterclockwise. (a) Current. (b) Induced EMF. (c) EMF and current directions.

the winding that is spanned by *N*-pole will be approached by *S*-pole. Hence, the EMF magnitude starts decreasing and at the point when *S*- and *N*-poles equally span the conductor segment "*c*" (when the magnetic neutral axis of the rotor is in middle of segment "*c*"), the EMF becomes zero. It takes 30° of rotation to get zero back EMF and the next 30° of rotation will reverse the sign of the induced EMF as conductor segment "*c*" gets completely covered by *S*-pole. Hence, the transition takes 60° of rotation. This instant is shown in the second row of Figure 5.16. The negative back EMF remains during the next 120° of rotation. Then the next transition from negative to positive maximum will take place during the further rotation of 60°. This positive maximum back EMF will remain for the next 120° of rotation. At the end, the cycle will repeat itself.

In summary, as the motor rotates, the back EMF patterns go through the following sequence:

a. Remain at positive value for 120° of rotation
b. Transition from positive to negative value during the next 60° of rotation
c. Remain at negative value for the next 120° of rotation
d. Transition from negative to positive value during the next 60° of rotation

The cycle occurs for all three windings. Since the back EMF transitions indicate the position of the rotor, we can decide when the switching of windings should be done, based on the back EMF observation. The switching-on and switching-off timings of the three windings have different phase relationships.

In practice, a combination of analog-sensing electronics and fast-computing devices are used to implement the switching. This will also involve some electronic design. There are a few problems in implementing a back EMF sensor-based switching scheme:

Usually, the neutral point may not be always available.
There is PWM switching going on all the time for speed and current control; therefore the waveforms are not easy to interpret.

A general schematic diagram for implementing such scheme is shown in Figure 5.27.

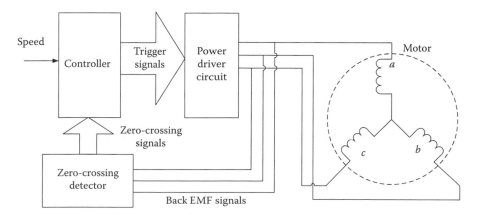

FIGURE 5.27   Generic scheme of back EMF sensing-based BLDC motor.

The general idea is that the controller processes the zero crossing signals and determines the appropriate instants of switching. At those appropriate instants of switching, the trigger signals are sent to the three-phase bridge inverter. There are many manufacturers who have solved the problem of determining the right instants of switching from zero crossings of the back EMF waveforms and provided off-the-shelf solutions. AVR444 is a good example of such a ready to use device (Atmel 2012). The method of switching is quite similar to the concept discussed above. Back EMF signal conditioning is done using a suitable filter due to the presence of switching noise. Since back EMF measurement is quite tricky at low speeds, a preprogrammed sensorless switching is implemented during start-up, and as speed picks up the back EMF-based switching is implemented.

5.4.3.2 Sensor-Based Switching

In earlier cases, we discussed the indirect way of locating the rotor position to determine the instants of commutation. However, if we have sensors fitted to sense the rotor position, then it is a straightforward task to do switching. Considering a two-pole machine, let us assume that Hall effect sensors are fitted around the stator spaced 120° apart. Then, we can see that *N*-pole, as well as *S*-pole, will hit the sensor three times, making the total hits six times in one rotation. Needless to say, if there are four poles in the system, they will be spaced 60° apart. By appropriately spacing the sensors, we can directly derive the instants of commutation. This method is a bit more expensive and requires more wiring, but completely eliminates the computational requirements. A block diagram of such a system is shown in Figure 5.28. For a four pole winding, the spread of the sensors need to be only 60° apart. As the rotor rotates, the poles "hit" the sensors and provide switching signals every 30° of rotation.

A commercial system to achieve the above control is available from Atmel with ICs ATA6832, ATmega88, and ATA6624. The system consists of three integrated circuits, Microcontroller ATmega88, Triple Half Bridge Driver ATA6832, and LIN System Basis Chip ATA6624 (Atmel 2012).

Ready pairs: There are also working pairs of a BLDC motor and its controller available on the market. Hobbywing Pentium-85A that drives a fan motor is an

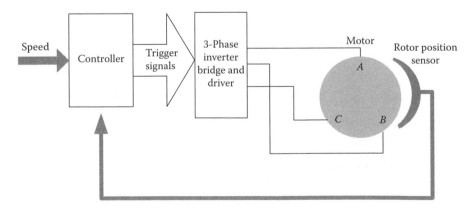

FIGURE 5.28   Block diagram of Hall effect sensor-based control of BLDC motors.

example (Hobbywing 2012). Such controller systems are capable of controlling the motor using RC signals or PWM signals connected physically through a standard three wire logic. These motor and controller pairs are mostly used in hobby model planes. The motors are capable of speeds of 50,000 rpm, and currents can reach up to 60 A.

5.5 CONCLUSION

In this chapter, we provided concise information regarding the electrical drive schemes that are commonly used in robot design and highlighted the practical issues. We presented basic DC electric motors and their operating principles using basic electrical laws that govern them. We presented a brief overview of servo motors. We furthered our discussion to describe more sophisticated actuators such as AC machines, stepping motors, and finally BLDC motors. Wherever necessary, we have provided information regarding the control of these machines, since accurate drive control is imperative to achieve precise robot motion.

REFERENCES

Allegro website. 2012. http://www.allegromicro.com/Products/Motor-Driver-And-Interface-ICs/Brush-DC-Motor-Drivers/A4973.aspx.

Atmel website. 2012. http://www.atmel.com.

Clayton, A.F. 1969. *The Performance and Design of Direct Current Machines*. London: Sir Isaac Pitman & Sons, Ltd.

Cotton, H. 1970. *Electrical Technology*. London: Sir Isaac Pitman & Sons, Ltd.

Edwards, J.D. 1991. *Electrical Machines and Drives*. Hampshire: Macmillan Education Ltd.

Faulhaber, 2011. *Miniature Drive Systems Catalogue*.

Fitzgerald, A.E., Kingsley, J.C. and Umans, S.D. 1990. *Electric Machinery*. New York: McGraw-Hill Publishing Company.

Hitec website. 2012. http://www.hitecrcd.com/products/servos/analog/standard-sport/hs-422.html.

Hobbywing website. 2012. http://www.hobbywing.com.

Kenjo, T. and Sugawara, A. 1994. *Stepping Motors and Their Microprocessor Controls*. Oxford: Oxford University Press.

Kuo, B.C. 1979. *Incremental Motion Control (Vol. II)—Step Motors and Control Systems*. Champaign, IL: SRL Publishing Company.

Langsdorf, A. 2001. *Theory of Alternating Current Machinery*. New Delhi: Tata McGraw-Hill.

McKenzie-Smith, I. and Hughes, E. 1995. *Hughes Electrical Technology*. Englewood Cliffs, NJ: Prentice-Hall.

Say, M.G. 1984. *Alternating Current Machines*. London: Pitman Publishing.

ST-Microelectronics. 2012. http://www.st.com/.

6 Motor Power Selection and Gear Ratio Design for Mobile Robots

6.1 GEAR RATIO FOR A MOBILE ROBOT

We have seen the various types of actuator motors and drive systems to power the robot motion. The most prominent form of drive used in robotics is electrical motors, which also come in various types such as DC motors (with commutator), brushless DC motors, DC servomotors, stepper motors, and so on. During the design phase, suitable motors have to be decided for the robot. This selection is usually done based on the experience and the specific needs of the robot. For example, if the desired robot motion is continuous, a DC motor can be selected and if there are motions in steps, then stepper motors can be chosen. Once the motor type is chosen, the next task is to decide power and torque requirements. In general, robots driven by such motors may need a high torque up to several newton-meters, even though they need to move relatively slow. We have seen the torque equation of motors in Chapter 5 on drives. The power developed by a motor is the product of the angular speed and the torque developed. If the motor develops a certain torque of τ Nm and runs at a speed of n revolution/s, then the equivalent power, in watts, is given as $2\pi n\tau$. To keep the robot at a reasonable weight, the motors should be light and small. Such small motors inherently develop low torque measured in milli-newton-meters. The motor needs to rotate faster, up to a few tens of thousands of revolution per minute (rpm), to achieve the high power required. This presents a conflicting situation where we have to use low-torque high-speed motors to power robot loads that move relatively slow, but require high torque. Therefore, the primary reason for using gears in any system is load matching since the high-speed low-torque motors have to drive low-speed high-inertia/friction loads requiring heavy torque. For example, it is not a good idea to drive a car up a slope in fourth gear. Car drive systems provide many selectable gear ratios, so that the driver can choose a ratio according to the circumstance. In robotic systems, it is quite difficult to have a gear-changing mechanism since it will make mechanical design cumbersome and complicated. In addition, limitations on robot size will preclude this approach. In robotics, only one gear ratio is used as shown in Figure 6.1, and a proper gear ratio is often decided by a trial-and-error method. This chapter aims to discuss the methods that may be useful in choosing the appropriate gear ratio (Kanniah, Ercan et al., 2004).

Inertia equivalent values reflected across a gear box: It often becomes necessary to calculate the reflected value of inertia across the gear box during the design process. We see below how the load inertia will appear at the motor side and motor

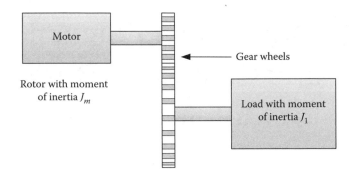

FIGURE 6.1 A typical use of speed reduction gear.

inertia will appear at the load side when there is a gear box between load and motor. However, the power is invariant from whatever side we see it due to the law of conservation of energy. Assume that the angular velocities at the load side and motor side are denoted by ω_l and ω_m, respectively. Let the motor moment of inertia be J_m and the equivalent reflected motor moment of inertia at the load side be J_{me}.

For a given angular acceleration of motor $d\omega_m/dt$, the torque required is given as $J_m(d\omega_m/dt)$.

The power at the motor side is defined by the product of the torque and the angular velocity. Hence, the power at the motor side is given by

$$P_m = J_m \frac{d\omega_m}{dt} \omega_m \tag{6.1}$$

Considering only the motor moment of inertia at the load side, the power at the load side can be written as

$$P_l = J_{me} \frac{d\omega_l}{dt} \omega_l \tag{6.2}$$

where J_{me} is the equivalent reflected value of the motor moment of inertia at the load side.

Using the law of conservation of energy, both the above power terms can be equated:

$$J_{me} \frac{d\omega_l}{dt} \omega_l = J_m \frac{d\omega_m}{dt} \omega_m \tag{6.3}$$

If the gear reduction ratio is N_g, we can write

$$\omega_m = N_g \omega_l$$

Substituting this into Equation 6.3, we get

$$J_{me} \frac{d\omega_l}{dt} \omega_l = J_m N_g^2 \frac{d\omega_l}{dt} \omega_l \qquad (6.4)$$

Hence

$$J_{me} = J_m N_g^2 \qquad (6.5)$$

Similarly, if the load inertia is J_l, the equivalent reflected value of the load inertia J_{le} on the motor side can be derived as

$$J_{le} = \frac{J_l}{N_g^2} \qquad (6.6)$$

6.2 POWER REQUIREMENT OF THE DRIVE MOTOR

Let us continue our discussion of motor power selection. After the type of motor is decided, the next task is to find the power rating of that motor. In the following discussions, commutator DC motors will be used as an example; however, the ideas developed can be easily modified and applied to other types of motors. The power requirement of the drive motor is a complex issue as it depends on the specific application. In robotics, speed and acceleration as well as accuracy are the major concerns. Any robot may have to achieve a velocity profile. It may be the change of angle of a joint or motion of a robot on a surface. The surface may be horizontal or inclined. Let us consider a mobile robot that has to adhere to a velocity profile to follow as shown in Figure 6.2. It is also assumed that the terrain is not horizontal and the robot is climbing on a slope as shown in Figure 6.3. Assume that the mass of the robot shown in these figures is M (kg), the required acceleration is a (m/s²), and the

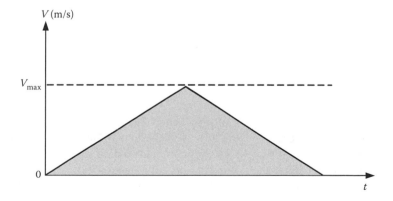

FIGURE 6.2 Desired velocity profile.

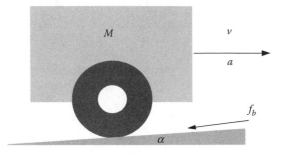

FIGURE 6.3 Robot moving on a mild slope.

maximum velocity required is V_{max} (m/s). The slope of climbing shown in Figure 6.3 is described by $\sin(\alpha)$, and the linear equivalent friction is B, which is measured in newton-second/meter.

To achieve the acceleration, the required force on the wheel contacts with the surface will be

$$f_a = Ma \tag{6.7}$$

The force required for overcoming gravity is

$$f_g = Mg\sin(\alpha) \tag{6.8}$$

and the force to overcome friction is

$$f_b = Bv \tag{6.9}$$

where v is the velocity of the robot. Hence, the total force can be given as

$$f = f_a + f_g + f_b \tag{6.10}$$

and the maximum power requirement in watts can be written as

$$P = SfV_{max} \tag{6.11}$$

In Equation 6.11, S is the factor of ignorance. The factor S has to be more than 1, while its actual value depends on how well the uncertainties in the system are estimated, such as rolling friction, gear friction, and so on. Depending on the robot design, forces acting upon the system that affect Equation 6.10 will be different, and hence the calculation of power in Equation 6.11 will also be different. The basic idea is to find out what maximum torque or force is required at the maximum angular velocity of the motor or linear speed of the robot. In other words, the worst loading condition should be tackled so that less severe conditions will be covered automatically.

EXAMPLE 6.1

Assume that for the robot shown in Figure 6.3 values for the parameters are given as follows:

$$M = 2 \, kg$$
$$B = 1.8 \, Ns/m$$
$$V_{max} = 2 \, m/s$$
$$a = 2 \, m/s^2$$
$$\alpha = 5.71°$$
$$S = 1.2$$

Using Equations 6.10 and 6.11, the power needed for the robot can be calculated as

$$P = 1.2 \times \left(2 \times 2 + 2 \times 9.81 \times \sin(5.71) + 1.8 \times 2\right) \times 2 = 22.95 \, W$$

6.2.1 Role of Motor Inertia and Friction

All motors have their own friction, mostly in their sleeve bearings and commutators. They do have some inertia as well. The fact is that most high-speed motors are made of air-core armatures, with very little inertia. Motor manufacturers usually include these figures as standard specifications in their manuals and data sheets. However, it is important to consider the effect of the motor friction and inertia when they are used in robots with appropriate gears. Let us consider an example to compare the relative effects of these quantities with respect to the overall system values, which include loads as well.

EXAMPLE 6.2

Motor inertia, J_m, for a typical 27 W motor is given as

$$J_m = 20 \times 10^{-7} \, kg \, m^2$$

Using Equation 6.5, if the gear ratio is 10, then the reflected motor inertia on the load side J_{me} is

$$J_{me} = 10^2 \times J_m = 20 \times 10^{-5} = 0.0002 \, kg \, m^2$$

This is the motor inertia reflected at the wheel. Now, let us assume that we are driving a load of 2 kg on wheels of 0.03 m radius mounted on an axis to which the output side of the gear system is attached. We need to compute the moment of inertia of the load at the drive wheel.

Assuming an acceleration of a, the linear force to be supplied by the drive wheel is given by

$$f_w = M a$$

Assuming that the drive wheel radius is R_w, the torque from the drive wheels is given by

$$t_{w1} = M a R_w \tag{6.12}$$

Looking at the wheel side, let the equivalent moment of inertia of the load be J_l; then, the drive wheel torque can be written as

$$t_{w2} = J_l \frac{d\omega}{dt} \tag{6.13}$$

Since the linear velocity is the product of the wheel radius and the angular velocity of the wheel

$$v = R_w \omega \tag{6.14}$$

Substituting for ω from above into the second torque equation, we get

$$t_{w2} = \frac{J_l}{R_w} \frac{dv}{dt} = \frac{J_l}{R_w} a \tag{6.15}$$

Since both torque values must be the same, let us equate the above two equations for the drive wheel torque

$$t_{w1} = t_{w2}$$
$$M a R_w = \frac{J_l}{R_w} a \tag{6.16}$$

This yields

$$J_l = M R_w^2 \tag{6.17}$$

Then, the moment of inertia of load "seen" on the driving wheel J_l can be calculated as

$$J_l = 2(0.03)^2 = 0.0018 \text{ kg m}^2$$

A comparison of J_l and J_{me} reveals that the motor inertia is quite small when compared to that of the load it is driving.

6.3 TYPICAL MOTOR CHARACTERISTICS DATA SHEET

Manufacturers list quite a number of specifications of their motors in data sheets. This information is valuable during the design process. An example data, based on Portescap minimotor manufacturer's information on ESCAP® 28 DT 12-222E DC motor, is shown in Table 6.1 (Portescap 2013).

There are other information and characteristics, though they are not critically important at this stage. We have seen in the chapter on drive systems that K_b and K_t must be numerically the same Chapter 5. However, they are different in Table 6.1, since the value for K_b is given in units of V/1000 rpm. For the analysis provided in the later sections, the values of K_b in Vs/rad are needed. Let us convert the units from V/1000 rpm to Vs/rad for this case as shown in Example 6.3.

EXAMPLE 6.3

$$K_b = 3.4 \text{ V/1000 rpm}$$
$$= 0.0034 \text{ V/rpm}$$
$$= 0.0034 \times 60 \text{ V/rps}$$
$$= 0.0034 \times 60/(2\pi) \text{ V/rad/s}$$

Hence

$$K_b = 0.0325 \text{ Vs/rad} = K_t$$

In summary, whenever we need K_b in the unit of Vs/rad in our calculations, we can take K_t from the catalog and readily use it.

Some of the motor parameters in the above list immediately help in the design process. For example, rotor inductance to rotor terminal resistance ratio (L/R) should be considered in deciding the PWM frequency. The L/R ratio is also the

---

TABLE 6.1
Typical Motor Data Sheet Information

Characteristics	Specification
Voltage	24 V
No-load speed (full voltage applied with no load on shaft)	6900 rpm
Stall torque (full voltage applied, but shaft arrested forcibly)	126 mNm
No load current	110 mA
Maximum continuous current	1.4 A
Maximum speed—recommended	9000 rpm
Maximum angular acceleration	91,000 rad/s²
Maximum continuous power	37 W
Back EMF constant, K_b	3.4 V/1000 rpm
Torque constant, K_t	32.5 mNm/A
Rotor inductance, L	0.75 mH
Rotor terminal resistance, r_a	6.2 ohms
Rotor moment of inertia	20×10^{-7} kg m²

time constant of the current path. If the L/R ratio is high and the PWM frequency is also too high, this will reduce the duration of the applied voltage and the current will have no time to rise.

Another instantly useful parameter is maximum acceleration. For example, the maximum acceleration given in Table 6.1 is 91,000 rad/s^2, which means that this motor can reach a speed of $(91,000 \times 60/2\pi) \times 0.01 = 8690$ rpm in 0.01 s. A higher acceleration rate will damage the motor mechanically. Apparently, the design should not push the motor beyond this rate.

6.4 FRICTION MEASUREMENT IN A LINEAR MOTION SYSTEM

A good robot design depends upon reasonable knowledge of robot parameters. Some of these parameters are obtained easily, such as the mass of a robot, which can be weighed with no trouble. However, measuring friction parameter is not very straightforward. It is not possible to use the values in specifications provided in data sheets to compute the overall friction coefficient. Moreover, when we are concerned about friction, there are so many friction coefficients involved, such as friction of motor bearings, commutators, friction of wheel bearings, and friction of the gear train, to name a few. To cloud the picture further, there are other loss-making elements as well, such as rolling friction of the rubber tire on the wheel of a mobile robot. It may be possible to measure all these individually, but it is not practical. What is needed is an approximation of the overall picture of friction interfering with the motion of the robot. It is important to note that the friction is a highly nonlinear phenomenon, even without considering static friction, which results in requiring a force to get the robot moving initially from standstill. However, the friction force can be approximately considered as a linear function of velocity during motion.

Any linear motion of a moving body may be considered to consist of a combination of a mass and an overall friction coefficient acted upon by a force. How do we get a reasonable idea about the linear motion friction coefficient? It is possible to devise a simple experiment to measure this friction as shown in Figure 6.4.

In this arrangement, the robot is allowed to slide down from the top to the lower end of the platform and we measure the time taken. Some precautions are necessary during the experiment. First, the slope angle θ cannot be very large. Second, the robot should travel down the slope in a straight line.

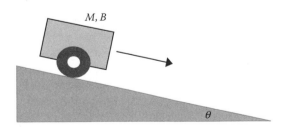

FIGURE 6.4   Measuring overall linear friction.

Let us perform a simple analysis. Assume that the distance traveled is X meters along the platform, and the time taken by the robot to travel is T seconds. We know that

$$M g \sin \theta = M \frac{d^2 x}{dt^2} + B \frac{dx}{dt} \qquad (6.18)$$

where x is the distance measured from the starting point. By assuming that all the initial conditions are zero and taking the Laplace transform, we get

$$\frac{Mg \sin \theta}{s} = [M s^2 + B s] X(s) \qquad (6.19)$$

Substituting $\beta = (B/M)$, which is the corner frequency of the system response and taking the inverse Laplace transform, we obtain

$$x(t) = \frac{g \sin \theta}{\beta} \left[t - \frac{(1 - e^{-\beta t})}{\beta} \right] \qquad (6.20)$$

As mentioned, assume that $t = T$, $x(t) = X$ are obtained conducting the above experiment.

Hence, we obtain

$$\beta = \sqrt{\frac{g \sin \theta}{X} \left[\beta T - (1 - e^{-\beta T}) \right]} \qquad (6.21)$$

It is important to note that in Equation 6.21, β appears in both sides of the equation. This equation can be solved easily by iterative techniques as shown with the simple MATLAB® program given in Figure 6.5.

EXAMPLE 6.4

Assume that in one such experiment described above, the following results were obtained:

Mass of the robot, $M = 2.2$ kg
Time taken by the robot to descend, $T = 3.2$ s
The distance traveled, $X = 2.13$ m
Slope of the platform, $\theta = 5.17°$

By running the MATLAB program shown in Figure 6.5, with these values, we can obtain the result for β. The program given in Figure 6.5 needs an initial guess for β and in this example it is taken as $\beta = 3$. Hence, we obtain

$$\beta = 0.8880 \quad \text{and} \quad B = 1.9537 \text{ Ns/m}$$

```
%Matlab script for calculating beta value

beta=3.0;   % initial guess for beta value

g=9.81;     % acceleration due to gravity

M=2.2;      % mass

T=3.2;      % time taken to descend

X=2.13;     % distance traveled

theta=5.173*3.14/180; % slope in radians

%iterate about 100 times to check convergence

%once converge beta value will repeat at later iterations

for i=1:100

beta= sqrt(g*sin(theta)*(T*beta-1+exp(-T*beta))/X)

end;

B=beta*M
```

FIGURE 6.5   MATLAB code for calculating β and B values.

6.5   FIRST APPROACH: GEAR RATIO DESIGN

The gear ratio design will vary according to the specific application. The robot may have to move on a horizontal surface, move on a slope, follow a velocity profile, or climb a wall carrying its own weight. In an industrial robot, requirements will vary from joint to joint, since some joints may be required to carry load vertically and other joints just move horizontally. This is also the case for a two-legged humanoid robot, where knees will be the fastest-moving joint and the hip pitch will be the heaviest-load-bearing joint. The question is for a given application how to design the suitable gear ratio. There could be many criteria used for this purpose. It is not possible to cover all the possible approaches, but some specific cases will be illustrated in the following discussions.

Let us assume that the robot has to follow the velocity profile shown in Figure 6.2 while traveling on an inclined plane as shown in Figure 6.3. This is a safe practice since it is not possible to assume that the terrain will always be horizontal. A sketch of the system is shown in Figure 6.6. The motor is designed for high-speed operation. The robot does not need to move that fast, but it needs to provide a high torque at the drive wheels. Therefore, we need a reduction gear to drive the robot. The discussion starts with an overall torque required on the drive wheel to move the robot. It is assumed that the proposed gear ratio is N_g and the drive wheel radius is R_w.

The force required on the drive wheel is given by the sum of Equations 6.7 through 6.9. Referring to the quantities shown in Figure 6.3 as well as Figure 6.6, the torque required on the drive wheel can be obtained easily as shown below.

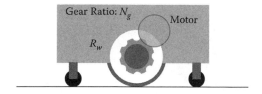

FIGURE 6.6   Block diagram of a gear-driven robot.

$$\tau_w = (f_a + f_b + f_g)R_w \qquad (6.22)$$

$$\tau_w = (M\,a + B\,v + Mg\sin\alpha)R_w \qquad (6.23)$$

Then, the torque to be developed by the motor is given by

$$\tau_m = \frac{(M\,a + B\,v + Mg\sin\alpha)R_w}{N_g} \qquad (6.24)$$

Some assumptions made regarding the motor parameters are as follows:

Supply voltage = V_s
Armature resistance = r_a
Torque constant (Nm/A) = K_t
Back EMF contant (Vs/rad) = K_b
Motor speed (rps) = n_m
Armature current (amp) = i_a

From the basic knowledge of the DC machine theory (Rosenblatt and Friedman 1984), the torque developed by the motor will be

$$\tau_m = \frac{V_s - K_b\,2\pi\,n_m}{r_a}\,K_t \qquad (6.25)$$

or

$$\tau_m = \frac{V_s K_t}{r_a} - \frac{K_b K_t\,2\pi\,n_m}{r_a} \qquad (6.26)$$

For any ground speed of v, the drive wheel speed can be obtained as

$$n_w = \frac{v}{2\pi R_w} \qquad (6.27)$$

or motor speed as

$$n_m = \frac{vN_g}{2\pi R_w} \tag{6.28}$$

Then, using Equation 6.28 in Equation 6.26, the torque developed by the motor at any ground speed v can be obtained as

$$\tau_m = \frac{V_s K_t}{r_a} - \frac{K_b K_t 2\pi}{r_a} \frac{vN_g}{2\pi R_w} \tag{6.29}$$

or

$$\tau_m = \frac{V_s K_t}{r_a} - \frac{K_b K_t vN_g}{r_a R_w} \tag{6.30}$$

The robot has to move with an acceleration of a and reach a velocity of v. Combining Equations 6.24 and 6.30, we can write

$$\frac{V_s K_t}{r_a} - \frac{K_b K_t vN_g}{r_a R_w} \geq \frac{(Ma + Bv + Mg\sin\alpha)R_w}{N_g} \tag{6.31}$$

At the limit, it becomes

$$\frac{V_s K_t}{r_a} - \frac{K_b K_t vN_g}{r_a R_w} = \frac{(Ma + Bv + Mg\sin\alpha)R_w}{N_g} \tag{6.32}$$

or

$$\frac{K_b K_t vN_g^2}{r_a R_w} - \frac{V_s K_t N_g}{r_a} + (Ma + Bv + Mg\sin\alpha)R_w = 0 \tag{6.33}$$

The above equation is quadratic in N_g, the solution of which yields two values N_{g1} and N_{g2}. For the above two gear ratios obtained, the current drawn i_{a1} and per-unit power efficiency η_1 can be calculated by ignoring iron and frictional losses of the DC motor as given below:

$$i_{a1} = \left[V_s - \frac{K_b vN_{g1}}{R_w} \right] \frac{1}{r_a} \tag{6.34}$$

$$\eta_1 = \frac{V_s i_{a1} - i_{a1}^2 r_a}{V_s i_{a1}} \tag{6.35}$$

and

$$i_{a2} = \left[V_s - \frac{K_b v N_{g2}}{R_w} \right] \frac{1}{r_a} \tag{6.36}$$

$$\eta_2 = \frac{V_s i_{a2} - i_{a2}^2 r_a}{V_s i_{a2}} \tag{6.37}$$

The above results created a dilemma. Which one of these two values should be used? The following numerical example throws some light on this problem.

EXAMPLE 6.5

Let us assume that we would like a robot to accelerate at 2 m/s² and reach a velocity of 2 m/s, while climbing a slope of 5.7°. The robot parameters are given as

$$V_s = 24 \text{ V}, \ m = 2 \text{ kg}, \ b = 2 \text{ Ns/m}, \ \sin(\theta) = 0.1, \ v = 2 \text{ m/s}, \ a = 2 \text{ m/s}^2,$$

$$K_b = 0.033 \text{ V–s/rad}, \ K_t = 0.033 \text{ Nm/rad}, \ r_a = 6.2 \ \Omega, \ R_w = 0.03 \text{ m}$$

Let us decide a suitable gear ratio using the above technique. For the solution, a simple MATLAB program can be utilized. By entering the above values and executing the MATLAB code shown in Figure 6.7, the two values of gear ratios are obtained.

The computation result produced two sets of solutions. Let us elaborate more by computing the motor speeds for both cases and decide on an acceptable solution.

For the higher gear ratio, the back EMF will be

$$E_{b1} = V_s - I_{a1} r_a$$

$$E_{b1} = 24 - 1.2047 \times 6.2 = 16.53 \text{ V}$$

Hence, the angular velocity of the motor is

$$\omega_{m1} = \frac{E_{b1}}{K_b} = \frac{16.53}{0.033} = 500.93 \text{ rad/s}$$

and then the motor shaft speed is

$$n_{m1} = \frac{\omega_{m1}}{2\pi} = \frac{500.93}{2\pi} = 79.77 \text{ rps} \quad \text{or} \quad 4786 \text{ rpm}$$

For the lower gear ratio, the back EMF is

$$E_{b2} = 24 - 2.6663 \times 6.2 = 7.469 \text{ V}$$

Hence, the angular velocity of the motor is

$$\omega_{m2} = \frac{7.469}{0.033} = 226.33 \text{ radian/s}$$

```
% Given parameters

vmax=2; acc=2; F=2.0; M=2.0; g=9.81; Kt=0.033;

Kb=0.033; %Kb in terms of volt-sec/rad.

ra=6.2; Rw=0.03; Vs=24;

slope=sin(5.7*3.14/180);

A=Kt*Kb*vmax/(ra*Rw);

B=-Vs*Kt/ra;

C=(M*acc+M*g*slope+F*vmax)*Rw;

%calculate gear ratio, motor current and efficiency

Ng1=(-B+sqrt(B*B-4*A*C))/(2*A)

Ng2=(-B-sqrt(B*B-4*A*C))/(2*A)

ia1=(Vs-(Kb*vmax*Ng1)/Rw)/ra

ia2=(Vs-(Kb*vmax*Ng2)/Rw)/ra

e1=(Vs*ia1-ia1*ia1*ra)/(Vs*ia1)

e2=(Vs*ia2-ia2*ia2*ra)/(Vs*ia2)
```

result:

```
Ng1 =      7.5140

Ng2 =      3.3950

ia1 =      1.2047

ia2 =      2.6663

e1 =       0.6888

e2 =       0.3112
```

FIGURE 6.7 MATLAB code and results for gear ratio, motor current, and efficiency.

and then the motor shaft speed is

$$n_{m2} = \frac{226.33}{2\pi} = 36.02 \, \text{rps} \quad \text{or} \quad 2161 \, \text{rpm}$$

Referring to the results obtained above, the following observations can be made. The higher gear ratio ($N_{g1} = 7.5140$) allows the motor to run at a higher speed of 4786 rpm, developing a higher back EMF of 16.53 V, thus drawing a lower current ($i_{a1} = 1.2047$). The lower current results in lower copper loss of $i_{a1}^2 r_a = 1.2047^2 \times 6.2 = 9$ W and delivers a higher efficiency of 0.6888 as shown above.

On the other hand, the lower gear ratio ($N_{g1} = 3.3950$) is in fact "strangling" the motor, although it will still do the job. This ratio makes the motor run at a lower speed of 2161 rpm, developing a lower back EMF of 7.469 V, thus drawing a higher current ($i_{a2} = 2.6663$) to develop a high torque. This current for the given example will generate a higher copper loss of $i_{a2}^2 r_a = 2.6663^2 \times 6.2 = 44$ W and result in a lower efficiency of 0.3112. In that case, a bigger problem will emerge, which is the heating of the motor and the drive system. Consequently, power dissipation has to be resolved. The above example provided one "good" and one "bad" solution. Apparently, for the robot in this example, a high gear ratio must be selected.

The case of inadequate power rating: If the motor power selection was not done properly, the given requirements will be impossible to achieve. It is interesting to find what happens if an impossible task is given to the system. Assume that the objective is to achieve a ground velocity of, say, 4 m/s and also a slightly higher acceleration. The following example highlights the consequence of such a situation.

EXAMPLE 6.6

Assume that the following requirements are given for the same system, $v = 4$ m/s, $a = 2.3$ m/s^2, while the other parameters remain the same as in Example 6.5. The program in Figure 6.6 can now be used just by entering these new values. Then, the results will be

$$N_{g1} = 4.7431 + j1.0114$$
$$N_{g1} = 4.7431 - j1.0114$$

The complex numbers obtained for gear ratios simply indicate that the task is impossible. The power check was not done properly to start with. If the power selection was marginally inadequate, we can have a quick-fix solution by raising the voltage a few volts. For instance, if we set $V_s = 27$ V and repeat the same calculations, keeping all other parameters unchanged, the results are

$$N_{g1} = 7.5615$$
$$N_{g2} = 3.1105$$
$$i_1 = 1.2693$$
$$i_2 = 3.0856$$
$$e_1 = 0.7085$$
$$e_2 = 0.2915$$

In many cases, it is not wrong to use this adjustment provided that motor is not driven with excessive voltages.

6.6 SECOND APPROACH: SYSTEM PERFORMANCE AS A FUNCTION OF GEAR RATIO

In the previous section, the optimum gear ratio was calculated by solving a quadratic equation, which yields two choices. We have no information as to what happens if the gear ratio is different from the two values obtained. Thus, we can analyze how the performance gets affected for a wide range of gear ratios. For this purpose, we can fix acceleration, and solve for maximum velocity obtainable as a function of gear ratio. We can develop a program to find the effect of gear ratio to find where the best performance occurs. From Equation 6.33, the limiting velocity can be obtained as

$$v = \frac{\dfrac{(V_s K_t N_g)}{r_a} - (Ma + Mg\sin\alpha)R_w}{\dfrac{K_b K_t N_g^2}{r_a R_w} + BR_w} \tag{6.38}$$

A sample program listing is provided in Figure 6.8. The program plots the maximum velocity obtainable, the current drawn from power source, and the efficiency of the motor. They are shown in Figures 6.9 through 6.11, respectively, as functions of the gear ratio selected.

```
% This program shows the effect of gear ratio on vmax

% given parameters

acc=2; B=2.0; M=2.0; g=9.81; Kt=0.033;

Kb=0.033; % volt-sec/rad

slope=sin(5.7*3.14/180); ra=6.2; Rw=0.03; Vs=24; n=25;

for i=2:n

Ng=i;

num=((Vs*Kt*Ng/ra)-(M*acc+M*g*slope)*Rw);

den=(Kb*Kt*Ng*Ng/(ra*Rw))+B*Rw;

vmax(i)=num/den;

i1(i)=(Vs-(Kb*vmax(i)*Ng)/Rw)/ra;

e1(i)=(Vs*i1(i)-i1(i)*i1(i)*ra)/(Vs*i1(i));

end;

figure(1); plot(vmax); xlabel('Gear Ratio');

ylabel('Maximum Velocity, m/s'); grid; axis([2,22,0,3]);

figure(2); plot(i1); xlabel('Gear Ratio');

ylabel('Current drawn by motor, Amps'); axis([2,22,0,4]);

grid;

figure(3); plot(e1); xlabel('Gear Ratio');

ylabel('Efficiency p.u');

axis([2,22,0,1]); grid;
```

FIGURE 6.8   MATLAB code that shows the effect of gear ratio on maximum velocity.

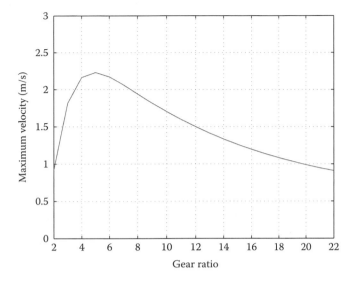

FIGURE 6.9   Effect of gear ratio on maximum velocity.

The aim of this exercise is not to obtain the maximum velocity, but to examine the effect of gear ratio on the system performance. Referring to Figure 6.9, we notice that for a desired velocity, two gear ratios are available. A higher gear ratio may be chosen as before. However, it is important to consider the current drawn and the efficiency before selecting a gear ratio. Referring to Figure 6.9, we notice that as we increase the gear ratio, the maximum velocity achievable by the robot increases and

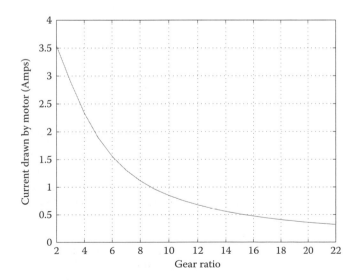

FIGURE 6.10   Effect of gear ratio on current drawn by motor.

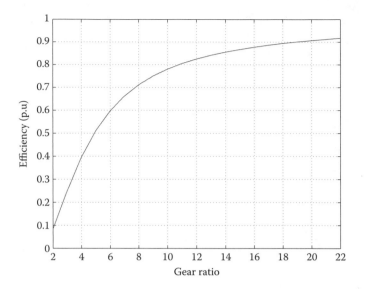

FIGURE 6.11   Effect of gear ratio on efficiency.

then decreases. Let us consider the point of maximum velocity (2.25 m/s) where the gear ratio is 5. Referring to Figure 6.10, the current drawn at the chosen gear ratio is 2 A. Furthermore, referring to Figure 6.11, the same gear ratio shows that the efficiency is only 0.5. It implies that half of the power drawn will be used to heat the motor. This may result in dissipation issues and the motor and the driver may get overheated. Considering all the three charts, a gear ratio of 8 may result in acceptable maximum velocity (1.9 m/s), current (1.1 A), and efficiency (0.71). Consulting all the three plots is essential when deciding upon a gear ratio.

6.7   GEAR RATIO DESIGN FOR STEPPER MOTORS

The working principle of stepper motors is different from that of DC motors. For these devices, the magnetic reluctance between the rotor and the stator of the stepper motor changes with respect to the position of the rotor. When a winding is excited, the rotor aligns with that winding. The power supply is switched to different windings in a sequence so that the rotor continues to rotate in the desirable direction and speed, which corresponds to the switching frequency. There are well-known switching circuits to achieve this. However, stepper motor-based design needs to be done cautiously. Let us evaluate this in detail. A typical stepper motor characteristic graph is shown in Figure 6.12 (for a typical stepper motor specification, see Portescap 2013). The motor can start at a given load condition from zero to any pulse rate (speed) as long as the point of operation falls within the region enclosed by "pull-in" line and two axes. Then, if the load torque increases gradually, from there the motor will continue to rotate. However, when the point of operation crosses the "pull-out" line upward, the motor will suddenly stop. Similarly, starting from the "pull-in" region, if the pulse rate (speed) is increased gradually, the motor will continue to accelerate. However, when

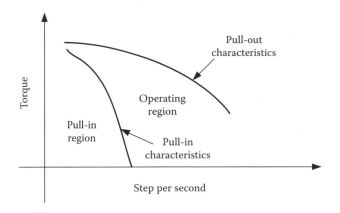

FIGURE 6.12 A typical stepper motor characteristic.

the point of operation crosses the "pull-out" line toward the right, the motor will suddenly stop. A fully analytical solution for gear ratio design is not strictly possible.

The following steps are useful in arriving at an optimum gear ratio for stepper motors:

1. Compute power requirement as before with a high *S* (ignorance factor) value.
2. Find the linear force and the driving wheel torque using Equation 6.23. The maximum speed of the robot is also known.
3. As a starting point, assume a large gear ratio.
4. Using the above gear ratio, calculate the required motor torque and stepping speed.
5. Place the point indicated by the torque and stepping speed on motor characteristic graph.
6. Check if this point falls on the left-hand side of the pull-in line on step motor characteristic graph with enough margin. If yes, then the design is complete!
7. If the margin is too low, or the point falls to the right-hand side of the pull-in line, lower the gear ratio in steps. Repeat the design from step 4.
8. If the margin is too large, raise the gear ratio in steps to achieve the desired margin.

The above design procedure starts with a high gear ratio and decides on the appropriate ratio by iteration. Alternatively, we can start with a gear ratio of 1 and increase it iteratively. That is, after steps 1 and 2, we can now proceed as follows:

1. Assume a gear ratio of 1 as a starting point.
2. Using the above gear ratio, find the torque and the stepping speed of the motor.
3. Place the point indicated by the torque and the stepping speed on the motor characteristic graph. By intuition, this point will fall too far to the left of the pull-in line. We will be underutilizing the motor.

4. If it is so, gradually increment the gear ratio until the torque versus speed point falls reasonably close but below the pull-in characteristics with the desired margin.

Using the "pull-in" line for design is conservative. In fact, this line gives the maximum stepping speed to which the motor can start from a standstill position for a given load torque on the motor shaft. In applications where the robot can accelerate slowly from standstill, the operating region may be used for the design.

6.8 DESIGN PROCEDURES FOR MOBILE ROBOT THAT ARE NOT GROUND BASED

The above discussions described procedures for gear ratio design for robots that are mobile and ground based. But there are many other applications where robots are not moving on a horizontal surface. For example, a wall-climbing robot moves on a vertical surface or even under the ceiling. A two-legged robot has many joints that will have different load–speed demands on them. For such special robots, the above techniques are not readily applicable, though the design procedure is still based on estimating the maximum torque and speed requirements. Since it is not possible to give a general procedure that is applicable to all, we will give some practical design examples with the following case studies.

Case Study I: Design of Robotic Arm Joints

The aim of the exercise is to rate the motors and gear ratios at the joints for a robot arm shown in Figure 6.13. We assume that the length of robot links are l_1, l_2, and l_3. The weights of the links are assumed to be negligible when compared to the load at the end point P_A. The design values are computed to cater to the worst-case situation. Let us find the maximum load variables of P_A in the worst-case situation. This will occur when the robot arm is horizontal and fully stretched as shown in Figure 6.14.

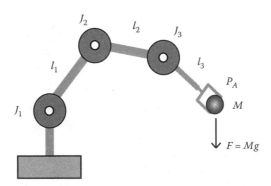

FIGURE 6.13   Block diagram of a three-joint robot arm.

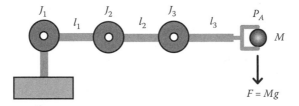

FIGURE 6.14   Robot joints having the maximum load.

Assume that the following parameters are given for the design process:

The load at the end effector: M
The acceleration with which the load has to be lifted: a
The required vertical velocity to be reached: v
Then, the downward load on the system: $M(a + g)$

For joint J_1: The torque on joint J_1 will be $\tau_{j_1} = (l_1 + l_2 + l_3)M(a + g)$, and the angular velocity will be $\omega_{j_1} = v/(l_1 + l_2 + l_3)$. Hence, the motor power at joint 1 can be estimated as $P_1 = \omega_{j_1}(l_1 + l_2 + l_3)M(a + g)$ or $P_1 = vM(a + g)$. Now, we can select a suitable motor with a power rating of P_1.

From the given angular velocity, the load shaft speed can be derived as $N_{S1} = (\omega_{j_1}/2\pi)$. Assume that the no-load speed of the selected motor is N_{M1} Hence, the required gear ratio is $G_1 = (N_{M1}/N_{S1})$. DC motors that will be used for this application are permanent magnet motors that have shunt motor characteristics and the speed drop at the loaded condition is minimal. Hence, the speed at full torque will not be different from the no-load speed. We can now pick a suitable ratio from a manufacturer's catalog. Let this selected ratio be G_{11} such that $G_{11} \leq G_1$ to guarantee the speed requirement of the arm. We now need to check the motor safety and performance. With the selected gear ratio, the torque on the motor shaft will be $\tau_{M1} = (\tau_{j_1}/G_{11})$. Assume that the torque constant given in the motor specifications is K_{t1}. The current drawn by the first joint motor will then be $I_{M1} = \tau_{j_1}/(G_{11} \cdot K_{t1})$. If I_{M1} is less than the maximum current rating of the selected motor, then the selected gear ratio is fine. Otherwise, a next higher gear ratio, say, G_{12} that is still less than G_1 should be considered and the motor current must also be recalculated accordingly. If such a gear ratio is not available, a higher motor power should be selected and the above calculations should be repeated.

For joint J_2: Let us assume that the vertical velocity, v, and acceleration, a, need to be achieved using J_2, while J_1 is fixed. The torque on joint, J_2, will be $\tau_{J_2} = (l_2 + l_3)M(a + g)$, and the angular velocity will be $\omega_{J_2} = v/(l_2 + l_3)$. The power of the motor at joint J_2 should be $P_2 = \omega_{J_2}(l_2 + l_3)M(a + g)$ or $P_2 = vM(a + g)$. Now, we select a suitable motor at the rating of P_2 watts. At the driving shaft of J_2, the load shaft speed is given by $N_{S2} = (\omega_{J_2}/2\pi)$.

Assume that the no-load speed of the selected motor is N_{M2}, then the required gear ratio will be $G_2 = (N_{M2}/N_{S2})$. We can now pick a suitable gear ratio G_{21} so that $G_{21} \leq G_2$. Again, we need to check the motor safety and performance. The torque on the motor shaft is given as $\tau_{M2} = (\tau_{j2}/G_{21})$; hence, the current drawn by the motor is $I_{M2} = \tau_{j2}/(G_{21} \cdot K_{t2})$ A, where K_{t2} is the torque constant of the selected motor. As discussed earlier, the current I_{M2} should be less than the maximum current rating of the motor.

For the rest of the joints, a similar procedure can be applied as discussed above. In the following, a numerical example of the design procedure using actual motors from a manufacturer's catalog will be given.

EXAMPLE 6.7

Assume that the basic design requirements of a robot arm are given as follows:

Load $(M) = 2$ kg
Length of link 1 $(l_1) = 0.2$ m
Length of link 2 $(l_2) = 0.2$ m
Length of link 3 $(l_3) = 0.2$ m
Maximum load velocity $(v) = 0.5$ m/s
Maximum load acceleration $(a) = 0.5$ m/s²

Let us design and select suitable drive motors with appropriate gear ratios.

For joint J_1, the downward load is equal to $M(a + g)$; hence, with the given parameters, it will be 20.62 N. The torque on J_1 is defined as $\tau_{j1} = (l_1 + l_2 + l_3) M(a + g)$ and with the given parameters this will yield 12.372 Nm. Similarly, the angular speed $\omega_{j1} = v/(l_1 + l_2 + l_3)$ is 0.833 rad/s and the load shaft speed is $N_{S1} = (\omega_{j1}/2\pi) = (0.833/2\pi) = 0.1326$ rps. The power required for J_1 is $P_1 = \omega_{j1}\tau_{j1}$ or 10.31 W.

Let us select type 2342024CR from the minimotor series offered by Faulhaber's minimotor catalog (Faulhaber 2013), which has the following specifications, power = 19 W, nominal speed = 8500 rpm, supply voltage = 24 V, torque constant = 26.1 mNm/A, and maximum current = 0.72 A. We find that the no-load speed of the motor is 141.7 rps and the ideal gear ratio is $(N_{m1}/N_{S1}) = (141.7/0.1326) = 1068$. The gears provided by the manufacturer, which can be factory fitted, are types 23/1, 26A, 26/1, 22/7, 30/1, and 38/3 (Faulhaber 2013). As we search for the gears, none of the above available gears is capable of giving this torque of 12.372 Nm.

Apparently, we need to change the motor, though it was satisfactory, since none of the factory-fitted gears will serve the purpose. Let us select the minimotor number 3242024CR (Faulhaber 2013), which has the following specifications: power = 26.3 W, nominal speed = 5300 rpm, supply voltage = 24 V, torque constant = 41.3 mNm/A, maximum current = 1.20 A. Then the no-load speed is 88.33 rps and the ideal gear ratio will be $(N_{m1}/N_{S1}) = (88.33/0.1326) = 666.16$. Gears provided by the manufacturer are types 32/3, 38/1, and 38/2 (Faulhaber 2013). Let us choose the gear heads series 38/1s and 38/2s ("s" indicates all steel gears), where the continuous output torque of the gear is 10 Nm and the intermittent maximum torque will be 15 Nm. We can pick that gear since our calculations for torque are for the worst-case scenarios and the torque of 12.372 Nm will not be continuous. We need to the select ratio that is <666.16. From the available ratios, let us choose a gear ratio for joint one, G_{11}, as 592 (Faulhaber 2013). In this case, the torque on the motor shaft will be $\tau_{m1} = (\tau_{J1}/G_{11}) = (12.372/592) = 0.02089$ Nm. The torque constant of the motor, K_{t1}, from the given specifications will be 0.0413 Nm/A. Hence, the current drawn can be calculated as $I_{M1} = (0.02089/0.0413) = 0.505$ A. For this motor, the maximum continuous current is specified as 1.20 A, so the estimated I_{M1} value is still acceptable in this case.

For joint J_2, the downward load will be the same, that is, 20.62 N. The torque on J_2 is given by $\tau_{J2} = (l_2 + l_3)M(a + g)$ or 8.248 Nm. The angular speed $\omega_{J2} = v/(l_2 + l_3)$ or 1.25 rad/s and the power required for J_2 is $P_2 = \omega_{J2}\tau_{J2}$ or 10.31 W. As the power will be the same as in joint 1, let us use the same motor which is Minimotor series 3242024CR (Faulhaber 2013). Similar to the above calculations, load shaft speed N_{S2} is calculated as 0.1989 rps for $\omega_{J2} = 1.25$ rad/s. Hence, the ideal gear ratio will be $(N_{m2}/N_{S2}) = (88.33/0.1989) \approx 444.1$. Let us choose the gear heads series 38/1 and 38/2 (Faulhaber 2013) in which a gear ratio of 415 is available, which is <444.1. The torque on the motor shaft will be $\tau_{m1} = (\tau_{J2}/G_{22}) = (8.248/415) = 0.01987$ Nm, and the motor current will be $I_{M2} = (0.01987/0.0413) = 0.482$ A. This is less than the maximum current of 1.2 A. Therefore, the motor and gear with a gear ratio of 415:1 is suitable.

Repeating the same calculations for the third joint and assuming the same velocity, mass, and acceleration will lead to the same motor power requirement, the torque on J_3 is given by $\tau_{J3} = l_3M(a + g)$ or 4.124 Nm. The angular speed $\omega_{J3} = (v/l_3)$ or 2.5 rad/s and the power required for J_3 is $P_3 = \omega_{J3}\tau_{J3}$ or 10.31 W. That is, we can still employ the same motor (3242024CR) for this joint. N_{S3} is calculated as 0.3981 rps for $\omega_{J3} = 2.5$ rad/s. Hence, the ideal gear ratio will be $(N_{m3}/N_{S3}) = (88.33/0.3981) \approx 221.9$. Let us choose the gear heads series 38/1 and 38/2 (Faulhaber 2013) in which gear ratios of 159:1 and 246:1 are available. Let us pick 159, which is <221.9. The torque on the motor shaft will be $\tau_{m3} = (\tau_{J3}/G_{31}) = (4.124/159) \approx 0.0259$ Nm and the motor current will be $I_{M3} = (0.0259/0.0413) = 0.627$ A, which is less than the maximum current of 1.2 A.

In summary, in the above example, we insisted on using the gears provided by the manufacturer and found out that the output torque provided by those gears meant for motors with suitable power ratings were inadequate. Hence, our motor selection was overshadowed by the availability of a suitable gear. At the end, the motors we chose were far more powerful than needed. An alternative for this kind of predicament is to use an external gear as an additional stage. However, this additional gear may cost more than the cost of a higher power-rated motor that we used in the above design. Furthermore, an external gear stage may occupy more space.

Case Study II: Motor Power and Gear Ratio Calculation for a Wall-Climbing Robot

In this second case study, we discuss another design example, which is a wall-climbing robot. It is possible to design a robot to climb a vertical wall, and there are many possible designs. A unique triangular structure design is shown in Figure 6.15. In this robot, A, B, and C are the pivotal joints that are powered by geared DC motors. Each joint is provided with a sticking pad that can be activated to stick to the surface. The problem is to design these joints so that the robot will be able to climb the vertical wall. Figure 6.15 shows the instant when the robot is climbing a vertical wall. We start this discussion considering the instant when the pads are stuck to the vertical wall. These pads, associated with joints A and C, are named

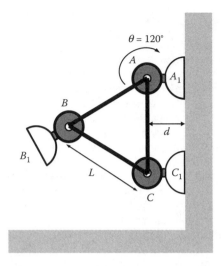

FIGURE 6.15   Block diagram of a wall-climbing robot.

A_1 and C_1, respectively. To climb, the robot needs to release the grip on pad C_1 and crank joint A so that the robot body rotates clockwise until the joint pad B_1 touches the vertical surface above and then gets stuck to it. After this motion, joint C will be the farthest away point of the robot from the climbing surface. Obviously, this sequence is continued to crank joint B and so on to achieve a climbing motion.

We need to design the joints in such a way that it will work when the load torque is maximum at the desired speed. In this case, the design approach used for a mobile robot discussed earlier is not strictly followed. Figure 6.16 shows the instant in which the load torque on joint A is nearly the maximum. This is only approximate since the actual position of the center of gravity will only be known when the exact positions of the motors and other components are known. Hence, only the approximate position is considered. Furthermore, joint A is not part of the load. We can argue that the major part of the load comes from these three joints and hence only two-third of the weight of the robot, W, needs to be cranked upward with a torque arm length of $L\cos(30°)$, which makes the maximum torque on joint A as $(2/3)WL\cos(30°) = (W L/\sqrt{3})$. This is not strictly correct, since there are additional masses of the frame and electronics and others on the robot. Hence, a factor of ignorance S can be included and the maximum torque on joint A is written as

$$T_{max} = S\left(W\frac{L}{\sqrt{3}}\right) \text{kgm}$$

or

$$\tau_{max} = S9.81\left(W\frac{L}{\sqrt{3}}\right) \text{Nm}$$

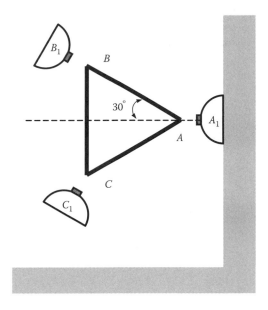

FIGURE 6.16 Robot position that has the maximum torque on the joint.

The other important parameter is the speed with which the joint has to rotate. Assume that a duration of t_s seconds is allowed for the robot to rotate 120°, which is needed for sticker pad C_1 to leave the wall and sticker pad B_1 to stick to the wall. Then, the angular velocity of joint A cranking up will be $\omega_0 = (120/t_s) \times (\pi/180) \approx (2.09/t_s)$ rad/s. The power developed by the motor, powering joint A, can be written as $P = \tau_{max}\omega_0$ and the power rating of the motor can be calculated as $P = \tau_{max} \times \omega_0 = S\,9.81(W\,L/\sqrt{3})(2.09/t_s)$. The shaft speed is given as $\omega_0/2\pi$, which is $N_0 = (60\omega_0/2\pi)$ rpm. We now have to choose a motor that has adequate power rating P_m. To simplify matters, we did not include the power required for the acceleration; therefore, we need to choose a slightly higher power rating than P. Furthermore, other than the instant shown in Figure 6.16, there will be excess torque available for acceleration.

Let the nominal speed of this motor, listed in the catalog, be N_m rpm; then, the ideal gear ratio is $G_1 = (N_m/N_0)$. As before, we need to find a gear ratio available for the gear head provided by the manufacturer. We pick the value G_{11} which is slightly less than G_1 such that $G_{11} \le G_1$. For this gear ratio, the torque to be developed by the motor will be $\tau_{m1} = (\tau_{max}/G_{11})$. If the torque constant of the motor is K_{t1}, it yields a motor current of $i_{M1} = (\tau_{m1}/K_{t1})$, and it should be less than the rated maximum current of the chosen motor. Otherwise, a higher gear ratio G_{21} may have to be chosen. If G_{21} is greater than G_1, the speed performance will be compromised. In an extreme case, the motor may have to be changed.

EXAMPLE 6.8

Assume that the following data are provided for the wall-climbing robot shown in Figure 6.15:

$$W = 4 \text{ kg}$$

$$L = 0.4 \text{ m}$$

$$t_s = 0.4 \text{ s}$$

Assuming $S = 1.3$, let us calculate the required motor power and gear ratio.

From the above definition, the motor power will be $P = 1.3 \times 9.81(4 \times 0.4/\sqrt{3})$ $\times (2.09/0.4) = 61.55$ W and the shaft speed will be $N_0 = (60/2\pi)(2.09/0.4) \approx 50$ rpm. Now, we select a suitable motor (such as minimotor 3257024C (Faulhaber 2013)) that has the following specifications: power = 83.2 W, nominal speed = 5900 rpm, supply voltage = 24 V, torque constant, $K_t = 37.7$ mNm/A, maximum current = 2.3 A. With these parameters, the ideal gear ratio will be $G_1 \approx (5900/50) \approx 118$ and the torque on the shaft will be $\tau_{max} = 1.3 \times 9.81(4 \times 0.4/\sqrt{3}) = 11.78$ Nm. From the available gears (Faulhaber 2013), all steel gears types 38/1s and 38/2s with maximum output torque option of 15 Nm, for intermittent peak loads, can be selected. From the catalog, the nearest gear ratios available are 66 or 134. We have to select 134, even though this value is higher than 118. For this gear ratio, the motor torque can be obtained as $\tau_{m1} = (11.78/134) \times 1000 = 87.9$ mNm, and the motor current is $(\tau_{m1}/K_t) = (87.9/37.7) = 2.33$ A. This compares well with the maximum current of 2.3 A, bearing in mind that this will only be a peak value coming on intermittently. Since the gear ratio is 134, which is higher than the required 118, the robot will climb at a slightly slower speed, which is a compromise.

6.9 CONCLUSION

In this chapter, we have described various ways of arriving at a suitable gear ratio for mobile robots. A low gear ratio demands a low-speed operation of a DC motor and needs more current, resulting in large copper losses and heating. It is necessary to check if more current is really needed. On the other hand, a very large gear ratio will result in too safe currents, but slows down the motion of the robot and its performance, which may not meet the objectives.

REFERENCES

Faulhaber. 2013. http://www.faulhaber.de.
Kanniah, J., Ercan, M.F. et al. 2004. *Bits and Bytes of Robotics*. Singapore: Prentice Hall.
Portescap. 2013. http://www.portescap.com.
Rosenblatt, J. and Friedman, M.H. 1984. *Direct and Alternating Current Machinery*. London: Charles E. Merrill Publishing Company.

7 Control Fundamentals

7.1 CONTROL THEORY FOR ROBOTICS

In the previous chapters, we looked into the various parts and components that go into building a robot. Nevertheless, these various parts need to be assembled and the ensemble must be controlled in a coordinated way to achieve the objective. There are many ways to build a robot, but typically in any robot design there will be a control system in place that is usually an onboard computing device. For example, a robotic arm is made of joints, links, and a grabber mechanism. If the robotic arm needs to pick up an object, all the joint motions must be coordinated so that the gripper moves to the target, opens the gripper, and picks up the object. This requires a close control of many actuators.

Let us take a micromouse robot as an example, where the constituent parts are sensors with relevant circuitry, motors, battery pack, motor-driving circuits, and a suitable microcontroller along with its support ICs. The assembled robot has to move appropriately. Typically, the onboard intelligence takes the decision as to where it should go and what it should do. This intelligence is handled by a program developed and stored in the memory of the onboard computing device. However, the very basic motions of the robot would be moving forward or backward, maintaining a certain speed, and making turns. Let us consider the situation that this two-wheeled robot starts from one point, accelerates to reach a certain speed, and decelerates and stops at the destination. This motion is simply achieved by making both wheels accelerate according to a profile until the top speed is reached, then decelerate and stop at the stipulated distance or target spot. Then, we have to fix the speed according to a plan or profile. The question is how to enforce those desired speeds? This is where the control theory comes in.

In fact, control theory can be applied to any robot joint, to a drive wheel motor, to an economy, or even to a population as long as the objective is clearly known. In this book, we are interested in applying control theory to robotics.

Over the years, control theory has grown immensely, and many techniques have been developed. Mathematical principles have been developed to facilitate better understanding of the "plants" to be controlled. For this, the first step is to understand the plant in terms of cause and effect or input and output. Then, controllers can be put in place to make the plants yield the desired results. Raven (1987), Philips and Harbor (1988), Ogata (1990), D'Azzo and Houpis (1995), Kuo (1987), and Astrom and Wittenmark (1990) are some of the valuable resources for further reading on this topic. Before proceeding any further, we will describe some of the basic terms as used in control.

Plant: A plant can be defined as a physical entity, which takes any form of energy as input (cause) and produces an output (effect). A DC motor driving the robot wheel

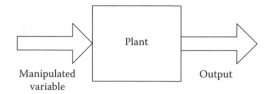

FIGURE 7.1   A plant with its input and output.

is a simple plant. A voltage applied to its terminals is the input, and the speed is the output. System and plant are two widely used terms in control engineering literature. At times, they are used interchangeably, which may be very confusing. A typical block diagram of a plant is shown in Figure 7.1.

Inputs and outputs: Input to the plant is the manipulated variable. For example, in a speed control system, the voltage applied to the DC motor is the *input* (manipulated variable), and the speed of rotation of the load shaft is the *output*.

Systems and subsystems: A system is more than a plant. It consists of a plant in its core together with other components around it. For example, a motor is the core for a "speed control system." There are instrumentation devices such as an encoder attached to the motor shaft. An encoder also needs additional devices such as a "decoder IC" to measure the position digitally. We can call the combination of the encoder and the decoder as an instrumentation "subsystem." The speed is measured and compared with the given desired value by the computer. The computer produces a manipulated variable to control the input voltage, hence the speed, of the motor. This action needs many intermediate stages. Each stage may be called a subsystem. The overall assembly of the plant (motor) and all the peripheral devices put together is called the "control system." In many cases, a control system may also comprise many control subsystems that are parts of it.

7.2   TYPES OF PLANTS

It is necessary to understand how plants and systems are classified. As a whole, controlled plants can be classified in several ways based on their nature of input and output relationship or the nature of their parameters.

7.2.1   LINEAR VERSUS NONLINEAR PLANTS

In linear plants, the input versus output relationship is linear and hence the plant obeys the superposition theorem. For example, the relationship between current and voltage in a resistance is linear as shown in Figure 7.2. The increase in the voltage across a resistor causes the current flowing through it to increase proportionally. On the other hand, for nonlinear plants, the input/output relationship is nonlinear. A good example is the relationship between the magnetizing current and flux density, which is nonlinear. As shown in Figure 7.2, by increasing the current further, we do not see a proportional increase in the flux density in a magnetic core.

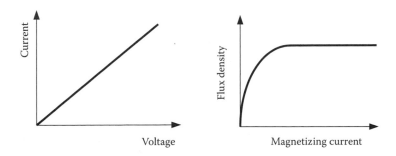

FIGURE 7.2   Examples of a linear plant (resistance) and a nonlinear plant (magnetic core).

7.2.2   TIME-INVARIANT VERSUS TIME-VARIANT PLANTS

In a time-invariant linear plant, all the plant parameters do not vary with time. A resistor circuit is a good example, as the voltage versus current relationship will not be affected by time. On the other hand, in time-variant linear plants, one or more plant parameters change with time.

EXAMPLE 7.1

The total mass of a rocket will continuously change as the fuel gets consumed along the way. This is a plant with a slowly time-varying parameter. But, when this rocket ejects a fuel stage, there will be a sudden change of weight and there will be a step change in weight parameter.

EXAMPLE 7.2

In the robot colony competition, a robot carries pellets on its basket, and the robot weight and speed response will not be the same as when the robot has dropped all the pellets in the goal location. If the pellet weight is substantial, then there should be a change in control strategy.

EXAMPLE 7.3

The load on the joint of a humanoid robot carrying an object in its arms will change as soon as it has placed the object in its destination.

In all the above examples, the controller gain may need adjustments at different time instances.

7.3   CLASSIFICATION BASED ON CONTROL SYSTEM

Another classification of the control system is based on the system implementations. The classification may depend on the type of control we use to make the system perform according to our requirements and specifications. These classifications are applicable to the entire system architecture.

7.3.1   ANALOG VERSUS DIGITAL SYSTEMS

Analog controllers control the plants directly using analog components such as amplifiers, pneumatic, or hydraulic controllers. However, the usage of analog control systems in robotics is almost obsolete.

Because of the availability of cheap computing hardware, which is more flexible and capable of handling sophisticated modern control methods, digital control has come to stay and dominate. In systems that use digital controllers, the computer reads the output of the plant, compares with the desired value, and computes the required control input or manipulated variable. As mentioned earlier, such systems are more flexible, cheaper, and more powerful. Furthermore, digital controllers can easily communicate with other systems both inside and outside. These characteristics make them fit very well into hierarchical systems as well as distributed systems. Robots are complex devices with many subsystems that respond or report to external systems. Therefore, it will not be wrong to assume that all robot controllers are digital as a rule.

EXAMPLE 7.4

An example of an analog controller is the voltage regulator. A typical voltage regulator, such as 7805, is an analog controller that regulates the output voltage for varying input voltage.

EXAMPLE 7.5

All off-the-shelf controllers sold now are digital controllers. Programmable logic controllers (PLCs) are a good example. They are flexible and reliable. Most controllers used in robots, such as humanoid robot, micromouse, and wall-climbing robots, are digital using onboard computers.

7.3.2   OPEN-LOOP VERSUS CLOSED-LOOP SYSTEMS

Open loop: The output of an open-loop system is neither measured nor used. So, the output does not influence the manipulated variable. A precise mathematical model must be evaluated and then the controller system should be carefully calibrated. Open-loop systems are hardly used in robotics.

EXAMPLE 7.6

Many humanoids move their hip and leg joints according to predetermined trajectories, without any gyro and foot sensor feedback. If the walking surface is a level ground with enough friction, the robot will walk smoothly. If there is level difference, the robot may take the next step before its swing foot has landed on the ground, since the controller will assume that the swing leg has landed based on joint angle values. Hence, the robot will topple. This is a typical example of an open-loop robotic system. For proper walking, there should be gyro and foot sensor feedback. What we have described above is a complex situation. Even though individual joints are activated by accurate servomotors, the overall walking control system is an open-loop system. In summary, any system without output monitoring is considered an open-loop system.

Closed loop: In closed-loop systems, the output is constantly monitored and fed back. According to the error, the manipulated variable is adjusted to achieve the objective. The main idea of a closed-loop system is a feedback-based control. The feedback must be adequate.

EXAMPLE 7.7

A pole-balancing robot control system has many control subsystems. Balancing the pole and moving the vehicle at the same time is done by one subsystem. Assume that the robot has to move through a distance of 1 m. A motor may be fitted with an encoder. But, if the program depends only upon this, the robot will not work due to the slippage of the drive wheel. Owing to inadequate feedback, the system behaves like an open-loop system and the distance moved will be inaccurate. To overcome this problem, we include ground feedback sensors that monitor the cross tapes placed on the platform and correct the errors in distance measurement. Here, we highlighted a case where the system has adequate feedback. In simple terms, a control system that uses adequate feedback to adjust the performance can be defined as a closed-loop system.

A popular closed-loop controller: So far, we have been using the term "controller" in general. Before we go any further, we discuss some basic ideas of what a "controller" is. Most closed-loop controllers are error based. The controller uses the error between the desired value and the actual value of the output to decide the magnitude and sign to be applied to the plant. The most popular error-based controller is called the PID controller, which means the proportional, integral, and derivative controllers together. Mathematically, we can write an expression for the output of such a controller that is fed to the plant as

$$ m(t) = K_p e + K_i \int e\, dt + K_d \frac{de}{dt} \qquad (7.1) $$

where e is the error between the desired value and the output, K_p is the proportional gain, K_i is the integral gain, and K_d is the derivative gain. After adjusting these three parameters, the output from a PID controller can be input to plant. We will discuss more about PID controllers in the following chapters.

7.4   NEED FOR INTELLIGENT ROBOT STRUCTURE

In control theory, we assume that the plant is already there to start with. However, in game robotics, the first task is to design the mechanical structure of the plant. This must be executed carefully, and the controller must be designed for that structure. Assume that we have a robot with two wheels, with unsymmetrical loading on them. Then, the robot will have a problem in moving straight. The controller will help, but the basic things such as load distribution must be done right. A badly designed robot cannot be forced to perform well just by using a good controller. One example is the wall-climbing robot. Most wall-climbing robots are event-driven systems. The term "event-driven" implies that when the robot has successfully completed one climbing step, then the next step should start. However, the task of finding when one step got safely completed is easier said than done. The sensor configuration used in robot design plays an important role. In another example, a good biped robot must be balanced and must have enough degrees of freedom. We can see some biped robots with intelligent structure that can walk down a slope without any external power or control. There are many such designs (see, e.g., Passive Walker 2009; Walking Robot 2010). It is easier to put a controller for robots with sound designs in which the control intelligence is built into the mechanical structure. Hence, the intelligence must be embedded in the mechanical structure of the robot.

7.5 A TYPICAL ROBOT CONTROL SYSTEM

The next task in game robotics is to design the robotics control system with many plants along with their control subsystems. For example, a biped robot has a master processor, which coordinates the control of many joints according to joint trajectories. For every joint, there is a control subsystem. There will be vision cameras, foot sensors, and gyroscopes, which are parts of the overall biped control system. The master coordinates all these subsystems. These coordinations must also be planned and programmed by the robot designer.

In any complex control system, the constituents are simple closed-loop controllers, which are the subsystems, and when assembled together they form the overall control system. We would like to look at one basic feedback controller.

Basic closed-loop controller and some terminologies used in control: Having provided basic ideas on controller classifications, we now consider a simple single-loop control system to familiarize ourselves with terminologies used in such systems (Raven 1987; Philips and Harbor 1988). A simplified block diagram of a typical single-loop control system is shown in Figure 7.3.

In Figure 7.3, the signal R is the reference signal or the set point and B is the feedback signal. Error E is obtained by subtracting B from R and fed to a PID controller. The controller can be any one of the controllers we mentioned earlier. The controller may also be one of the many other types. This controller produces an appropriate control signal U, which is fed to the final control element. This final control element

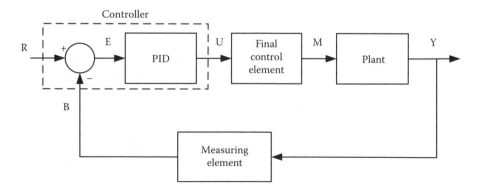

FIGURE 7.3 Simplified block diagram of a closed-loop control system.

can be a PWM-based H-bridge driver or a pneumatic valve controller. There are many other possibilities. The output M of the final control element may produce a power voltage, pressure, or heat to the controlled plant. The output Y is measured by measurement element whose output B is fed back.

7.6 TRENDS IN CONTROL

In addition to the above classifications, a number of new controller types are emerging in modern control systems. Some of them are very useful in robotics. This has been made possible by advances in technology, control theory, instrumentation technology, and computing power. All these new control techniques are digitally implemented. Therefore, they also fall under the general category of digital controls. The list is growing as research progresses in this field. We briefly mention some relevant ones below.

In the case of a biped robot, there are many joints to be controlled to make a humanoid robot walk without falling. Here, we may have a master controller and many other subcontrollers controlling the joints, which act like slaves taking commands from the master. This type of control is usually called a hierarchical control.

In a complex industrial environment, there may be many control systems acting independently, but cooperating with each other. Actually, in swarm robotics, each robot has its own controller, but they constantly communicate with each other. This can be broadly classified as a distributed control system.

In adaptive controllers, the parameters of the controller will be adapted according to the plant parameters that are identified by an iterative identifier (Mendel 1973). Model reference adaptive controllers are also used in robot control (Astrom and Wittenmark 1989).

In systems that cannot be modeled mathematically with the desired ease, fuzzy logic-based controllers are used. Nowadays, they are everywhere, starting from washing machines to pole balancing robots and biped robots. Similar difficulties are tackled by neural network-based controllers that mimic the human brain model of functioning (Kosko 1992).

In systems where the plant parameters change as the operating point changes, the controller structure is changed to suit the operating point. They are called variable structure controllers. In such systems, fast adaptation is also useful.

7.7 CONCLUSION

Typically, all feedback control systems are error based, which implies "no error–no action" with a few exceptions. Open-loop systems do not use error or feedback to produce control action. If there is no disturbance, they work well. When a disturbance occurs, we need error feedback to take care of it.

We have provided a brief introduction to the concept of control in the above sections. This has been an exercise in generality. We have not provided any detailed discussion on any specific system here. However, we have provided a brief description of the objectives of control, types of control, and their relevance. We have also introduced some terminologies used in control. We have also indicated that error is an important factor in control. In many systems, the error is the driving force in taking corrective action. In the subsequent chapters, we will discuss more quantitative aspects of controllers, their mathematical modeling, and their time domain analysis and synthesis.

REFERENCES

Astrom, K.J. and Wittenmark, B. 1990. *Computer Controlled Systems: Theory and Design*, 2nd Edition. Englewood cliffs, NJ: Prentice-Hall International Editions.

Astrom, K.J. and Wittenmark, B. 1989. *Adaptive Control*. New York: Addison-Wesley.

D'Azzo, J.J. and Houpis, C.H. 1995. *Linear Control System Analysis and Design: Conventional and Modern*, 3rd Edition. New York: McGraw-Hill Book Company.

Kosko, B. 1992. *Neural Networks and Fuzzy Systems: A Dynamical Systems Approach to Machine Intelligence*. Englewood Cliffs, NJ: Prentice-Hall.

Kuo, B.C. 1987. *Automatic Control Systems*, 5th Edition. Englewood Cliffs, NJ: Prentice-Hall.

Mendel, J.M. 1973. *Discrete Techniques of Parameter Estimation: The Equation Error Formulation*. New York: Marcel Dekker, Inc.

Ogata, K. 1990. *Modern Control Engineering*. Englewood Cliffs, NJ: Prentice-Hall.

Passive Walker. 2009. http://www.youtube.com/watch?NR=1&feature=endscreen&v=N64K OQkbyiI

Philips, C.L. and Harbor, R. 1988. *Feedback Control Systems*. Englewood Cliffs, NJ: Prentice-Hall International Editions.

Raven, F.H. 1987. *Automatic Control Engineering*. New York: McGraw-Hill International Editions.

Walking Robot. 2010. http://www.youtube.com/watch?v=UJLH5GYyVhY

8 Review of Mathematical Modeling, Transfer Functions, State Equations, and Controllers

8.1 INTRODUCTION

Every part or subsystem of a control system has some input to output relationship. This relationship is implicitly contained in its transfer function. A transfer function is nothing but the ratio of Laplace transforms of the output and the input. Transfer functions provide good insight into the subsystems they represent. For example, this can be the relationship between the speed and input voltage to the armature of the robot drive motor or can be the output to input relationship of a transducer used for feedback. However, such transfer functions in bits and pieces themselves will not be very useful for understanding the overall system. We should be able to obtain the ratio of output of any part to the input to any part of the system in terms of Laplace transforms ratio. That leads to the conclusion that there can be many transfer functions for a single system depending on the objective. If we look at such a ratio while the feedback is absent (open loop), then we call it the open-loop transfer function. Usually, an important aim of modeling is to obtain the transfer function of the open-loop plant. If the feedback is included (close loop), then we call it the closed-loop transfer function. These ideas are well discussed in the control literature (Ogata 1990; D'Azzo and Houpis 1995; Kuo 1987; Palani 1997; Nagrath and Gopal 1985), and we only wish to highlight the basic concepts for a robotics engineer so that it provides a starting point in forming system equations to design the controllers for the robot.

A state equation is another variant of a transfer function, which still represents the system dynamics. The main distinction between transfer functions and state equations is that where a transfer function has only one input and one output, a state equation is capable of representing more than one input and more than one output. Where the controller specifications focus on only one output of the plant, transfer functions are used to design controllers using classical control theory. However, when the specifications involve a few outputs of the plant, the state space approach is more convenient. Hence, depending on the complexity of the problem, either the transfer function-based design or the state equation-based design is chosen. In modern practice, whatever method is chosen, during the design of controllers, software tools such as MATLAB® are actively used (Cavallo et al. 1996).

8.2 IMPORTANCE OF MODELING

The understanding of the plant dynamics is the first step in designing a suitable controller. Once we know the dynamics, we can choose a suitable controller structure. The controller has to cater to the nature of the plant. For robotics application, we can summarize the major benefits of modeling as follows.

When we use transfer functions of plants with known components, this knowledge leads us to the order and the possible behavioral patterns of the plant. This plays an important part in deciding the controller. Even if the components are not known, we may have to use some identification techniques to understand the plant dynamics. In robotics, most components are with known dynamics, and this should be used to our advantage.

To study the stability of a system, we need to examine the open-loop transfer function. Some types of plants may become unstable when controlled in a closed loop. If the model is known, it provides an opportunity for offline testing of the controller. Before actually testing the controller in real time, it is important to predict to some extent how the closed-loop system will behave in terms of performance and safety. Once we see some discrepancy between the model performance and the expected performance, we can correct the mistakes in design. By the same reasoning, we can say that a mathematical model also helps in designing various controllers rapidly. Using the model, we can generate different designs of controllers. Then, they can be tried in simulation, and the time response can be readily computed. Different controller responses can be compared to choose the most suitable one for possible adoption in the actual implementation. In the following sections, the term "model" is used for a transfer function model as well as a state model.

8.3 TRANSFER FUNCTION MODELS

Transfer functions give a better understanding of the system behavior. The transfer function concept is applicable to only "linear and time-invariant" plants, which can be described by linear differential equations with constant coefficients. For solutions of such systems, Laplace transform techniques are very useful. The transfer function for a plant is defined as the ratio of the Laplace transform of the output to the Laplace transform of the input, where the initial conditions are assumed to be zeros.

Consider the following nth-order differential equation:

$$a_n \frac{d^n y}{dt^n} + a_{n-1} \frac{d^{n-1} y}{dt^{n-1}} + \cdots + a_1 \frac{dy}{dt} + a_0 y = b_m \frac{d^m x}{dt^m} + b_{m-1} \frac{d^{m-1} x}{dt^{m-1}} + \cdots + b_1 \frac{dx}{dt} + b_0 x$$

for $m < n$

$$(8.1)$$

In the above equation, $y(t)$ is the output and $x(t)$ is the input.

By taking the Laplace transform of Equation 8.1, we can obtain

$$\left[a_n s^n + a_{n-1} s^{n-1} + \cdots + a_1 s + a_0 \right] Y(s) = \left[b_m s^m + b_{m-1} s^{m-1} + \cdots + b_1 s^1 + b_0 \right] X(s)$$
(8.2)

We have ignored the initial conditions completely. The transfer function is then

$$G(s) = \frac{Y(s)}{X(s)} = \frac{b_m s^m + b_{m-1} s^{m-1} + b_{m-2} s^{m-2} + \cdots + b_2 s^2 + b_1 s + b_0}{a_n s^n + a_{n-1} s^{n-1} + a_{n-2} s^{n-2} + \cdots + a_2 s^2 + a_1 s + a_0}$$
(8.3)

We note from the above equation that the transfer function of a system is an operational method of expressing the differential equation that relates the output variable to the input variable. We add a few thoughts on transfer functions below.

Even though the applicability of the concept of the transfer function is limited to systems that can be described by linear differential equations with constant, time-invariant coefficients, some nonlinear systems can be approximated to their linear equivalent, and this approach is extensively used in the analysis and design of controllers for such nonlinear systems as well. While we say that the transfer function is a property of a system itself, independent of the magnitude and nature of the input or driving function, if saturation occurs, the linearity may not be applicable. Then, the model becomes an inadequate representation. We will highlight this later in our case studies.

If the plant representation or differential equation is known, the transfer function can be derived as shown above. However, even if the constituent components of a plant are unknown, the overall transfer function may be established experimentally by introducing known inputs and studying the output of the system. Once established, a transfer function gives a full description of the dynamic characteristics of the system, as distinct from its physical description.

8.3.1 DIFFERENT FORMS OF TRANSFER FUNCTIONS

The transfer function given in Equation 8.3 is in a polynomial form, since we start from differential equations. There are a few different forms of transfer functions; depending on the purpose, it will be used for.

Polynomial form: This polynomial form is the result of taking the Laplace transform of the differential equation. An example of it is given in Equation 8.4.

$$G(s) = \frac{b_2 s^2 + b_1 s + b_0}{s^3 + a_2 s^2 + a_1 s + a_0}$$
(8.4)

Pole/zero form: Factorizing and rewriting, we get the form

$$G(s) = \frac{K_1 [s - \alpha_1][s - \alpha_2]}{[s - \beta_1][s - \beta_2][s - \beta_3]}$$
(8.5)

which is in the pole/zero form. The roots of the numerator polynomial are called zeros. For the above system, it is obvious that $s = \alpha_1$, $s = \alpha_2$ are the zeros of the plant; similarly, the β values are the poles of the plant. These poles represent stability characteristics of the open-loop system. The negative real parts of the poles indicate an open-loop stable system. But, considering the numerator, if the real parts of zeros are negative, the plant is called a minimum phase system. If they have positive real parts, they are called nonmiminum phase systems, and it will be difficult to control them in a closed loop.

Time-constant form: Again, rearranging the above equation differently, we get another form:

$$G(s) = \frac{K_2(1 + T_{z1}s)(1 + T_{z2}s)}{(1 + T_{p1}s)(1 + T_{p2}s)(1 + T_{p3}s)} \tag{8.6}$$

We can observe that the DC gain is K_2 and the T_{zi} terms are numerator time constants and the T_{pi} terms are denominator time constants. This information may be helpful in assessing the speed of the response of the open-loop system. The shorter time constant indicates a fast response of the open-loop system.

Corner frequency form: We can reorganize the above to reveal more information about the frequency response of the system:

$$G(s) = K_3 \frac{(1 + (s/\omega_{z1}))\,(1 + (s/\omega_{z2}))}{(1 + (s/\omega_{p1}))\,(1 + (s/\omega_{p2}))\,(1 + (s/\omega_{p3}))} \tag{8.7}$$

In the above equation, K_3 is DC gain and the numerator ω terms are the corner frequencies related to the zero terms and denominator ω terms are the corner frequencies related to the pole terms. The corner frequency form is useful in sketching Bode diagrams for analyzing the frequency response and stability criteria.

8.4   STEPS IN MODELING

We can arrive from the discussion in the previous section that the primary aim of modeling is to obtain the open-loop transfer function or the state equation of the physical subsystem of the robot for which we want to design a controller. Then, the steps followed in modeling can be listed as follows.

The first thing to do is to identify the output of the plant and the possible input that will influence that output. Then, we find out if the plant has a linear input to output relationship. If that relationship is not linear, we decide the operating point around which we want to use the plant and find the linear approximation of the plant around that point. In case of electrical systems, we write the differential equations connecting the inputs and outputs by applying the circuit laws. For mechanical systems, we apply Newton's second law and write the applied forces and the reactive forces. Again, we write the differential equations connecting the input and the output.

If we are focusing on transfer functions, we take the Laplace transform of the relevant differential equations and deduce the ratio of the output to the input. If we need to formulate state equations, we define the states and write the state equations directly from the differential equations. If the transfer functions are readily available, state space equations can also be written from them directly. Similarly, transfer functions can also be obtained from the state space model. Once the model, either in the transfer function or in the state equation, has been obtained, we can design a viable controller.

8.5   SOME BASIC COMPONENTS OFTEN ENCOUNTERED IN CONTROL SYSTEMS

Control is an interdisciplinary subject. A control system may include electrical, mechanical, hydraulic, and pneumatic plants. Hence, we need to familiarize ourselves with components found in all of them. For example, some two-legged robots use hydraulic drives while some wall-climbing robots use pneumatic drives. To implement an effective controller, we need to understand them. On the other hand, feedback devices may also involve many disciplines. For example, simple encoders and gyroscopes are integral parts of the feedback control in many robotic systems. In the same robot, thermistors may be needed for monitoring temperature for safety. Here, we will mainly focus on electrical and some mechanical systems as a starting point.

8.5.1   Electrical Components

The main electrical components are resistances, inductances, and capacitors, and they are described below.

Ohm's law is the system equation of a resistor, where R is in ohms, v_R is in volts, and i is in amperes as shown in Figure 8.1a.

$$v_R = R\,i \tag{8.8}$$

The ideal inductance is shown in Figure 8.1b. The input–output relationship for an inductance is given by Faraday's law, where L is in henries, i is in amperes, and v_L is in volts.

$$v_L = L\,\frac{di}{dt} \tag{8.9}$$

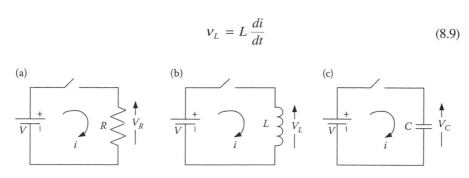

FIGURE 8.1   Common electrical components: (a) Resistance, (b) inductance, (c) capacitance.

A capacitor is shown in Figure 8.1c. The relationship between the input and the output of a capacitor is given in Equation 8.10, where C is in farads, i is in amperes, and v_C is the voltage across the capacitor in volts.

$$v_c = \frac{1}{C}\int i\,dt \quad \text{or} \quad i = C\frac{dv_c}{dt} \tag{8.10}$$

Note that in Figure 8.1, the diagrams are marked with uppercase letters, emphasizing the Laplace transform relationship. Note that we have ignored "s" terms within brackets for simplicity. But, Equations 8.8 through 8.10 are written for instantaneous values.

8.5.2 MECHANICAL COMPONENTS

The main mechanical components are mass, damper, and spring. Masses are distributed everywhere on robots and form an integral part of it. A mass placed on "frictionless" wheels is shown in Figure 8.2a. The input–output relationship of a mass is nothing but Newton's second law as given in Equation 8.11, where f is the applied force in newtons, x is the translation in meters, and M is the mass in kilograms.

$$f = M\frac{d^2x}{dt^2} \tag{8.11}$$

Dampers are also important in robotics. Dampers may be introduced on purpose to stabilize the system; on the other hand, friction in the system may act as a damper. The symbolic representation is shown in Figure 8.2b. The input–output relationship of a damper is given in Equation 8.12, where D is the damping coefficient in newtons-seconds/meter.

$$f = D\frac{dx}{dt} \tag{8.12}$$

Springs are used in many applications of control theory, including robotics. Springs are devices that produce a restraining force against pushing or pulling. For example, a spring can be used to press an encoder wheel toward the platform for

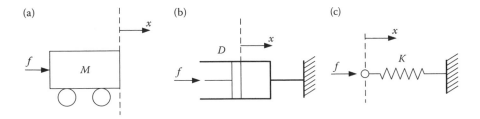

FIGURE 8.2 Common mechanical components: (a) mass, (b) damper, (c) spring.

achieving a proper contact. Springs are also an integral part of elastic actuators. A symbolic representation is shown in Figure 8.2c. The applied force, f, is in newtons, the linear translation, x, is in meters, and K is the spring constant in newtons/meter, the relationship is described in Equation 8.13.

$$f = Kx \qquad (8.13)$$

In addition to the above, we will encounter pneumatic, hydraulic, thermal, and other types of systems in robotics. We are not dealing with all such systems here, since the list of such components is very long. More details about them can be found in D'Azzo and Houpis (1995), Kuo (1987), and Nagrath and Gopal (1985).

8.6 BLOCK DIAGRAM CONCEPTS

We saw earlier some transfer function concepts. The plant transfer function is written inside the block that represents the plant. Usually, a complex control system has many components represented by blocks with their own respective transfer functions. They are interconnected in such a way that one block's output is the input of another block and so on. Hence, an overall block diagram provides information regarding the interconnections and functional relationships among the various constituents that form the controlled system. The basic components of a block diagram are blocks representing transfer functions, summing points, take-off points, and arrows indicating the direction of signal flow. When we start deciphering the relationship among the blocks, our aim is to obtain a simplified diagram with the forward system, a feedback loop, and a summing point, clearly marking the input and the output of the system. This is generally called the canonical form and is shown in Figure 8.3. In the figure, $G(s)$ represents the overall forward transfer function and $H(s)$ is the overall feedback transfer function. We emphasis the term "overall" because they are usually derived by simplifying many constituent blocks.

The following notations are typically used in the literature:

The system output is $C(s)$.
The reference input is $R(s)$.
The feedback signal is $B(s)$.

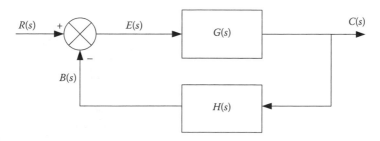

FIGURE 8.3 Canonical form.

The open-loop transfer function is

$$G(s)H(s) = B(s)/E(s) \qquad (8.14)$$

The feedback transfer function is

$$H(s) = B(s)/C(s) \qquad (8.15)$$

The closed-loop transfer function is $C(s)/R(s)$, and it is derived in the following steps:

$$E(s) = R(s) - H(s)C(s) \qquad (8.16)$$

Then

$$C(s) = G(s)E(s) = G(s)[R(s) - H(s)C(s)]$$

$$C(s)[1 + G(s)H(s)] = G(s)R(s)$$

$$\frac{C(s)}{R(s)} = \frac{G(s)}{1 + G(s)H(s)} = \frac{1}{H(s) + \dfrac{1}{G(s)}} \qquad (8.17)$$

8.6.1 BLOCK DIAGRAM REDUCTIONS

We mentioned earlier that $G(s)$ and $H(s)$ may not represent one block diagram, and the overall transfer function may have to be derived. Hence, when a block diagram is very complicated, it may be necessary to use some reduction techniques to obtain the overall transfer function. There are a number of guidelines for achieving this reduction, and it is well described in the literature. Some sample types are shown in Table 8.1. For a complete list of such possibilities, refer to Ogata (1990), D'Azzo and Houpis (1995), Kuo (1987), and Nagrath and Gopal (1985).

8.7 SOME SYSTEM EXAMPLES

Having learned the concepts of transfer functions and block diagrams, let us apply them to derive and simplify the transfer functions of systems we would often encounter in robotics. The devices described in Section 8.5 are seldom used in isolation. They are used in many forms of interconnections in real-world systems. The control designer should be able to write transfer functions for all types of interconnections. Let us see a few examples. In the notations used in the following sections, variables indicated with lowercase are functions of time and for simplicity, we may ignore showing (*t*).

EXAMPLE 8.1: SIMPLE ELECTRICAL SYSTEM

Let us start with a simple example. Consider the system shown in Figure 8.4 consisting of resistive, inductive, and capacitive components. Using Kirchoff's law, we can write

TABLE 8.1
Block Diagram Reductions

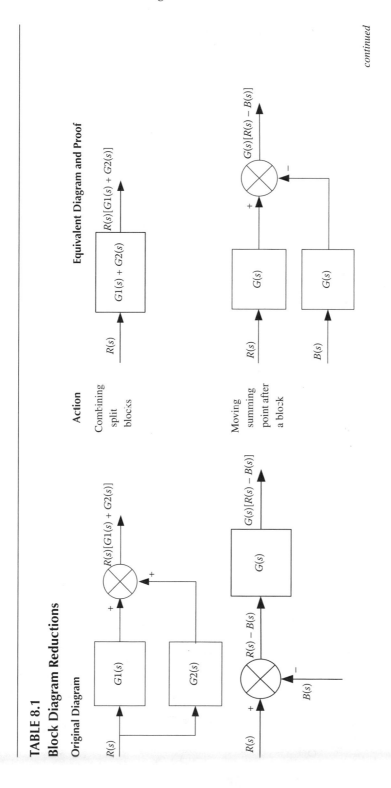

continued

TABLE 8.1 (continued)
Block Diagram Reductions

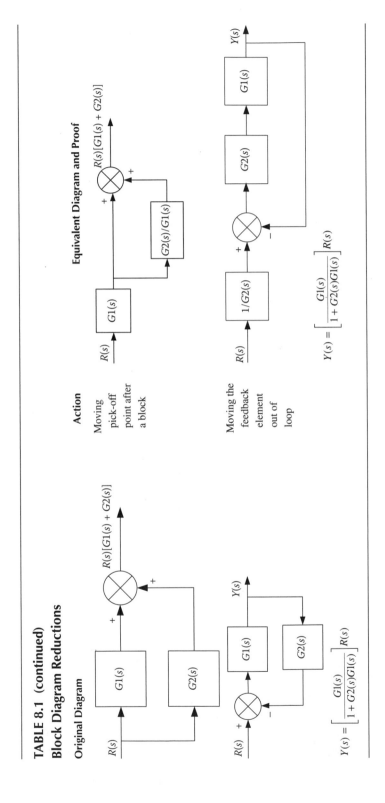

Original Diagram

Action

Equivalent Diagram and Proof

Moving pick-off point after a block

Moving the feedback element out of loop

$$Y(s) = \left[\frac{G1(s)}{1 + G2(s)G1(s)} \right] R(s)$$

$$Y(s) = \left[\frac{G1(s)}{1 + G2(s)G1(s)} \right] R(s)$$

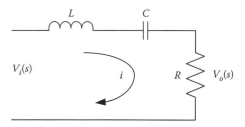

FIGURE 8.4   Simple electrical system with resistive, inductive, and capacitive components.

$$Ri + \frac{1}{C}\int idt + L\frac{di}{dt} = v_i(t) \qquad (8.18)$$

Taking the Laplace transform, while ignoring initial conditions, we can obtain

$$RI(s) + \frac{1}{Cs}I(s) + LsI(s) = V_i(s) \qquad (8.19)$$

$$I(s)[R + \frac{1}{Cs} + Ls] = V_i(s) \qquad (8.20)$$

Multiply both sides with R and using the fact that $V_o(s) = RI(s)$, we get

$$\frac{V_o(s)}{V_i(s)} = \frac{R}{[R + (1/Cs) + Ls]} = \frac{RCs}{[RCs + 1 + LCs^2]} = \frac{(R/L)s}{s^2 + (R/L)s + (1/LC)} \qquad (8.21)$$

EXAMPLE 8.2: TRANSFER FUNCTION OF A PERMANENT MAGNET DC MOTOR DRIVE SYSTEM (ELECTROMECHANICAL SYSTEM)

The basic components of a permanent magnet DC motor are shown in Figure 8.5. There is one electrical system which is a motor armature circuit. We also have the rotating mechanical system with its inherent mass and friction. We can use the following steps to derive a transfer function:

i. Write the equation describing each system and its transfer function.
ii. Simplify the equation to obtain the transfer function relating angular velocity to the input voltage.

The relevant quantities are marked on the diagram in Figure 8.5. Given that the back EMF constant is K_b, the torque constant is K_m, the outputs of the system are $\omega(s)$ and $\theta(s)$, the input of the system is $V_a(s)$, and the armature current is i_a, our objective is to derive the transfer functions $\omega(s)/V_a(s)$ and $\theta(s)/V_a(s)$.

Consider the electrical circuit

$$v_a(t) = R_a i_a + L_a \frac{di_a}{dt} + e_b \qquad (8.22)$$

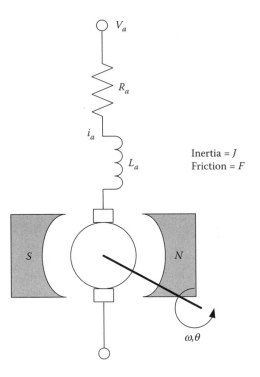

FIGURE 8.5   Servomotor system.

where the back EMF is e_b. Then, using the back EMF constant

$$e_b = K_b \omega \qquad (8.23)$$

Using the torque constant, the motor torque can be written as

$$t_m = K_m i_a \qquad (8.24)$$

Since there is no other load, the motor torque generated is equal to the load torque due to the motor inertia and friction. Then, the torque balance equation is written as

$$t_m = J\frac{d\omega}{dt} + F\omega \qquad (8.25)$$

Taking the Laplace transforms of Equations 8.22 through 8.25, respectively, we can write

$$V_a(s) = [R_a + L_a s]I_a(s) + E_b(s) \qquad (8.26)$$

$$E_b(s) = K_b \omega(s) \qquad (8.27)$$

$$T_m(s) = K_m I_a(s) \qquad (8.28)$$

$$T_m(s) = [Js + F]\omega(s) \qquad (8.29)$$

Our first objective is to derive $\omega(s)/V_a(s)$, which can be done using mathematical manipulations, since the numerator term and the denominator term are found in the above equations. Substituting for T_m and I_a, we can write

$$[Js + F]\omega(s) = K_m \frac{[V_a(s) - E_b(s)]}{R_a + L_a s} = K_m \frac{[V_a(s) - K_b \omega(s)]}{R_a + L_a s}$$

$$(R_a + L_a s)[Js + F]\omega(s) = K_m V_a(s) - K_m K_b \omega(s) \qquad (8.30)$$

Assembling these terms, we get

$$G_1(s) = \frac{\omega(s)}{V_a(s)} = \frac{K_m}{(R_a + L_a s)[Js + F] + K_m K_b} \qquad (8.31)$$

Since, we know

$$\omega(t) = \frac{d\theta}{dt} \text{ and } \omega(s) = s\theta(s) \qquad (8.32)$$

We get

$$G_2(s) = \frac{\theta(s)}{V_a(s)} = \frac{K_m}{(R_a + L_a s)[Js^2 + Fs] + K_m K_b s} \qquad (8.33)$$

Intuitive approach: Referring to Figure 8.6, we can draw a block diagram, first, by obtaining the driving voltage $E(s)$ as the difference between $V_a(s)$ and $E_b(s)$. After that, we put a block of impedance to get current $I_a(s)$. Following this, adding a block with K_m leads to torque $T_m(s)$. From there, we add a block using the inertia and friction terms to get the angular speed $\omega(s)$. From the angular speed, we obtain the back EMF, E_b, by including a block with K_b and feed it back to the summing point. We now obtain the angle as an integral of the angular speed. Figure 8.6 represents the overall block diagram of the system.

From the above diagram, we can obtain the closed-loop transfer function quite easily.

The forward transfer function is

$$G(s) = \frac{\omega(s)}{E(s)} = \frac{K_m}{R_a + L_a s} \frac{1}{Js + F} \qquad (8.34)$$

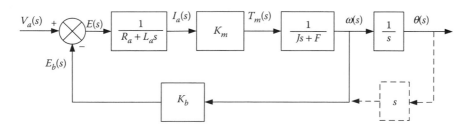

FIGURE 8.6 Overall block diagram.

The feedback transfer function is

$$H(s) = \frac{E_b(s)}{\omega(s)} = K_b \qquad (8.35)$$

The open-loop transfer function is

$$G(s)H(s) = \frac{E_b(s)}{E(s)} = \frac{K_m}{R_a + L_a s} \frac{K_b}{Js + F} \qquad (8.36)$$

Then, the closed-loop transfer function is

$$G_1(s) = \frac{\omega(s)}{V_a(s)} = \frac{G(s)}{1+G(s)H(s)} = \frac{K_m}{(Js + F)(R_a + L_a s) + K_m K_b} \qquad (8.37)$$

Then, by using the relationship between $\omega(s)$ and $\theta(s)$, we write

$$G_2(s) = \frac{\theta(s)}{V_a(s)} = \frac{K_m}{(Js^2 + Fs)(R_a + L_a s) + K_m K_b s} \qquad (8.38)$$

The above transfer function is of the third order; however, if the motor leakage inductance can be ignored, then it is reduced to the second order.

EXAMPLE 8.3: MASS DAMPER AND SPRING ASSEMBLY

Let us find the transfer function, $Y(s)/F(s)$ of the mechanical translation system shown in Figure 8.7, which is another popular mention in the literature.

The quantities are marked in the figure, and the mass is supported with frictionless wheels.

Considering only the two ends of the spring, we know that the force is transmitted through the mass partly to the spring and partly to the damper or dash-pot. Hence, we can write

$$f(t) = M\frac{d^2y}{dt^2} + D\frac{dy}{dt} + Ky \qquad (8.39)$$

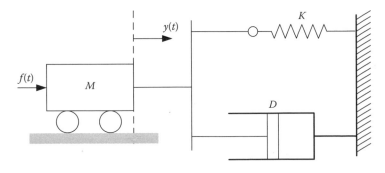

FIGURE 8.7   A mechanical translation system.

Now, taking the Laplace transform, we obtain

$$F(s) = \left[Ms^2 + Ds + Ky\right]Y(s) \tag{8.40}$$

$$\frac{Y(s)}{F(s)} = \frac{1}{Ms^2 + Ds + K} \tag{8.41}$$

Usually, in practical systems, the damper gets connected in parallel to a spring, and this combination is used in tandem with a mass. In fact, if we turn the system clockwise, it becomes a setup of a shock absorber of an automobile or some autonomous robot.

EXAMPLE 8.4: VEHICLE INSIDE A VEHICLE

This is a popular example that can be found in the literature (Palani 1997). We will analyze this system since it is a bit more complex for the transfer function as well as the state equation. Figure 8.8 shows such an arrangement of a system in which a vehicle is placed inside another vehicle. The outer vehicle is a container of mass M_2, which can move on wheels on a platform with no friction. This is attached to the wall on the right by a spring with stiffness of K_2 and a damper of damping coefficient B_2. Inside this vehicle, there is another mass M_1 moving without friction. This mass is attached to the wall of the outer vessel (vehicle) wall through a spring of constant, K_1. Assume that a force $f(t)$ is applied to the outer vehicle as shown in Figure 8.8. The displacements of the inner and outer masses are marked as y_1 and y_2, respectively, and are measured with respect to the reference wall on the right. We are interested in displacements $y_1(t)$ and $y_2(t)$ in response to the input force $f(t)$.

This example needs a careful observation. Let us first look at the forces that act upon mass M_2. The forces can be listed as follows:

1. The applied force $f(t)$ acting to the right.
2. The reaction force, due to acceleration of M_2 to the right, is $M_2(d^2y_2/dt^2)$ acting to the left.

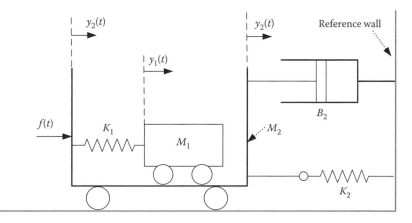

FIGURE 8.8   Vehicle inside a vehicle.

3. The restraining spring force K_2y_2, which is acting to the left.
4. The damping force $B_2(dy_2/dt)$ which is acting to the left.
5. As the inner vehicle moves, the spring K_1 will generate a force $K_1(y_2 - y_1)$, which is also acting to the left. This can be understood by visualizing the situation if M_1 does not move ($y_1 = 0$), y_2 motion will compress the spring K_1, and hence it will push back the mass M_2 to the left.

We can write a force balance equation as

$$M_2\frac{d^2y_2}{dt^2} + B_2\frac{dy_2}{dt} + K_2y_2 + K_1(y_2 - y_1) = f(t) \tag{8.42}$$

and taking the Laplace transform, we get

$$\left[M_2s^2 + B_2s + (K_2 + K_1)\right]Y_2(s) - K_1Y_1(s) = F(s) \tag{8.43}$$

Next, let us list the forces acting on mass M_1 for the positive displacement of y_1:

1. The inertial reaction of mass M_1 equal to $M_1(d^2y^1/dt^2)$, acting to the left
2. The restraining force of spring K_1 equal to $K_1(y_1 - y_2)$, acting to the left

There is no other directly applied force. We can write the force balance equation as

$$M_1\frac{d^2y_1}{dt^2} + K_1(y_1 - y_2) = 0 \tag{8.44}$$

Taking the Laplace transform, we have

$$\left[M_1s^2 + K_1\right]Y_1(s) = K_1Y_2(s) \tag{8.45}$$

Substituting for $Y_2(s)$ in Equation 8.43, we get

$$\left[M_2s^2 + B_2s + (K_2 + K_1)\right]\frac{\left[M_1s^2 + K_1\right]}{K_1}Y_1(s) - K_1Y_1(s) = F(s)$$

$$\left[M_2s^2 + B_2s + (K_2 + K_1)\right]\left[M_1s^2 + K_1\right]Y_1(s) - K_1^2Y_1(s) = K_1F(s) \tag{8.46}$$

This yields the transfer function

$$\frac{Y_1(s)}{F(s)} = \frac{K_1}{M_2M_1s^4 + M_1B_2s^3 + \left[M_1(K_2 + K_1) + K_1M_2\right]s^2 + K_1B_2s + K_1K_2} \tag{8.47}$$

Alternative approach: We now attempt to solve the same problem using the block diagram reduction approach. We can get a block diagram for Equation 8.45 as shown in Figure 8.9.

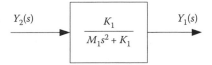

FIGURE 8.9   Block diagram from Equation 8.45.

We can get a block diagram for Equation 8.43 as shown in Figure 8.10.

There is a positive feed-in of $Y_1(s)$. This has to be dealt with later. Let us now combine the two block diagrams of Figures 8.9 and 8.10 as shown in Figure 8.11.

Note that we have introduced two "-" signs in the feedback loop, which changes nothing to make it an usual negative feedback system. Then, the overall transfer function can be written as

$$\frac{Y_1(s)}{F(s)} = \frac{1}{H + (1/G)} = \frac{1}{-K_1 + ((M_2 s^2 + B_2 s + K_1 + K_2)(M_1 s^2 + K_1)/K_1)} \tag{8.48}$$

$$\frac{Y_1(s)}{F(s)} = \frac{K_1}{M_2 M_1 s^4 + M_1 B_2 s^3 + \left[M_1(K_2 + K_1) + K_1 M_2\right]s^2 + K_1 B_2 s + K_1 K_2} \tag{8.49}$$

This is exactly the same result that we got earlier in Equation 8.47.

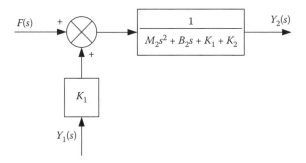

FIGURE 8.10   Block diagram from Equation 8.43.

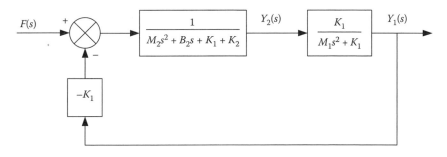

FIGURE 8.11   Overall block diagram from Figures 8.9 and 8.10.

EXAMPLE 8.5: A MECHANICAL ACCELEROMETER

Now, take a look at another popular example found in the literature (Nagrath and Gopal 1985). We will study accelerometers that are often used in robotics, especially in humanoid and mobile robots. We present one version of such a device here and derive the transfer function as well as the state equation. The physical arrangement of an accelerometer is shown in Figure 8.12. It shows a simplified accelerometer fitted to a moving vehicle the displacement of which as the vehicle moves is $y_2(t)$.

The device is shown as mounted on a robot body. It has a box consisting of a mass spring and a damper fitted to it. The encoder reading at standstill is $y_1(t) = 0$. We want to show that the linear encoder reading is a measure of the acceleration of the vehicle. We assume that the mass M moves without any friction. Obviously, $y_1(t)$ is measured with respect to the frame of the accelerometer, while $y_2(t)$ is measured with respect to an external stationary absolute frame. We further assume that the positive movement of the robot, $y_2(t)$, is to the left.

As marked, $y_1(t)$ moving to the right of the scale is positive. Then, the absolute displacement of M toward the left is $(y_2(t) - y_1(t))$ since the positive $y_1(t)$ motion is opposite to the vehicle motion $y_2(t)$.

For the positive displacement of $y_1(t)$ on the scale, the forces acting on the mass, M, are

1. The reactive force $M(d^2(y_2 - y_1)/dt^2)$ acting to the right
2. The reactive damping force of B, $B(dy_1/dt)$ acting to the left
3. The restraining force of spring $K_1 y_1$ acting to the left

Since there is no other force, the total of these three should be zero.

$$M\frac{d^2(y_1 - y_2)}{dt^2} + B\frac{dy_1}{dt} + K y_1 = 0$$

$$M\frac{d^2 y_1}{dt^2} + B\frac{dy_1}{dt} + K y_1 = M\frac{d^2 y_2}{dt^2} \tag{8.50}$$

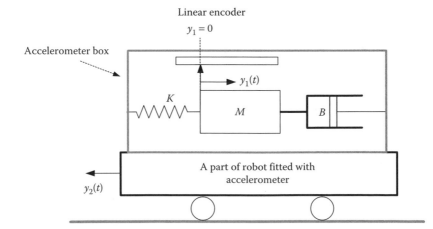

FIGURE 8.12 Simplified arrangement of an accelerometer.

For a steady acceleration of $(d^2y_2/dt^2) = a$, $(dy_1/dt) = (d^2y_1/dt^2) = 0$.

Hence, $a = (K/M)y_1$ and the encoder reads the acceleration value of the outer vehicle. Moreover, if we are interested in finding the response of $Y_1(s)/Y_2(s)$, then we can take the Laplace transform and obtain

$$\frac{Y_1(s)}{Y_2(s)} = \frac{Ms^2}{Ms^2 + Bs + K} \tag{8.51}$$

8.8   STATE EQUATIONS

We will introduce the concept of state equations purely with an objective of designing controllers for state control. At the outset, one can say that the state equations are used to represent an nth-order differential equation of a system as a set of n first-order equations and assemble them in a matrix form by defining the state vectors, output vectors, and system matrices.

Where the transfer functions can represent one output and one input in a single transfer function, the state equations can take care of many inputs and many outputs. State equations are a well-developed concept and hence quite a few control tools are available for designing the controllers for the systems, such as pole-balancing robots and biped robots represented by state equations (Ogata 1990; Kuo 1987; Nagrath and Gopal 1985; Wilburg 1971).

8.8.1   BASIC CONCEPTS OF STATE EQUATIONS FROM DIFFERENTIAL EQUATIONS

Let us consider the differential equation in Equation 8.52, which is of the nth order.

$$a_n \frac{d^n y}{dt^n} + a_{n-1} \frac{d^{n-1} y}{dt^{n-1}} + a_{n-2} \frac{d^{n-2} y}{dt^{n-2}} + \cdots a_1 \frac{dy}{dt} + a_0 y = b_1 u_1 + b_0 u_0 \tag{8.52}$$

We have included two inputs to make it a general system with more than one input. Let us define the following quantities as states:

$$
\begin{aligned}
x_1 &= y \\
x_2 &= \frac{dy}{dt} = \dot{x}_1 \\
x_3 &= \frac{d^2 y}{dt^2} = \dot{x}_2 \\
&\cdots \\
x_{n-1} &= \frac{d^{n-2} y}{dt^{n-2}} = \dot{x}_{n-2} \\
x_n &= \frac{d^{n-1} y}{dt^{n-1}} = \dot{x}_{n-1}
\end{aligned}
\tag{8.53}
$$

We can rewrite Equation 8.52 as

$$a_n \frac{d^n y}{dt^n} = -a_{n-1}\frac{d^{n-1}y}{dt^{n-1}} - a_{n-2}\frac{d^{n-2}y}{dt^{n-2}} - \cdots - a_1 \frac{dy}{dt} - a_0 y + b_0 u_0 + b_1 u_1$$

or   (8.54)

$$\frac{d^n y}{dt^n} = -\frac{a_{n-1}}{a_n}\frac{d^{n-1}y}{dt^{n-1}} - \frac{a_{n-2}}{a_n}\frac{d^{n-2}y}{dt^{n-2}} - \cdots - \frac{a_1}{a_n}\frac{dy}{dt} - \frac{a_0}{a_n}y + \frac{b_0}{a_n}u_0 + \frac{b_1}{a_n}u_1$$

Using the definitions in Equation 8.53, we can write

$$\dot{x}_n = \frac{d^n y}{dt^n} = -\frac{a_{n-1}}{a_n}x_n - \frac{a_{n-2}}{a_n}x_{n-1} - \cdots - \frac{a_1}{a_n}x_2 - \frac{a_0}{a_n}x_1 + \frac{b_0}{a_n}u_0 + \frac{b_1}{a_n}u_1 \quad (8.55)$$

We can assemble Equations 8.53 and 8.55 into a matrix form as

$$\begin{bmatrix} \dot{x}_1 \\ \dot{x}_2 \\ \dot{x}_3 \\ \cdots \\ \cdots \\ \dot{x}_{n-1} \\ \dot{x}_n \end{bmatrix} = \begin{bmatrix} 0 & 1 & 0 & \cdots & 0 \\ 0 & 0 & 1 & \cdots & 0 \\ \cdots & & & & \\ 0 & 0 & \cdots & & 1 \\ -\frac{a_0}{a_n} & -\frac{a_1}{a_n} & -\frac{a_2}{a_n} & \cdots & -\frac{a_{n-1}}{a_n} \end{bmatrix} \begin{bmatrix} x_1 \\ x_2 \\ x_3 \\ \cdots \\ x_{n-1} \\ x_n \end{bmatrix} + \begin{bmatrix} 0 & 0 \\ 0 & 0 \\ 0 & 0 \\ \cdots & \cdots \\ 0 & 0 \\ \frac{b_0}{a_n} & \frac{b_1}{a_n} \end{bmatrix}\begin{bmatrix} u_0 \\ u_1 \end{bmatrix} \quad (8.56)$$

Equation 8.56 is called the state equation and is usually abbreviated as

$$\dot{X} = [A]\bar{X} + [B]\bar{U} \quad (8.57)$$

where A is the system matrix, B is the control matrix, $\bar{X}$ is the sate vector, and $\bar{U}$ is the control vector. If y is the output, then the output can be expressed as a function of states as follows:

$$y = [C]\bar{X} + D\bar{U} \quad (8.58)$$

Here, D is the transmission matrix. Usually, input is seldom transmitted directly to the output, except in rare cases where the orders of numerator polynomial and denominator polynomial of the system transfer function are equal. The "bars" above the vectors $\bar{X}$ and $\bar{U}$ are dropped for convenience.

Thus far, we have provided some introduction to the state space representation of linear systems. What we must note is that states consist of higher derivatives, if we follow the above method of forming state equations. In noisy situations, only the first two states can be measured or estimated, but other states involve taking derivatives of second or higher order and it is almost impossible. Even estimates of second-order derivatives will be inaccurate in a noisy situation. When we implement state feedback control techniques, we may need most of the states. This presents a tricky situation in which we need state observers, and they are costly in terms of computation time. A good practice is to select state variables that can either be measured or estimated easily with good accuracy.

8.8.2 STATE EQUATIONS FROM PLANT KNOWLEDGE

In many robotic systems, we have good knowledge regarding the dynamics of the constituent plants. Hence, it is easy to write the differential equations and hence derive the state equations directly without the need for transfer functions. In many such cases, the state measurement or estimation is quite straightforward. We see such examples below.

EXAMPLE 8.6: MASS, SPRING, AND DAMPER SYSTEM

Here, we would like to work on the example of a mechanical plant and derive the state equations for it. Earlier in Example 8.3, we derived the transfer function for a mechanical plant. Here, we will take up the same example and derive state equations for that plant. The mechanical structure is shown in Figure 8.7. The differential equations are as follows:

$$f = M\frac{d^2y}{dt^2} + D\frac{dy}{dt} + Ky \qquad (8.59)$$

Let us define states

$$x_1 = y$$
$$x_2 = \frac{dy}{dt} = \dot{x}_1 \qquad (8.60)$$
$$u = f$$

We can rewrite the differential equation as

$$u = M\dot{x}_2 + Dx_2 + Kx_1, \text{ hence}$$

$$\dot{x}_2 = -\frac{K}{M}x_1 - \frac{D}{M}x_2 + \frac{1}{M}u \qquad (8.61)$$

Combining, we can obtain

$$\begin{bmatrix} \dot{x}_1 \\ \dot{x}_2 \end{bmatrix} = \begin{bmatrix} 0 & 1 \\ -\dfrac{K}{M} & -\dfrac{D}{M} \end{bmatrix} \begin{bmatrix} x_1 \\ x_2 \end{bmatrix} + \begin{bmatrix} 0 \\ \dfrac{1}{M} \end{bmatrix} u \qquad (8.62)$$

EXAMPLE 8.7: VEHICLE INSIDE A VEHICLE
PROBLEM USING STATE EQUATIONS

Consider the system in Figure 8.8. The differential equations are given as follows:

$$M_2\frac{d^2y_2}{dt^2} + B_2\frac{dy_2}{dt} + K_2y_2 + K_1(y_2 - y_1) = f(t) \tag{8.63}$$

Also, we have

$$M_1\frac{d^2y_1}{dt^2} + K_1(y_1 - y_2) = 0 \tag{8.64}$$

Let us define the following states:

$$
\begin{aligned}
x_1 &= y_1 \\
x_2 &= \frac{dy_1}{dt} = \dot{x}_1 \\
x_3 &= y_2 \\
x_4 &= \frac{dy_2}{dt} = \dot{x}_3
\end{aligned}
\tag{8.65}
$$

Substituting into above equations, we get

$$M_2\dot{x}_4 + B_2x_4 + K_2x_3 + K_1(x_3 - x_1) = u(t)$$

$$\dot{x}_4 = -\frac{B_2}{M_2}x_4 - \left(\frac{K_2}{M_2} + \frac{K_1}{M_2}\right)x_3 + \frac{K_1}{M_2}x_1 + \frac{1}{M_2}u(t) \tag{8.66}$$

and

$$M_1\dot{x}_2 + K_1(x_1 - x_3) = 0$$

$$\dot{x}_2 = -\frac{K_1}{M_1}x_1 + \frac{K_1}{M_1}x_3 \tag{8.67}$$

Assembling Equations 8.65, 8.66, and 8.67, we get

$$
\begin{bmatrix} \dot{x}_1 \\ \dot{x}_2 \\ \dot{x}_3 \\ \dot{x}_4 \end{bmatrix}
=
\begin{bmatrix}
0 & 1 & 0 & 0 \\
-\dfrac{K_1}{M_1} & 0 & \dfrac{K_1}{M_1} & 0 \\
0 & 0 & 0 & 1 \\
\dfrac{K_1}{M_2} & 0 & -\dfrac{K_1 + K_2}{M_2} & -\dfrac{B_2}{M_2}
\end{bmatrix}
\begin{bmatrix} x_1 \\ x_2 \\ x_3 \\ x_4 \end{bmatrix}
+
\begin{bmatrix} 0 \\ 0 \\ 0 \\ \dfrac{1}{M_2} \end{bmatrix} u
\tag{8.68}
$$

In this example, all states can be easily measured or estimated.

EXAMPLE 8.8: MECHANICAL ACCELEROMETER

We considered the case of a mechanical accelerometer in Example 8.5. We derived a transfer function describing the relationship between $Y_1(s)/Y_2(s)$ in Equation 8.51. Here, we would like to define the states for that system and explore how those states are affected by the acceleration of the vehicle. If one of the states chosen is the scale reading y_1, we get a better picture of the response of the acceleration measurement of the system. For Figure 8.12, the differential equation is given as

$$M\frac{d^2 y_1}{dt^2} + B\frac{dy_1}{dt} + K y_1 = M\frac{d^2 y_2}{dt^2} = M a \qquad (8.69)$$

Let us define the states as

$$x_1 = y_1$$
$$x_2 = \frac{dy_1}{dt} = \dot{x}_1 \qquad (8.70)$$

And substituting for x_2

$$\dot{x}_2 = -\frac{B}{M}x_2 - \frac{K}{M}x_1 + a \qquad (8.71)$$

Hence, the state and output equations are

$$\begin{bmatrix} \dot{x}_1 \\ \dot{x}_2 \end{bmatrix} = \begin{bmatrix} 0 & 1 \\ -\dfrac{K}{M} & -\dfrac{B}{M} \end{bmatrix}\begin{bmatrix} x_1 \\ x_2 \end{bmatrix} + \begin{bmatrix} 0 \\ 1 \end{bmatrix} a$$
$$y = \begin{bmatrix} 1 & 0 \end{bmatrix}\begin{bmatrix} x_1 \\ x_2 \end{bmatrix} \qquad (8.72)$$

Here, the input is the acceleration, a, and the output is the reading y_1 on the accelerometer scale. In this example, all states can be easily measured or estimated as well.

8.8.3 STATE EQUATIONS DIRECTLY FROM TRANSFER FUNCTIONS

In many cases, transfer functions are obtained by running a frequency response test of the systems. In other words, only transfer functions are available for getting the system insight.

Cross-multiplying the transfer function equation relating the output to the input and taking the inverse Laplace transform, we can write the differential equation. Then, we can define the states and write the state equations. Also, state equations can be formed from transfer functions directly even without writing any differential equations. We provide some examples to describe methods widely discussed in the control literature (Ogata 1990; Kuo 1987; Palani 1997; Nagrath and Gopal 1985).

EXAMPLE 8.9

Consider the transfer function

$$\frac{Y(s)}{U(s)} = \frac{5}{s^4 + 5s^3 + 4s^2 + 4s + 7} \tag{8.73}$$

This example has no zero term, which makes things straightforward. The numerator polynomial order, m, is 0 and the denominator polynomial order, n, is 4. We define the following states:

$$x_1 = y$$

$$x_2 = \dot{x}_1 = \frac{dy}{dt}$$

$$x_3 = \dot{x}_2 = \frac{d^2y}{dt^2} \tag{8.74}$$

$$x_4 = \dot{x}_3 = \frac{d^3y}{dt^3}$$

Using the definition of the states, the state equation can be written as

$$\begin{bmatrix} \dot{x}_1 \\ \dot{x}_2 \\ \dot{x}_3 \\ \dot{x}_4 \end{bmatrix} = \begin{bmatrix} 0 & 1 & 0 & 0 \\ 0 & 0 & 1 & 0 \\ 0 & 0 & 0 & 1 \\ -7 & -4 & -4 & -5 \end{bmatrix} \begin{bmatrix} x_1 \\ x_2 \\ x_3 \\ x_4 \end{bmatrix} + \begin{bmatrix} 0 \\ 0 \\ 0 \\ 5 \end{bmatrix} u \tag{8.75}$$

The output equation becomes

$$y = \begin{bmatrix} 1 & 0 & 0 & 0 \end{bmatrix} \begin{bmatrix} x_1 \\ x_2 \\ x_3 \\ x_4 \end{bmatrix} \tag{8.76}$$

Again, the drawback is that the states involve higher derivatives of the output.

EXAMPLE 8.10: ALTERNATIVE APPROACH FOR EXAMPLE 8.9

As it is obvious from the example given above, the solution to the state feedback becomes difficult to tackle, since states involve higher-order derivatives and cannot be evaluated. However, in the following, we employ an alternative approach

(Palani 1997). In this method, the states are defined in such a way that they do not involve higher-order derivatives in state definitions. Here, the numerator order $m = 0$, and the denominator order $n = 4$. We repeat Example 8.9 for some comparison of state equations. Given that

$$\frac{Y(s)}{U(s)} = \frac{5}{s^4 + 5s^3 + 4s^2 + 4s + 7} \tag{8.77}$$

Cross-multiplying and taking the inverse Laplace transform, we have

$$\ddddot{y} = -5\dddot{y} - 4\ddot{y} - 4\dot{y} - 7y + 5u$$

Then, integrating three times, we get

$$\dot{y} = -5y - 4\int y - 4\iint y - 7\iiint y + 5\iiint u \tag{8.78}$$

Defining $x_1 = y$ as before

$$\dot{x}_1 = -5x_1 + x_2$$

where x_2 is tacitly defined as

$$x_2 = -4\int y - 4\iint y - 7\iiint y + 5\iiint u \tag{8.79}$$

Now, differentiating Equation 8.79 once

$$\dot{x}_2 = -4y - 4\int y - 7\iint y + 5\iint u$$

$$\dot{x}_2 = -4x_1 + x_3 \tag{8.80}$$

where

$$x_3 = -4\int y - 7\iint y + 5\iint u \tag{8.81}$$

Again, differentiating Equation 8.81 once

$$\dot{x}_3 = -4y - 7\int y + 5\int u$$

$$\dot{x}_3 = -4x_1 + x_4 \qquad (8.82)$$

where

$$x_4 = -7\int y + 5\int u \qquad (8.83)$$

Again, differentiating Equation 8.83 once

$$\dot{x}_4 = -7y + 5u = -7x_1 + 5u \qquad (8.84)$$

Let us assemble the above equations into the state equations

$$\begin{bmatrix} \dot{x}_1 \\ \dot{x}_2 \\ \dot{x}_3 \\ \dot{x}_4 \end{bmatrix} = \begin{bmatrix} -5 & 1 & 0 & 0 \\ -4 & 0 & 1 & 0 \\ -4 & 0 & 0 & 1 \\ -7 & 0 & 0 & 0 \end{bmatrix} \begin{bmatrix} x_1 \\ x_2 \\ x_3 \\ x_4 \end{bmatrix} + \begin{bmatrix} 0 \\ 0 \\ 0 \\ 5 \end{bmatrix} u \qquad (8.85)$$

and the output is given by

$$y = \begin{bmatrix} 1 & 0 & 0 & 0 \end{bmatrix} \begin{bmatrix} x_1 \\ x_2 \\ x_3 \\ x_4 \end{bmatrix} \qquad (8.86)$$

When we compare this set (Equations 8.85 and 8.86) with the state equation set obtained earlier (Equations 8.75 and 8.76), we note that the output equations are the same, but system matrices look different. We know that both state models originate from the same system represented by the transfer function given by Equation 8.73 and hence should represent the same dynamics as that of the transfer function. In state equations, the eigenvalues represent the system dynamics. Since both state models were derived from the same transfer function, their eigenvalues cannot be different and must be the same as the poles of the original transfer function. Let us use MATLAB to check the poles of the original transfer function and the eigenvalues of both state equations, for comparison. The MATLAB dialog is listed in Figure 8.13 (Cavallo et al. 1996).

From Figure 8.13, it is clear that both state matrices obtained by different methods yield the same eigenvalues, which are also the same as the roots of the characteristic equation. It is important to note that in Example 8.9, the state variables are not measurable since the states' estimation involves higher-order derivatives as it can be seen in Equation 8.74. Hence, states are not available for feedback. But, in Example 8.10, the state variables can be estimated with some efforts of integration as it can be seen from Equations 8.79, 8.81, and 8.83.

```
% define polynomial
C = [1 5 4 4 7];
% define state 1
A1=[0 1 0 0;0 0 1 0;0 0 0 1;-7 -4 -4 -5];
% define state 2
A2=[-5 1 0 0;-4 0 1 0;-4 0 0 1;-7 0 0 0];
rts=roots(C)
Lambda1=eig(A1)
Lambda2=eig(A2)

Results:

rts =
  -4.1753
  -1.3415
   0.2584 + 1.0876i
   0.2584 - 1.0876i
Lambda1 =
  -4.1753
  -1.3415
   0.2584 + 1.0876i
   0.2584 - 1.0876i
Lambda2 =
  -4.1753
   0.2584 + 1.0876i
   0.2584 - 1.0876i
  -1.3415
```

FIGURE 8.13 MATLAB code and results for comparing the eigenvalues of the two state systems.

EXAMPLE 8.11: CASE WHERE PLANT HAS ZERO TERMS WITH $M = 2$ AND $N = 4$

Consider

$$\frac{Y(s)}{U(s)} = \frac{24s^2 + 32s + 48}{4s^4 + 5s^3 + 8s^2 + 12s + 120} \tag{8.87}$$

In this case, $m < n$ and there will be no direct transmission term D (i.e., $D = 0$). This can be directly dealt without factorizing it. Cross-multiplying and taking the inverse Laplace transform, we have

$$4\ddddot{y} + 5\dddot{y} + 8\ddot{y} + 12\dot{y} + 120y = 24\ddot{u} + 32\dot{u} + 48u$$

Solving for the highest derivative

$$\ddddot{y} = -1.25\dddot{y} - 2\ddot{y} - 3\dot{y} - 30y + 6\ddot{u} + 8\dot{u} + 12u$$

and integrating both sides four times, we obtain

$$y = -1.25\int y - 2\iint y - 3\iiint y - 30\iiiint y + 6\iint u + 8\iiint u + 12\iiiint u \tag{8.88}$$

The above step is somewhat redundant, but written to streamline the thinking process. Now, let us define the first state variable, which is also the output:

$$x_1 = y$$

At this step, if there is a nonintegral term u on the right-hand side, it must be included in the state variable x_1. Since there was no such term, the definition of x_1 was equated to a simple measurement. Let us differentiate Equation 8.88 once:

$$\dot{y} = -1.25y - 2\int y - 3\iint y - 30\iiint y + 6\int u + 8\iint u + 12\iiint u$$

$$\dot{x}_1 = -1.25x_1 + x_2 \qquad (8.89)$$

We have tacitly equated the integral terms to a new state variable as

$$x_2 = -2\int y - 3\iint y - 30\iiint y + 6\int u + 8\iint u + 12\iiint u \qquad (8.90)$$

Let us differentiate above equation once:

$$\dot{x}_2 = -2y - 3\int y - 30\iint y + 6u + 8\int u + 12\iint u$$

$$\dot{x}_2 = -2x_1 + x_3 + 6u \qquad (8.91)$$

where, as before, we have equated all integral terms to a new state variable as

$$x_3 = -3\int y - 30\iint y + 8\int u + 12\iint u \qquad (8.92)$$

Differentiating Equation 8.92 once

$$\dot{x}_3 = -3y - 30\int y + 8u + 12\int u$$

$$\dot{x}_3 = -3x_1 + x_4 + 8u \qquad (8.93)$$

Again, we have equated all integral terms to the fourth state variable as

$$x_4 = -30\int y + 12\int u \qquad (8.94)$$

Differentiating Equation 8.94 once again, we get

$$\dot{x}_4 = -30y + 12u$$

$$\dot{x}_4 = -30x_1 + 12u \qquad (8.95)$$

Then, the state equation becomes

$$
\begin{bmatrix} \dot{x}_1 \\ \dot{x}_2 \\ \dot{x}_3 \\ \dot{x}_4 \end{bmatrix} = \begin{bmatrix} -1.25 & 1 & 0 & 0 \\ -2 & 0 & 1 & 0 \\ -3 & 0 & 0 & 1 \\ -30 & 1 & 1 & 1 \end{bmatrix} \begin{bmatrix} x_1 \\ x_2 \\ x_3 \\ x_4 \end{bmatrix} + \begin{bmatrix} 0 \\ 6 \\ 8 \\ 12 \end{bmatrix} u \tag{8.96}
$$

$$
y = \begin{bmatrix} 1 & 0 & 0 & 0 \end{bmatrix} \begin{bmatrix} x_1 \\ x_2 \\ x_3 \\ x_4 \end{bmatrix} \tag{8.97}
$$

Note: In both Examples 8.10 and 8.11, the approach is very similar.

EXAMPLE 8.12: CASE WHERE NUMERATOR AND DENOMINATOR ORDERS ARE EQUAL WITH $M = N = 3$

Let us consider

$$
\frac{Y(s)}{U(s)} = \frac{s^3 + 13s^2 + 50s + 62}{s^3 + 12s^2 + 47s + 60} \tag{8.98}
$$

In this case, $m = n$ and we will see that it will result in the direct transmission term D.

We will also realize that the state definition needs to be done a bit more cautiously. Cross-multiplying the original Equation 8.98 and taking the inverse Laplace transform, we have

$$
\dddot{y} + 12\ddot{y} + 47\dot{y} + 60y = \dddot{u} + 13\ddot{u} + 50\dot{u} + 62u
$$

Solving for the highest derivative, we get

$$
\dddot{y} = -12\ddot{y} - 47\dot{y} - 60y + \dddot{u} + 13\ddot{u} + 50\dot{u} + 62u \tag{8.99}
$$

By integrating both sides two times, we have

$$
\dot{y} = -12y - 47\int y - 60\iint y + \dot{u} + 13u + 50\int u + 62\iint u \tag{8.100}
$$

Now, we need to define the state variable. But, if we define $x_1 = y$, we will get one state equation as

$$
\dot{x}_1 = -12y - 47\int y - 60\iint y + \dot{u} + 13u + 50\int u + 62\iint u \tag{8.101}
$$

Note that we have $\dot{u}$ on the right-hand side of the state equation we wrote above. This is not acceptable in the standard form of state equations, where u is expected to assume piecewise constant values from sample to sample. Hence, its derivatives cannot be dealt with in the solution methods employed. So, $x_1 = y$ is not a valid state definition. So, we need to include u in the definition of the first state variable. Let us redefine the first state variable as

$$x_1 = y - u \tag{8.102}$$

With the above definition in mind, Equation 8.100 can be written as

$$\dot{y} - \dot{u} = \dot{x}_1 = -12(y - u) - 12u - 47\int y - 60\iint y + 13u + 50\int u + 62\iint u$$

$$\dot{x}_1 = -12x_1 - 12u + 13u - 47\int y - 60\iint y + 50\int u + 62\iint u = -12x_1 + x_2 + u \tag{8.103}$$

where we have tacitly equated the integral terms to a new state variable as

$$x_2 = -47\int y - 60\iint y + 50\int u + 62\iint u \tag{8.104}$$

Let us differentiate the above equation once:

$$\dot{x}_2 = -47y - 60\int y + 50u + 62\int u \tag{8.105}$$

$$\dot{x}_2 = -47(y - u) - 47u - 60\int y + 50u + 62\int u$$

$$= -47x_1 - 60\int y + 62\int u + 3u$$

$$= -47x_1 + x_3 + 3u \tag{8.106}$$

where, as before, we have equated all integral terms to a new state variable as

$$x_3 = -60\int y + 62\int u \tag{8.107}$$

Differentiating equation once

$$\dot{x}_3 = -60y + 62u = -60(y - u) - 60u + 62u \Rightarrow \dot{x}_3 = -60x_1 + 2u \tag{8.108}$$

$$\begin{bmatrix} \dot{x}_1 \\ \dot{x}_2 \\ \dot{x}_3 \end{bmatrix} = \begin{bmatrix} -12 & 1 & 0 \\ -47 & 0 & 1 \\ -60 & 0 & 0 \end{bmatrix} \begin{bmatrix} x_1 \\ x_2 \\ x_3 \end{bmatrix} + \begin{bmatrix} 1 \\ 3 \\ 2 \end{bmatrix} u \tag{8.109}$$

From Equation 8.102, the output can be written as

$$y = \begin{bmatrix} 1 & 0 & 0 \end{bmatrix} \begin{bmatrix} x_1 \\ x_2 \\ x_3 \end{bmatrix} + 1u \qquad (8.110)$$

There is a direct transmission term in the output equation because $m = n$. In Examples 8.10 through 8.12, we see that the states do not involve higher-order derivative terms. If the initial values of u and y are zeros, the states can be estimated with relative ease. We have seen a few methods of forming state equations with examples. We have only focused on those methods that yield state equations where states can be easily measured or estimated. We may conclude that the modeling must keep the actual measurable or easy-to-estimate variables as the state variables so that they are available for feedback control. Whether the state equations are derived from the actual dynamic system knowledge, or by factorizing transfer functions, or from polynomials directly, this point must be kept in mind. When we have a distributed control system where we encounter many subsystems, it is easy to pick measurable variables as state variables. This is true in the case of robotic systems.

8.9   TIME DOMAIN SOLUTIONS USING TRANSFER FUNCTIONS APPROACH

In this section, we will deal with the closed system along with a controller of some sort. We will exclude open-loop systems, since open-loop system response or open-loop stability matters are usually trivial once the system poles are determined. Therefore, we will focus on closed-loop performance with a controller in place.

For a system with a frequency domain transfer function, the steps involved in getting a system with desirable performance are rather straightforward. The first thing we need to do is to decide on the controller configuration. In this section, we will see closed-system responses for some elementary controllers such as proportional controller (P-only) and proportional, derivative, and integral controllers (PID), to get familiar with time domain solutions with controllers. For more controller types and further reading on this issue, refer to Ogata (1990), D'Azzo and Houpis (1995), and Kuo (1987). Once the controller is decided, the feedback system must be analyzed and the overall transfer function must be computed. Then, by applying partial fraction techniques, such complex systems can be broken down, and the inverse Laplace transform solutions can be applied to obtain time domain solutions. We can also analyze the system performance by using software simulation tools such as MATLAB. If necessary, the design can be further improved iteratively. Here, we will demonstrate one example using a few different techniques.

EXAMPLE 8.13: MASS, SPRING, AND DAMPER SYSTEM IN CLOSED LOOP

We have seen a mechanical system in Example 8.3, and the plant configuration is shown in Figure 8.7. We derived the open-loop transfer function of the

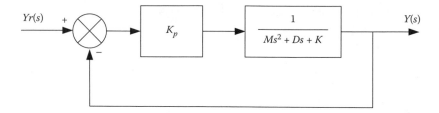

FIGURE 8.14 Closed-loop system used for simulation.

plant and the output versus input relationship is given by Equation 8.41, which is repeated here:

$$\frac{Y(s)}{F(s)} = \frac{1}{Ms^2 + Ds + K} \tag{8.111}$$

We would like to consider the performance of the plant when controlled in a closed loop. We may consider a few types of controllers. Let us assume that we want to control y to follow a reference input, y_r, which may be a step input. To do this, we need to select a controller and create a feedback control system. Let us choose a "P-only controller," the action of which is proportional to the error. Such a controller uses a gain factor, denoted by K_p, to amplify the error between the desired value and the actual value of the output, and feeds the resulting value as the input to the plant. Then, the feedback system can be formulated as shown in Figure 8.14.

Let us assume that the following arbitrary parameters for the above system are given:

$$M = 1 \text{ kg}, D = 1.5 \text{ Ns/m}, K = 6 \text{ kg/m, and } K_p = 20$$

8.9.1 ANALYTICAL SOLUTION FOR MASS, SPRING, AND DAMPER SYSTEM IN CLOSED LOOP

By referring to Figure 8.14, the closed-loop transfer function is obtained as

$$\frac{Y(s)}{Y_r(s)} = \frac{1}{1 + (Ms^2 + Ds + K)/K_p} = \frac{K_p}{Ms^2 + Ds + (K + K_p)}$$

$$= \frac{K_p/M}{s^2 + (D/M)s + ((K + K_p)/M)} \tag{8.112}$$

$$\frac{Y(s)}{Y_r(s)} = \frac{K_p}{(K + K_p)} \frac{((K + K_p)/M)}{s^2 + (D/M)s + ((K + K_p)/M)} = \left(\frac{20}{26}\right) \frac{26}{s^2 + 1.5s + 26} \tag{8.113}$$

For a step input

$$Y(s) = \left(\frac{20}{26}\right) \frac{26}{s^2 + 1.5s + 26} \frac{1}{s} = \frac{20}{26} \frac{\omega_n^2}{s^2 + 2\xi\omega_n s + \omega_n^2} \frac{1}{s} \tag{8.114}$$

```
zeta=0.1471;
omegan=5.1;
omegad=omegan*sqrt(1-zeta*zeta);
phi=81.54*pi/180
delt=0.01
i=1;
for i=1:1500
y(i)=(20/26)*(1-(exp(-zeta*omegan*i*delt)/sqrt(1-
zeta*zeta))*sin(omegad*delt*i+phi));
t(i)=i*delt;
i=i+1;
end;
figure(1)
plot(t,y);
ylabel('position')
xlabel('time, seconds')
grid;
```

FIGURE 8.15 Code for plotting response yielded by Equation 8.116.

For finding solution, we match it with a suitable entry in the standard Laplace transform table. The comparison yields

$$\omega_n^2 = 26 \Rightarrow \omega_n = \sqrt{26} = 5.1$$
$$2\xi\omega_n = 1.5 \Rightarrow \varsigma = 0.1471 \tag{8.115}$$

Then, the solution is

$$y(t) = \frac{20}{26}\left\{1 - \frac{e^{-\varsigma\omega_n t}}{\sqrt{1-\xi^2}}\sin(5.1\sqrt{1-0.1471^2}\,t + \phi)\right\} \tag{8.116}$$

where $\phi = \cos^{-1} 0.1471 = 81.54°$.

The simple code to plot this is shown in Figure 8.15. The response obtained using MATLAB is shown in Figure 8.16.

8.9.2 SIMULATION SOLUTION FOR MASS, SPRING, AND DAMPER SYSTEM IN CLOSED LOOP

In the MATLAB environment, Simulink® can be easily used to test various controllers. We do not need to do any closed-loop system calculations or deal with Laplace transforms. A Simulink setup for this simple proportional controller is shown in Figure 8.17.

The MATLAB code that calls this Simulink model is given in Figure 8.18. Using this code, we can pass the system parameters to the model. This gives some flexibility to the programmer. In all the codes in which we have used sim("xyz"), xyz is the file name of the Simulink model shown in the figure referred.

The response we obtain is shown in Figure 8.19. We can see that the response is quite oscillatory, and it has a large steady-state error as well. We may try to improve this by adding a few additional terms in the controller function.

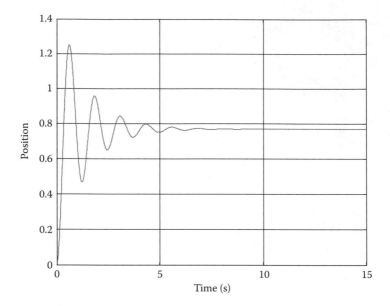

FIGURE 8.16   Time response of the system.

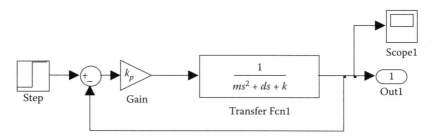

FIGURE 8.17   Simulation setup.

```
%This is the code for second order system solution
%controller parameters
kp=20;
%system parameters
m=1.0;
d=1.50;
k=6.0;
%Simulation
[t,x,Out1]=sim('figure817mdl');
figure(1);
plot(t,Out1)
ylabel('output')
xlabel('time,seconds')
grid;
```

FIGURE 8.18   MATLAB code for P-only controller.

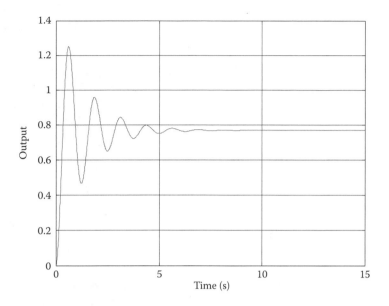

FIGURE 8.19 Time response from Simulink P-only controller.

8.9.3 PID CONTROLLER RESPONSE

We presented a basic PID controller equation in Chapter 7 as

$$m(t) = k_p e(t) + k_i \int e(t)dt + k_d \frac{de(t)}{dt} \tag{8.117}$$

where k_p is the proportional gain, k_i is the integral gain, k_d is the derivative gain, and $e(t) = y_r - y$.

In terms of the Laplace transform, the controller output can be written as

$$M(s) = k_p E(s) + k_i \frac{E(s)}{s} + k_d s E(s) \tag{8.118}$$

We would like to apply such a controller to the plant and see how the system responds. Before we do that, we would like to discuss some fundamental concepts regarding PID controllers.

Controller actions: As it can be seen from Equation 8.117, the controller output is the sum of three terms. The first term $k_p e(t)$ represents the proportional control action, which produces an output proportional to the error. This is usually used as the base controller. The second term $k_i \int e(t)dt$ represents the integral control action. As long as the error is present, this term keeps growing. For example, if the output is less than the desired value, the error will be positive and the integral term's value grows and increases the input to the plant, thus pushing the output closer toward the desired value. In general, an integral controller eliminates the error.

The third term, $k_d(de/dt)$, represents the derivative control action. When the derivative of the error is positive, implying a tendency for the error to increase,

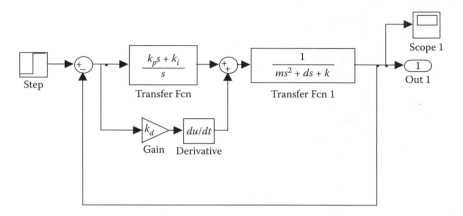

FIGURE 8.20 Simulink setup for PID implementation.

this controller produces an output to reduce the error, thus producing an anticipatory/preemptive action. The process of adjusting the proportional, integral, and derivative gains to get a good performance is called "'tuning." Tuning a PID controller is usually a trial-and-error process, which itself is a vast area of research. At first, the proportional gain is adjusted to get a reasonably fast response without much oscillations of the output. Then, the integral gain is adjusted to eliminate the steady-state error. Finally, the derivative gain is adjusted to reduce the oscillations. These adjustments need to be repeated a few times, until response is satisfactory.

We can see that the PID controller is implemented as a sum of "PI" block $(k_p s + k_i)/s$ and a derivative block $k_d(de/dt)$ in Figure 8.20. It takes considerable effort of tuning by trial-and-error to get the appropriate parameters $k_p = 50$, $k_d = 8$, and $k_i = 50$ for the controller. Now, we will apply this PID controller to the same mass, spring, damper problem, and compare its performance.

The MATLAB code for implementing it is given in Figure 8.21. The response can be viewed in Figure 8.22.

```
%This is the code for second order system solution
%controller parameters
kp=50;
kd=8.0;
ki=30.0;
%system parameters
m=1.0;
d=1.50;
k=6.0;
[t,x,Out1]=sim('figure820mdl');
figure(1);
plot(t,Out1)
ylabel('output (PID control)');
xlabel('time, seconds');
grid
```

FIGURE 8.21 MATLAB code for PID controller simulation.

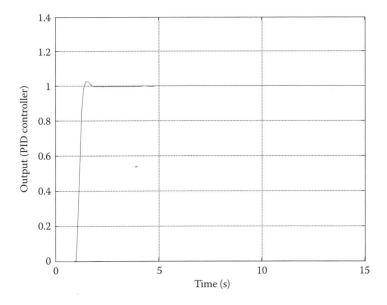

FIGURE 8.22   Time response of PID controller implemented in Simulink.

Figure 8.22 shows the system response. See that the overshoot above the reference value is <5%, and the final steady-state value is reached in <1 s. Comparing these values to the response obtained using the P-only controller, we can observe that the performance has tremendously improved in terms of overshoot and settling time.

8.10   TIME DOMAIN SOLUTIONS OF STATE EQUATIONS

As mentioned earlier, in robotics, we prefer state space representation because that will be more suitable to represent multi-input multi-output (MIMO) systems, which we often encounter in this field. Furthermore, there are well-established methods of controller design for such representation. For state space equations, the time domain solutions can be obtained using computational methods, analytical methods, and simulation methods. We discuss them in the following sections.

8.10.1   TIME DOMAIN SOLUTIONS USING ANALYTICAL METHODS

Computer solution: A system is represented by state equations in the form

$$\dot{\bar{X}} = A\bar{X} + B\bar{U}$$
$$\bar{Y} = C\bar{X} + D\bar{U}$$

(8.119)

We can drop the "bars" above the vectors for convenience. If there are n states, r inputs, and m outputs, then X is a vector of size n, A is an $n \times n$ matrix, U is a vector of size r, B is an $n \times r$ matrix, C is an $m \times n$ matrix, and D is $m \times r$ matrix. In most physical systems, the direct transmission term may not be present, hence, $D = 0$.

The solution is of the form

$$X(t) = e^{At}X(0) + \int_{0}^{t} e^{A(t-\tau)}BU(\tau)\,d\tau \tag{8.120}$$

and in general, if the initial time is t_0, then the solution consists of two terms a free solution plus a forced solution (Ogata 1990; Nagrath and Gopal 1985; Wilburg, 1971):

$$X(t) = e^{A(t-t_0)}X(t_0) + \int_{t_0}^{t} e^{A(t-\tau)}BU(\tau)\,d\tau \tag{8.121}$$

In the above equations, e^{At} and $e^{A(t-t_0)}$ are called the state transition matrices. The free solution implies the state variable response when there is no input. We imply that system states will drift according to the free solution pattern. When we have an input U that will create an additional response, it is called the forced solution. The total solution is the sum of these two.

Hence, it becomes necessary to compute e^{At} to obtain the solution. It can in fact be computed as a series given below:

$$e^{At} = I + At + \frac{A^2 t^2}{2!} + \frac{A^3 t^3}{3!} + \frac{A^4 t^4}{4!} + \cdots = \varphi(t,0)$$

or

$$e^{A(t-t_0)} = I + A(t-t_0) + \frac{A^2(t-t_0)^2}{2!} + \frac{A^3(t-t_0)^3}{3!} + \frac{A^4(t-t_0)^4}{4!} + \cdots = \varphi(t,t_0) \tag{8.122}$$

The solution of the systems can be easily programmed in computers. In addition to the free (undriven) solution, even the forced solution can be computed in steps using a suitable computer program. We will now proceed to see some analytical solutions in the following sections. We are not providing an example for this since we will provide a simulation procedure later.

Laplace transform technique: Let us take the Laplace transform of the state equation, which is straightforward and write

$$sX(s) - X(0) = AX(s) + BU(s) \tag{8.123}$$

Rearranging

$$[sI - A]X(s) = X(0) + BU(s) \tag{8.124}$$

Premultiply by $[sI - A]^{-1}$

$$X(s) = [sI - A]^{-1}X(0) + [sI - A]^{-1}BU(s) \tag{8.125}$$

By taking the inverse Laplace transform, we get

$$X(t) = L^{-1}\left\{[sI - A]^{-1}X(0)\right\} + L^{-1}\left\{[sI - A]^{-1}BU(s)\right\} \tag{8.126}$$

In the above step, $X(0)$ is constant, so we can rewrite Equation 8.126 as

$$X(t) = L^{-1}\left\{[sI - A]^{-1}\right\}X(0) + L^{-1}\left\{[sI - A]^{-1}BU(s)\right\} \tag{8.127}$$

The output vector becomes

$$Y(t) = CX(t) = CL^{-1}\left\{[sI - A]^{-1}\right\}X(0) + CL^{-1}\left\{[sI - A]^{-1}BU(s)\right\} \tag{8.128}$$

Comparing Equations 8.120 and 8.127, we can see that the term $L^{-1}\{[sI - A]^{-1}\}$ is in fact the state transition matrix. However, this procedure is quite hard to compute, which we will show in the example below. We will use the same example for all the methods in this section, and we are not attempting any controller design yet.

EXAMPLE 8.14: TYPICAL STATE EQUATION SOLUTION

Evaluate the time domain solution for the following system:

$$\dot{X} = \begin{bmatrix} -6 & 8 \\ -1 & 0 \end{bmatrix}\bar{X} + \begin{bmatrix} 1 \\ 0 \end{bmatrix}u \quad \text{and} \quad y = \begin{bmatrix} 1 & 0 \\ 0 & 1 \end{bmatrix}\bar{X} + Du$$

$$\text{Given that} \quad \bar{X}(0) = \begin{bmatrix} 1 \\ -1 \end{bmatrix} \text{ and } D = 0 \tag{8.129}$$

SOLUTION

We write the characteristic equation as

$$[sI - A] = \begin{bmatrix} s + 6 & -8 \\ 1 & s \end{bmatrix} \tag{8.130}$$

$$s(s + 6) + 8 = s^2 + 6s + 8 = (s + 2)(s + 4) \tag{8.131}$$

$$[sI - A]^{-1} = \frac{1}{(s + 2)(s + 4)}\begin{bmatrix} s & +8 \\ -1 & s + 6 \end{bmatrix} \tag{8.132}$$

The free solution in the Laplace form is represented by

$$\left[sI - A\right]^{-1} \bar{X}(0) = \frac{1}{(s+2)(s+4)}\begin{bmatrix} s & +8 \\ -1 & s+6 \end{bmatrix}\begin{bmatrix} 1 \\ -1 \end{bmatrix} \tag{8.133}$$

The forced solution in terms of the Laplace transform is represented by

$$\left[sI - A\right]^{-1} BU(s) = \frac{1}{(s+2)(s+4)}\begin{bmatrix} s & +8 \\ -1 & s+6 \end{bmatrix}\begin{bmatrix} 1 \\ 0 \end{bmatrix}\frac{1}{s}$$

$$= \frac{1}{(s+2)(s+4)}\begin{bmatrix} s \\ -1 \end{bmatrix}\frac{1}{s} = \frac{1}{(s+2)(s+4)}\begin{bmatrix} 1 \\ -\frac{1}{s} \end{bmatrix} \tag{8.134}$$

Hence, the solution is

$$X(t) = L^{-1}\begin{bmatrix} \dfrac{s-8}{(s+2)(s+4)} \\[2mm] \dfrac{-(s+7)}{(s+2)(s+4)} \end{bmatrix} + L^{-1}\begin{bmatrix} \dfrac{1}{(s+2)(s+4)} \\[2mm] \dfrac{-1}{s(s+2)(s+4)} \end{bmatrix} \tag{8.135}$$

Referring to any standard Laplace transform table in the literature, we can write the solution as

$$\bar{X}(t) = \begin{bmatrix} 0.5(4e^{-4t} - 2e^{-2t}) - 4(e^{-2t} - e^{-4t}) \\ -0.5(4e^{-4t} - 2e^{-2t}) - 3.5(e^{-2t} - e^{-4t}) \end{bmatrix} + \begin{bmatrix} 0.5(e^{-2t} - e^{-4t}) \\ -\dfrac{1}{8}\{1 - 0.5(4e^{-2t} - 2e^{-4t})\} \end{bmatrix} \tag{8.136}$$

Simplifying, we get

$$\bar{X}(t) = \begin{bmatrix} -5e^{-2t} + 6e^{-4t} \\ -2.5e^{-2t} + 1.5e^{-4t} \end{bmatrix} + \begin{bmatrix} 0.5e^{-2t} - 0.5e^{-4t} \\ -0.125 + 0.25e^{-2t} - 0.125e^{-4t} \end{bmatrix} \tag{8.137}$$

Note that the initial values of states (at $t = 0$) are 1 and –1, respectively, in the first term, while the forced solution in the second term is zero for both states. The total solution is

$$\bar{X}(t) = \begin{bmatrix} -4.5e^{-2t} + 5.5e^{-4t} \\ -0.125 - 2.25e^{-2t} + 1.375e^{-4t} \end{bmatrix} \quad \text{for } t > 0 \tag{8.138}$$

Let us plot the above states using the MATLAB code in Figure 8.23 for future comparison.
The plots are shown in Figure 8.24.

```
%Matlab calculation of analytical solution
delt=0.01
i=1;
for i=1:1000
x1(i)=-4.5*exp(-2*i*delt)+ 5.5*exp(-4*i*delt);
x2(i)=-0.125-2.25*exp(-2*i*delt)+ 1.375*exp(-4*i*delt);
t(i)=i*delt;
i=i+1;
end;
figure(1);
subplot(2,1,1)
plot(t,x1)
ylabel('state X1')
grid;
subplot(2,1,2)
plot(t,x2)
ylabel('state X2')
xlabel('time, seconds')
grid;
```

FIGURE 8.23   Code for calculation for plotting Equation 8.138.

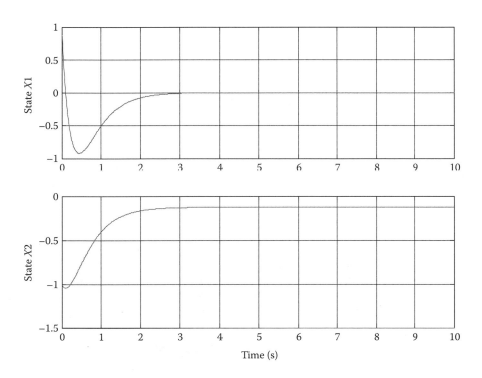

FIGURE 8.24   Time response of regulator output from calculations.

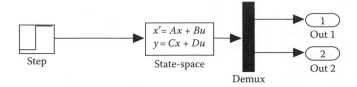

FIGURE 8.25 Simulink setup for state space system.

```
%state1 model;
A=[-6 8;-1 0];
B=[1;0];
C=[1 0;0 1];
D=[0;0];
X0=[1 -1];
%controllers
%simulation
[t,x,Out1,Out2]=sim('figure825mdl');
figure(1);
subplot(2,1,1)
plot(t,Out1)
ylabel('state X1')
grid;
subplot(2,1,2)
plot(t,Out2)
ylabel('state X2')
xlabel('time, seconds')
grid;
```

FIGURE 8.26 MATLAB code for state model simulation.

Time domain solutions using simulation: Here, we would like to take the same example for continuity of discussion and get the outputs using MATLAB simulation of state space system and compare the results.

$$\dot{\bar{X}} = \begin{bmatrix} -6 & 8 \\ -1 & 0 \end{bmatrix} \bar{X} + \begin{bmatrix} 1 \\ 0 \end{bmatrix} u \quad \text{and} \quad y = \begin{bmatrix} 1 & 0 \\ 0 & 1 \end{bmatrix} [\bar{X}] + Du$$

$$\text{Given that } \bar{X}(0) = \begin{bmatrix} 1 \\ -1 \end{bmatrix}, \text{ where } D = 0 \qquad (8.139)$$

The simulation diagram is quite routine, and it is shown in Figure 8.25. A MATLAB program that defines the parameters and invokes the Simulink model is listed in Figure 8.26. The simulation results in Figure 8.27 tally with the results of earlier methods shown in Figure 8.24.

8.11 REGULATOR AND SERVO CONTROLLERS

Earlier, we have seen simple types of P-only and PID controllers. For many systems in robotics, such simple controllers may not work. We may need more sophisticated

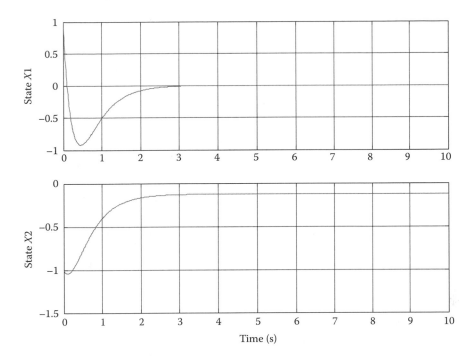

FIGURE 8.27 Time response from simulation.

controllers. The most popular types of controllers used in robotics are pole placement regulators, servos with integrator using pole placement concept, linear quadratic criteria-based regulators, linear quadratic criteria-based servo controllers, adaptive regulators, and adaptive servo controllers.

At the outset, in a day-to-day industrial control application, the objectives of the control and the actual specifications for performance are not really clear. However, to some extent, the picture gets better in robotics. In most of the cases in robotics, the objectives are clearly known in the beginning of the design process. On the other hand, there may be little knowledge of some parameters that are difficult to estimate. For example, in game robotics, we may not know the parameters such as friction of the robotic vehicle or reflectivity of a maze wall.

The design is going to be challenging and iterative mainly in two stages. We derive a model and evaluate its parameters. The next step is to check the model accuracy by experimenting. If our plant response is not according to the expectation of the model, then we have to revise the model. This has to be repeated until the model response and the plant response match.

The next stage is controller design. Once we design the controller, we simulate the response; if the result is satisfactory, then we test the actual closed-loop system taking utmost care. If the response matches the expectations, then we can conclude that the design is complete. Otherwise, controller structure or parameters may have to change. We can improve the performance to match the specifications still following the process of "simulate first and test later." This process has to go on through a few cycles until the result is satisfactory.

So far, we have discussed analog controllers to introduce the concepts in controller design. However, in robotics, controllers are hardly implemented in analog form. Therefore, we will present the design exercise in the next chapter where we describe digital controllers.

8.12 CONCLUSION

In this chapter, we have reviewed the basics of modeling of physical systems in terms of transfer functions and state variable analysis. Even though the transfer function models are sufficient to design and implement controllers of different types, we skipped detailed designs of controllers in favor of state models since all transfer function model-based designs are by trial and error. Another problem is that as the systems get complicated as in the case of robotics, such trial-and-error-based design is very time consuming and at times becomes impractical. We only presented a P-only controller and a PID controller as examples. Then, we moved on to state modeling. We went through some analytical methods of response calculations and dwelled upon some simulation techniques as well. We skipped the design methods for regulator problem and servo control problem, since digital controllers are used in robotics. We have focused on analog techniques so far, since we believe that control systems are learned first from analog systems. Since the actual implementations are done using digital computers, it is necessary to learn discrete system concepts. In the next chapter, we will discuss the discrete systems concepts before we go through some case studies.

REFERENCES

Cavallo, A., Setola, R., and Vasca, F. 1996. *Using MATLAB, Simulink, and Control System Toolbox: A Practical Approach*. Hertfordshire, UK: Prentice Hall Europe.

D'Azzo, J.J. and Houpis, C.H. 1995. *Linear Control System Analysis and Design: Conventional and Modern*. New York: McGraw-Hill Book Company.

Kuo, B.C. 1987. *Automatic Control Systems*. Englewood Cliffs, NJ: Prentice-Hall.

Nagrath, I.J. and Gopal, M. 1985. *Control Systems Engineering*. New Delhi: New Age International Limited Publishers.

Ogata, K. 1990. *Modern Control Engineering*. Englewood Cliffs, NJ: Prentice-Hall.

Palani, S. 1997. *Control Systems*. Regional Engineering College, Tiruchirappalli, India: Shanmuga Priya Publishers.

Wilburg, D.M. 1971. *Schaum's Outline Series*: *State Space and Linear Systems*. New York: McGraw-Hill.

9 Digital Control Fundamentals and Controller Design

9.1 INTRODUCTION

In earlier chapters, we focused on systems working in continuous mode, which implies that the system is monitored and controlled at each and every instant of time.

While analog control is quite important to understand control theory, implementing the designed controller using analog components would pose a number of problems. We will describe these issues in the following sections. In fact, as we have pointed out earlier, analog controllers are seldom implemented in modern control systems especially in robotics. They are expensive in the long run, since they need more calibration and maintenance than a digital system, due to aging of analog components. The falling price of computing hardware also makes digital controllers more attractive. They can work in most of the harsh environments where some robots are deployed. Furthermore, the distributed processing, which is very useful in robotics, is also possible to implement with digital controllers. They can fit in to small spaces and communicate with systems in the vicinity or far away. The necessary hardware or protocols for such communication network such as IEEE-488 are readily available. The instrumentation field also is moving toward digital systems. Therefore, digital controllers fit well with digital instrumentation.

We will present some ideas related to digital implementation of controllers in the next section. These concepts can be learned from a whole body of control literature (Astrom and Wittenmark 1990; Cadzow 1973; Nagrath and Gopal 1996; Ogata 1995). We have provided a glimpse of what we feel is essential knowledge for a robot designer from a large collection of literature. We have provided some examples in this chapter and some case studies in Chapter 10, where we show how this knowledge is used in robotics.

9.2 DIGITAL CONTROL OVERVIEW

A robot operates in a continuous world, and all the processes involved in robotics are also continuous by nature. Hence, digital controllers cannot directly deal with such continuous systems. Therefore, we need to introduce some intermediate steps to adapt digital systems to control analog plants. Figure 9.1 shows a block diagram

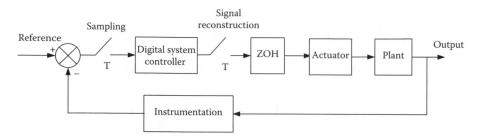

FIGURE 9.1   General block diagram of a digital controlled system.

of a typical single-input single-output digital control system. We explain briefly the basic elements involved in digital control below.

9.2.1   SIGNAL SAMPLER

This is usually denoted by a switch in all diagrams, though there is more to it. The computer represents the controller, and it only works with numbers. Hence, to take decisions and produce a control output, it needs to sample the system outputs periodically and receive them as numbers. On the other hand, an analog controller acts on the signal values every instant, which is not possible for digital controllers. The computer has to take one value of error at a time and compute the output. The function of a sampler is to catch the samples of the system variables and keep them until they can be converted into a digital value. A typical sampler is shown in Figure 9.2.

Even though Figure 9.1 shows only one input, there may be more than one channel to be sampled. Figure 9.2 shows a case where there are four signals to be sampled simultaneously. The digital computer takes control action at periodic intervals called sampling time. At the instant of sampling time, sample/hold signal is activated and all the signals are captured simultaneously at the current values and are available at the multiplexer inputs. Then multiplexer passes the captured signals to the ADC one by one for conversion based on the select lines provided by the computer. The

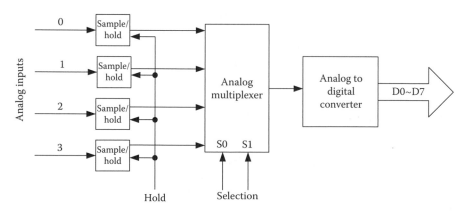

FIGURE 9.2   A typical synchronous sampling system.

program should allow enough settling time for the output of the multiplexer to be stabilized before conversion begins. In the example given in Figure 9.2, ADC is 8 bits; hence, captured samples are converted into 8-bit length digital numbers. The number of bits of conversion depends upon the resolution required. These operations are usually done under software control of the microcontroller or microprocessor. Some devices, such as HCTL 2016 decoder with on-board processor, provide data directly in digital form. For such input signals, there is no need for conversion.

When the system variables change quite slowly, compared to the ADC conversion speed, sample/hold devices are not needed. In such cases, the total time taken by ADC to convert all channels may be far less than the shortest time constant of the closed-loop system.

9.2.2 Digital Controller

The digital controller is nothing but a computing device equipped with software stored in its memory. In the case of a microcontroller, the software is stored in its EEPROM or "flash" memory. Sometimes, programmable logic controller (PLC) devices play the role of digital controllers. In many cases, the designers tend to use readily available motherboards (such as eZdsp 2407), which come with serial and parallel ports for communication and adopt them as digital controllers. Such boards also come with usable software development tools. These motherboards do need additional sister boards for power driving, and it is the task of the robotics engineer to develop sister boards that suit his needs. This ensemble described above as a whole can be called a digital controller.

9.2.3 Zero-Order Hold

As we mentioned at the outset, physical systems are continuous in nature, such as DC motors, hydraulic systems, or pneumatic systems, to name a few. They need continuous signal inputs. In contrast, digital controllers can only provide a set of numbers at each and every control instant. These numbers need to be converted to analog signals to be used by the plant. Since a microprocessor is used as a controller, the output number from the controller stays on the data bus for a short period of time. We need to keep this number for the duration of sampling time (control interval) and convert it to analog value so that it can be used by the plant. The dual function of holding the number and converting into analog value is achieved by the combination of zero-order hold (ZOH) and digital-to-analog converter (DAC). Figure 9.3 shows

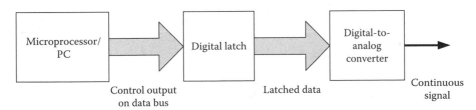

FIGURE 9.3   Implementation of zero-order hold.

the basic structure of ZOH implementation. After this, the continuous control signal can be processed as in any other analog control system.

When a microprocessor is used as controller, a separate digital latch circuit is necessary as shown in Figure 9.3. Nowadays, many microcontrollers come with built-in output ports with latches. Even the DACs are integrated to microcontrollers, which simplifies the circuit design significantly.

After considering all three important parts of a digital controller, we can incorporate all the constituent devices and redraw the block diagram as a general guide line as shown in Figure 9.4.

Since digital controllers are popular these days, it will be useful to compare them with analog controllers. Is it worth-learning microcontrollers, encoders, H-bridges, and so on, instead of simply implementing an analog controller which only requires the knowledge of power electronics and analog control theory? In practice, it is necessary to make this additional effort to implement a reliable and cheaper system, which avoids expensive precision analog hardware. Even though digital controllers are more complex, their advantages largely outweigh their complexity. Since they need less maintenance and calibration, digital controllers are far more reliable than analog control systems.

Digital computers are very flexible because the controller parameters can be changed by changing a value in the program without doing any hardware changes. One digital controller can take care of many control loops simultaneously due to the possibility of fast computation. Moreover, sophisticated control techniques such as adaptive control can be implemented using digital controllers. Such controllers include adaptive prefiltering following an anti-aliasing filter in data acquisition. They also use a postfiltering stage to smoothen the control output to avoid exciting hidden resonance especially in robotic systems. These features cannot be implemented in analog systems. For the examples of fast digital adaptive control, using prefiltering and postfiltering with sampling time of 0.5 ms, refer to Astrom et al. (1994) and Astrom and Kanniah (1993).

Warning messages, alarms, and other safety systems along with a user-friendly interface can be implemented in a digital system. The instrumentation technologies are also moving fast into digital types. It is difficult to make use of digital information obtained from such devices in an analog control system. Recently, hierarchical and distributed control systems are employed in robotics as well as in many other complex systems. Obviously, such controllers can be implemented within the framework of a digital control system easily.

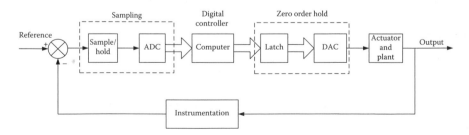

FIGURE 9.4   General outline of a direct digital control system.

9.3   SIGNAL REPRESENTATION IN DIGITAL SYSTEMS

When analog plants are controlled digitally, they are called sampled data systems. For this, we need to sample the analog signals and represent them digitally. Plants also need to be represented in discrete domain. In this section, we will consider the representation of sampled signals and the analog plants, in a computer-controlled environment.

9.3.1   SAMPLING PROCESS

A standard representation of a sampler switch is shown in Figure 9.5a. The switch closes every T seconds to sample input signal. Input signal is represented by the continuous function $f(t)$ as shown in Figure 9.5b. The outputs of the sampler are shown by a train of impulses $f^*(t)$ in Figure 9.5c. Sampling is equivalent to multiplying the signal by an infinite sequence of impulses spaced at constant intervals.

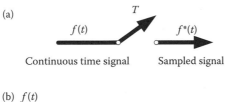

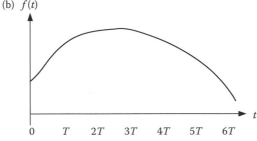

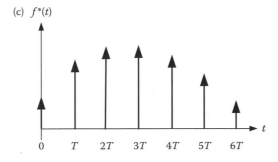

FIGURE 9.5   (a) Sampler switch, (b) continuous signal $f(t)$, (c) impulse approximation of switch output $f^*(t)$.

Summing up the unit impulse train, we can represent them mathematically as

$$\delta_T(t) = \sum_{k=-\infty}^{\infty} \delta(t - kT) \tag{9.1}$$

We multiply $f(t)$ using the sequence of impulses $\delta_T(t)$ and define it as $f^*(t)$ which represents the sampled signal. We can now write

$$f^*(t) = f(t)\delta_T(t) = \sum_{k=-\infty}^{\infty} f(t)\delta(t - kT)dt$$

$$f^*(t) = f(0)\delta(t) + f(T)\delta(t - T) + f(2T)\delta(t - 2T) + \cdots$$

$$f^*(t) = \sum_{n=0}^{\infty} f(nT)\delta(t - nT) \tag{9.2}$$

Figure 9.6 illustrates a part of the above function.

In Figure 9.6, $\delta(t)$ is a unit impulse at $t = 0$ and $\delta(t - nT)$ is a unit impulse at $t = nT$. Since

$$L\{\delta(t - nT)\} = e^{-nTs}$$

$$L\{f(nT)\delta(t - nT)\} = f(nT)e^{-nTs} \tag{9.3}$$

then taking the Laplace transform of Equation 9.2 and using Equation 9.3, we can obtain the Laplace transform of the sampled signal as

$$F^*(s) = L[f^*(t)] = L\{f(0)\delta(t) + f(T)\delta(t - T) + f(T)\delta(t - T) + \cdots\}$$
$$= f(0) + f(T)e^{-Ts} + f(2T)e^{-2Ts} + \cdots$$

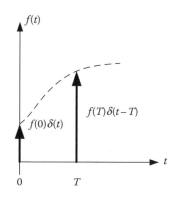

FIGURE 9.6   Sampled signal as weighted impulses.

$$F^*(s) = \sum_{n=0}^{\infty} f(nT)\, e^{-nTs} \tag{9.4}$$

A major question in digital systems is that how often should we sample. It depends upon the purpose of sampling and we will elaborate below.

9.3.1.1   Sampling for Reconstruction

Sampling theorem states that a sampled continuous signal may be reconstructed from the samples if and only if the sampling frequency is more than twice the highest-frequency content of the signal. Otherwise, the high-frequency component of the signal may pass as a low-frequency signal in the output.

9.3.1.2   Sampling for the Purpose of Control

The sampling frequency described above is not sufficient for control purposes. The sampling frequency should be more than 12 times the closed-loop system bandwidth. That leads to the conclusion that the sampling interval should be less than half of the smallest time constant of the closed-loop system. We have to bear in mind that the closed-loop system bandwidth will be much higher than the open-loop system bandwidth depending upon the gain of the controller.

9.3.2   Z-Transform of Signals

The Z-transform is an important mathematical tool for understanding and analyzing discrete time systems. Similar to the Laplace transform analysis, which is used for analyzing continuous time systems, we need another tool to analyze digital systems. Z-transform analysis serves that purpose. We have three possible scenarios:

1. Taking the Z-transform of continuous signals if they are sampled at regular intervals and represent them in the Z-domain. Then the continuous signal of the system output is recoverable from samples, if the sampling interval had been chosen appropriately.
2. If only signal description is available as sample values in terms of sample count k, they can still be represented in the Z-domain. But in this case, sampling time is transparent since it is not explicitly stated. This situation happens when one system receives data from another system in digital form only and the receiving system has to process the data further. The receiving system can be a filter or control computer. For design purposes, we need to have the Z-transform of such signals as well.
3. Z-domain representation can be obtained from s-domain transfer functions, by introducing a ZOH block as shown in Figure 9.4. We will deal with such a case in a later section. For now, we will discuss the first two cases only.

9.3.2.1 Z-Transform of Continuous Signals

Equation 9.4 gives the generic description of the Laplace transform of the sampled signal.

The relationship between the Z-transform and the Laplace transform is defined as

$$z = e^{sT} \tag{9.5}$$

Substituting Equation 9.5 in Equation 9.4, it can be rewritten as

$$F^*(s) = \sum_{n=0}^{\infty} f(nT)z^{-n} = F(z) \tag{9.6}$$

Equation 9.6 defines the Z-transform of the sampled signal $f*(t)$. We note that the Laplace transform of the sampled signal is nothing but the Z-transform of that signal. This concept is the foundation of sampled system analysis. Now, let us see some example signals to get familiarized with the idea.

EXAMPLE 9.1

A unit impulse function is given in Equation 9.7. It has a unity value at $k = 0$ and zero at all other instants of sampling.

$$
\begin{aligned}
f(kT) &= 0 \quad \text{for} \quad k \neq 0 \\
&= 1 \quad \text{for} \quad k = 0
\end{aligned}
\tag{9.7}
$$

The graphical representation of an impulse is shown in Figure 9.7. Applying Equation 9.4 to the signal shown in Figure 9.7

$$
\begin{aligned}
F^*(s) &= L[f^*(t)] = f(0) + f(T)e^{-Ts} + f(2T)e^{-2Ts} + \cdots \\
F(z) &= 1
\end{aligned}
\tag{9.8}
$$

Hence, we find that the Z-transform of a unit impulse is just 1.

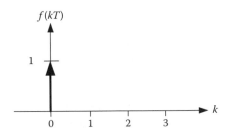

FIGURE 9.7   A unit impulse function.

EXAMPLE 9.2

A unit step function is defined as

$$f(kT) = 0 \quad \text{for} \quad k < 0$$
$$= 1 \quad \text{for} \quad k \geq 0 \tag{9.9}$$

Figure 9.8 shows the sampled unit step function.
By using Equation 9.6 and following the definition below

$$F(z) = \sum_{n=0}^{\infty} f(nT)z^{-n} = f(0)z^0 + f(T)z^{-1} + f(2T)z^{-2} + \cdots$$

the Z-transform of unit step function becomes

$$F(z) = 1 + z^{-1} + z^{-2} + \cdots = \frac{z}{z-1} \tag{9.10}$$

EXAMPLE 9.3

Find the Z-transform of the sampled signal

$$f(t) = t \quad \text{for} \quad t \geq 0 \tag{9.11}$$

Using the definition given in Equation 9.6, we can write

$$Z[f(t)] = F(z) = \sum_{k=0}^{\infty} kTz^{-k} = Tz^{-1} + 2Tz^{-2} + 3Tz^{-3} + \cdots = \frac{Tz}{(z-1)^2} \tag{9.12}$$

EXAMPLE 9.4

To further illustrate the preceding concepts, consider the continuous input

$$f(t) = e^{-at} \tag{9.13}$$

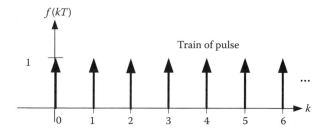

FIGURE 9.8 A unit step function.

The Z-transform of the sampled signal by definition is

$$F(z) = \sum_{n=0}^{\infty} e^{-nTa} z^{-n} = 1 + e^{-Ta} z^{-1} + e^{-2Ta} z^{-2} + \cdots \qquad (9.14)$$

$$F(z) = \sum_{n=0}^{\infty} e^{-naT} z^{-n} = \frac{z}{z - e^{-aT}} \qquad (9.15)$$

EXAMPLE 9.5

Find the Z-transform of the time function

$$f(t) = 1 - e^{-at}, \quad \text{for} \quad t \geq 0 \qquad (9.16)$$

Using the definition in Equation 9.6

$$F(z) = 0 + (1 - e^{-Ta}) z^{-1} + (1 - e^{-2Ta}) z^{-2} + \cdots \qquad (9.17)$$

Adding and subtracting a 1

$$F(z) = [1 + z^{-1} + z^{-2} + \cdots] - [1 + e^{-aT} z^{-1} + e^{-2aT} z^{-2} + e^{-3aT} z^{-3} + \cdots] \quad (9.18)$$

$$F(z) = \sum_{n=0}^{\infty} z^{-n} - \sum_{n=0}^{\infty} e^{-naT} z^{-n} = \frac{z}{z - 1} - \frac{z}{z \ c^{-aT}} = \frac{(1 - e^{-aT})z}{(z - 1)(z - e^{-aT})} \qquad (9.19)$$

The above derivations illustrate the methods used for calculating Z-transforms. Normally, Z-transform tables provide these transform values for sampled signals in terms of sampling time.

9.3.2.2 Z-Transform of Signals Represented Only as Sample Count, *k*

In all the previous five examples, we took continuous signals and sampled them at constant interval, say, T seconds. Then, term T appeared explicitly in the transform equations. Under some circumstances, the signal values are provided only at samples as a function of sample count, k, without indicating the sampling time explicitly. Then in that case, the sampling time is not known, and it cannot appear in the Z-transform. This kind of insight about the sampling time is not relevant, but we still need the Z-transform representation to analyze the overall system. To get a better picture, let us consider some examples here.

EXAMPLE 9.6

Find the Z-transform of the sampled signal

$$f(k) = k \quad \text{for} \quad k \geq 0 \qquad (9.20)$$

Let us invoke the definition in Equation 9.6. By dropping T from it, we can write

$$Z[f(k)] = F(z) = \sum_{k=0}^{\infty} kz^{-k} = z^{-1} + 2z^{-2} + 3z^{-3} + \cdots = \frac{z}{(z-1)^2} \qquad (9.21)$$

On comparing Equation 9.21 with Equation 9.12, we note that T is not explicit in Equation 9.21.

EXAMPLE 9.7

Find the Z-transform of the sampled signal

$$f(k) = a^k, \quad \text{for} \quad k \geq 0 \qquad (9.22)$$

Let us invoke the definition in Equation 9.6. By dropping T from it, we can write

$$Z[f(k)] = F(z) = \sum_{k=0}^{\infty} a^k z^{-k} = 1 + az^{-1} + a^2 z^{-2} + a^3 z^{-3} + \cdots$$

This is being a geometric progression, we can write

$$F(z) = \frac{1}{1 - az^{-1}} = \frac{z}{z - a} \qquad (9.23)$$

Alternatively, let us choose an arbitrary sampling time T and a constant α such that

$$a = e^{-\alpha T} \qquad (9.24)$$

then

$$f(k) = a^k \quad \text{for} \quad k \geq 0$$

can be written as

$$f(kT) = (e^{-\alpha T})^k \quad \text{for} \quad k \geq 0$$
$$= e^{-\alpha kT} \quad \text{for} \quad k \geq 0 \qquad (9.25)$$

This is similar to signal $e^{-\alpha t}$ sampled at arbitrary intervals of T seconds. Since

$$Z[e^{-\alpha t}] = \frac{z}{z - e^{-\alpha T}}$$

we can conclude

$$Z[e^{-\alpha kT}] = Z[a^k] = \frac{z}{z - e^{-\alpha T}} = \frac{z}{z - a} \qquad (9.26)$$

EXAMPLE 9.8

Find the Z-transform of the sampled signal

$$f(k) = 1 - a^k \quad \text{for} \quad k \geq 0 \qquad (9.27)$$

Let us choose as before an arbitrary sampling time T and a constant α such that

$$a = e^{-\alpha T} \qquad (9.28)$$

then $f(k) = 1 - a^k$ for $k \geq 0$ can be written as

$$\begin{aligned} f(kT) &= 1 - (e^{-\alpha T})^k \quad \text{for} \quad k \geq 0 \\ &= 1 - e^{-\alpha kT} \quad \text{for} \quad k \geq 0 \end{aligned} \qquad (9.29)$$

This is similar to signal $1 - e^{-\alpha t}$ sampled at arbitrary intervals of T seconds. Since

$$Z[e^{-\alpha t}] = \frac{z}{z - e^{-\alpha T}} \qquad (9.30)$$

we can conclude

$$Z[1 - e^{-\alpha kT}] = Z[1 - a^k] = \frac{z}{z - 1} - \frac{z}{z - e^{-\alpha T}} = \frac{z}{z - 1} - \frac{z}{z - a} = \frac{(1 - a)z}{(z - 1)(z - a)} \qquad (9.31)$$

9.4   PLANT REPRESENTATION IN DIGITAL SYSTEMS

So far we have seen how the sampled signals and signals that are represented in terms of sample count can be written in a discrete domain. We note that in Figure 9.4 there is a latch and a DAC just before the process under digital control that operates as ZOH. The control computer uses a program and produces appropriate values as manipulated variables. They are nothing more than numbers. These numbers have to be used to activate power control devices in robot systems. As we have seen earlier, there are two processes involved. One process is to hold the numerical value until it is changed, and the other is to convert it to analog form.

We will ignore the conversion process during our analysis. Actually, it does not affect the basic understanding of the process of reconstruction and implementation.

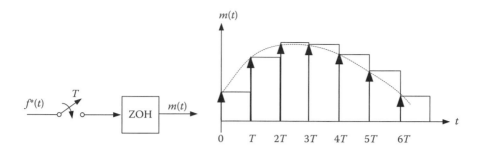

FIGURE 9.9   Signal reconstruction for control manipulation.

In practice, the holding process is done by a digital latch and the overall system as illustrated in Figure 9.4. The controller output signal $m(t)$ is shown in Figure 9.9.

Our aim is to find a mathematical representation for the ZOH shown in Figure 9.9 starting from computer output $f^*(t)$ to the ZOH output of the plant $m(t)$. This representation can be used to calculate the overall system transfer function. For this, we need to consider the transfer function of ZOH first.

9.4.1   TRANSFER FUNCTION OF ZOH

The ZOH is very important to understand and be accounted for when the physical system is being controlled from the computer. We will derive a transfer function for the ZOH. Referring to Figure 9.9, the reconstructed output can be written as

$$m(t) = f(0)\big[u(t) - u(t - T)\big] + f(T)\big[u(t - T) - u(t - 2T)\big]$$
$$+ f(2T)\big[u(t - 2T) - u(t - 3T)\big] + \cdots \qquad (9.32)$$

The Laplace transform of Equation 9.32 is

$$M(s) = f(0)\frac{1 - e^{-Ts}}{s} + f(T)\frac{e^{-Ts} - e^{-2Ts}}{s} + f(2T)\frac{e^{-2Ts} - e^{-3Ts}}{s} + \cdots$$

$$M(s) = \frac{1 - e^{-Ts}}{s}\Big[f(0) + f(T)e^{-Ts} + f(2T)e^{-2Ts} + \cdots\Big]$$

$$M(s) = \frac{1 - e^{-Ts}}{s}F^*(s) \qquad (9.33)$$

Hence, the Laplace transform for a ZOH is

$$\frac{1 - e^{-Ts}}{s} = \frac{1 - z^{-1}}{s} \quad \text{since } z = e^{Ts} \qquad (9.34)$$

9.4.2 Z-TRANSFORM OF PLANT FED FROM ZOH

Combining the above ideas, the discrete domain transfer function of a plant can be written as

$$G(z) = (1 - z^{-1})Z\left\{\frac{G(s)}{s}\right\} \tag{9.35}$$

EXAMPLE 9.9

Let us find the Z-transform of the following computer-controlled plant: $K/(s + a)$. It is a first-order plant fed through a ZOH from a computer as shown in Figure 9.10. Hence, it becomes necessary to find the Z-transform representation of this plant. The discrete transfer function can be obtained as

$$G(z) = \frac{C(z)}{M(z)} = (1 - z^{-1})Z\left[\frac{K}{s(s + a)}\right] \tag{9.36}$$

Referring to the standard Z-transform tables found in the control literature (see for instance Ogata 1995), we can write the Z-transform as

$$G(z) = \frac{C(z)}{M(z)} = 1 - z^{-1}\left[\frac{Kz(1 - e^{-aT})}{(z - 1)(z - e^{-aT})}\right] \tag{9.37}$$

Hence

$$G(z) = \frac{C(z)}{M(z)} = \frac{K(1 - e^{-aT})}{(z - e^{-aT})} \tag{9.38}$$

Then, the Z-transform of the output is

$$C(z) = G(z)M(z) \tag{9.39}$$

9.4.3 TUSTIN'S APPROXIMATION

At times, the evaluation of the Z-transform becomes a lengthy process. It is possible to use some approximate methods to obtain Z-domain representation of plants and still achieve reasonable accuracy of results in discrete domain analysis and simulations. Tustin's formula is one of such methods available and provides a good approximation of the Z-transform (Astrom and Wittenmark 1990). Tustin's formula is given as

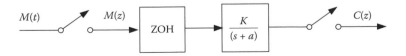

FIGURE 9.10 Manipulated variable acting on the plant through ZOH.

$$s \rightarrow \frac{2}{T} \frac{1 - z^{-1}}{1 + z^{-1}} = \frac{2}{T} \frac{z - 1}{z + 1} \qquad (9.40)$$

It simply implies that the discrete domain approximation can be obtained by replacing s terms by the above function in Equation 9.40 in the s-domain transfer function. A ZOH is built into this approximation. Therefore, there is no need to include a separate ZOH.

EXAMPLE 9.10

Let us rework Example 9.9 using Tustin's approximation. All we need to do is substitute Equation 9.40 into the system transfer function.

$$Z\left[\frac{K}{s + a}\right]_{Tustin} = \frac{K}{\dfrac{2}{T} \dfrac{z - 1}{z + 1} + a} = \frac{KT(z + 1)}{2z - 2 + aTz + aT} = \frac{K(z + 1)}{z(aT + 2) + (aT - 2)}$$

$$(9.41)$$

We note that Equations 9.38 and 9.41 are different. However, if we compute the responses, we will see that they yield the same results for any arbitrary input function.

9.5 CLOSED-LOOP SYSTEM TRANSFER FUNCTIONS

We have learned to compute the overall transfer functions of digital sampled data systems. Since there are rarely any open-loop control systems, we need to understand calculating closed-loop transfer functions. Even though fundamentally there is no major difference between analog and digital systems in this aspect, there are some minor differences depending upon how and at which points the sampling is done. We are deriving some useful formulas similar to what we saw in Chapter 8. Let us consider a typical closed-loop control system and see how the closed-loop transfer function can be derived.

In Figure 9.11, $G_c(z)$ is the controller implemented by a digital system such as a computer. There is an equivalent controller transfer function as $G_c(s)$ in frequency domain. We prefer that the controller be in Z-domain, since such a controller exists

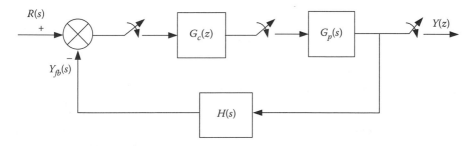

FIGURE 9.11 Block diagram of a closed-loop computer-controlled system.

only in the discrete domain. Furthermore, the plant transfer function $G_p(s)$ includes a ZOH. Also note that the output measurement, broadly called instrumentation, happens in the continuous domain through the feedback transfer function $H(s)$. In some cases, instrumentation can be digital. We will see such a scenario later. In fact in real systems, we can replace one sampler before $G_c(z)$ with two samplers, one on the path of $R(s)$ and another on the path of $Y_{fb}(s)$ without changing anything. After all, we are aware that the reference $R(s)$ may be even generated by the computer itself or in response to an external command link, say, from another computer. Let us derive the closed-loop transfer function for this system. However, before we proceed, it is important to note that

$$Z[G_p(s)H(s)] \neq G_p(z)H(z)$$

The left-hand side of the equation implies "multiply the transfer functions and find the Z-transform of the product" and the right-hand side means "find the Z-transforms individually and then multiply." These two are not the same. The typical notation used for $Z[G_p(s)H(s)]$ is $G_pH(z)$.

Now, referring to Figure 9.11

$$e = R - Y_{fb}$$
$$e(z) = R(z) - G_pH(z)m(z) \tag{9.42}$$
$$= R(z) - G_c(z)G_pH(z)e(z)$$

then we have

$$[1 + G_pH(z)G_c(z)]e(z) = R(z)$$
$$e(z) = \frac{R(z)}{1 + G_pH(z)G_c(z)} \tag{9.43}$$

We see that

$$Y(z) = G_c(z)G_p(z)e(z) \tag{9.44}$$

Substituting for $e(z)$, we get

$$\frac{Y(z)}{R(z)} = \frac{G_c(z)G_p(z)}{1 + G_c(z)G_pH(z)} \tag{9.45}$$

9.5.1 SYSTEMS WITH DIGITAL INSTRUMENTATION

What happens if the instrumentation is done digitally by taking in the sampled signal, $Y(z)$? We can redraw the block diagram by shifting the take off point for feedback path after the sampler as shown in Figure 9.12. We may have an awkward situation where the error detector deals with analog $R(s)$ and sampled feedback from $H(z)$. To avoid

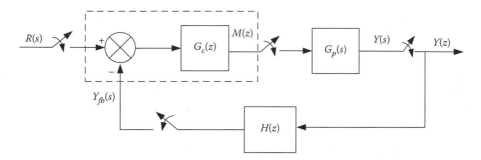

FIGURE 9.12   System with digital instrumentation.

this, we replace one sampler after the subtraction by two samplers before it. Shifting the samplers does not make any difference in the analysis, except that it indicates the control computer now does the subtraction of feedback from the reference. The new transfer function can be derived by changing the steps of analog instrumentation described earlier. We have rewritten $H(s)$ as $H(z)$ since it represents digital instrumentation. The difference finder is a part of the digital controller program, which is marked by dotted lines in Figure 9.12. We can now write the error equation as

$$e = R - Y_{fb}$$
$$e(z) = R(z) - G_p(z)H(z)m(z)$$

From Figure 9.12, $m(z) = G_c(z)e(z)$; hence

$$e(z) = R(z) - G_c(z)G_p(z)H(z)e(z) \tag{9.46}$$

We note that the term $G_pH(z)$ in Equation 9.42 has been replaced by $G_p(z)H(z)$ in Equation 9.46, since there is a sampler between $Y(s)$ and $Y(z)$ as shown in Figure 9.12. Then, Equation 9.43 gets modified as

$$[1 + G_p(z)H(z)G_c(z)]e(z) = R(z)$$
$$e(z) = \frac{R(z)}{[1 + G_p(z)H(z)G_c(z)]} \tag{9.47}$$

Following the same logic, we can proceed

$$Y(z) = G_c(z)G_p(z)e(z) \tag{9.48}$$

Substituting for $e(z)$ from Equation 9.47, we get

$$\frac{Y(z)}{R(z)} = \frac{G_c(z)G_p(z)}{1 + G_c(z)G_p(z)H(z)} \tag{9.49}$$

In complex systems, instrumentation may be happening in more than one point, making the block diagram much more complex than what we discussed in Figure 9.12. This may result in more complex interconnections.

9.6 RESPONSE OF DISCRETE TIME SYSTEMS, INVERSE Z-TRANSFORMS

We have so far presented the techniques to obtain the pulse transfer functions of signals and systems. If we know the pulse transfer functions of control systems, then it is easy to obtain pulse transfer functions of the output. Inverse Z-transform provides methods of finding the output as sequence of numbers in sampled instances, say, $y(k)$ from $Y(z)$. There are a few methods of obtaining such time responses. We will present two techniques and give some examples to provide insight into the techniques. One method is by splitting the signal into partial fractions and using Z-transform tables. The other method uses the "difference equation" concept.

9.6.1 PARTIAL FRACTION TECHNIQUE

EXAMPLE 9.11

Assume that Z-transform of a signal is given by

$$C(z) = \frac{z}{(z-1)(z-2)} \tag{9.50}$$

We would like to determine the inverse Z-transform using partial fractions and apply the transform tables to obtain signal description in terms of sample count k. Partial faction expansion of Equation 9.50 yields

$$C(z) = \frac{z}{(z-1)(z-2)} = -\frac{z}{(z-1)} + \frac{z}{(z-2)} \tag{9.51}$$

The inverse Z-transform using the standard Z-transform table found in the literature is

$$x(k) = -1 + (2)^k \quad \text{for} \quad k \geq 0 \tag{9.52}$$

9.6.2 DIFFERENCE EQUATION TECHNIQUES

Another example shows how time domain solution can be obtained by difference equations.

EXAMPLE 9.12

The transfer function of a closed-loop system is given below, and we would like to compute the solution in terms of sample count.

$$\frac{C(z)}{R(z)} = \frac{z^2 + 2z + 1}{10z^2 + 8z + 1} \tag{9.53}$$

We can cross multiply Equation 9.56 and get the difference equation from there:

$$C(z)(10z^2 + 8z + 1) = (z^2 + 2z + 1)R(z) \qquad (9.54)$$

$$C(z)(10 + 8z^{-1} + z^{-2}) = (1 + 2z^{-1} + z^{-2})R(z)$$

$$10C(z) = -8z^{-1}C(z) - z^{-2}C(z) + R(z) + 2z^{-1}R(z) + z^{-2}R(z) \qquad (9.55)$$

We now take the inverse Z-transform using time shift theorem

$$c(k) = -0.8c(k-1) - 0.1c(k-2) + 0.1r(k) + 0.2r(k-1) + 0.1r(k-2) \qquad (9.56)$$

We know that the input $r(k)$ is one for all $k \geq 0$ and let us assume that output $c(k) = 0$ for $k < 0$, which implies that $c(-1)$ and $c(-2)$ are zeros.

Now, we are ready to compute the $c(k)$ by substituting values for $k = 0,...,n$ in Equation 9.56

$$c(0) = -0.8c(-1) - 0.1c(-2) + 0.1r(0) + 0.2r(-1) + 0.1r(-2) = 0.1 \qquad (9.57)$$

$$\begin{aligned} c(1) &= -0.8c(0) - 0.1c(-1) + 0.1r(1) + 0.2r(0) + 0.1r(-1) \\ &= -0.8 \times 0.1 + 0.1 + 0.2 = 0.22 \end{aligned} \qquad (9.58)$$

$$\begin{aligned} c(2) &= -0.8c(1) - 0.1c(0) + 0.1r(2) + 0.2r(1) + 0.1r(0) \\ &= -0.8 \times 0.22 - 0.1 \times 0.1 + 0.1 + 0.2 + 0.1 = 0.214 \end{aligned} \qquad (9.59)$$

This calculation continues for all the values of k.

9.6.3 Time Domain Solution by MATLAB®

It is possible to obtain time responses using MATLAB simulation for digital systems (Cavallo et al. 1996; Ogata 1994). If we compute the closed-loop transfer function, then we can use the "filter" command to obtain the response. We will see a few different MATLAB techniques to achieve the same objective for the sake of diversity.

EXAMPLE 9.13

Consider an open-loop plant with transfer function $G(s) = (2\alpha/s + \alpha)$ controlled in a closed loop digitally with ZOH. It will have an open-loop transfer function of

$$G(z) = \frac{2(1-a)}{z-a} \qquad (9.60)$$

where a is the system constant dependent on sampling time and defined as

$$a = e^{-\alpha T} \qquad (9.61)$$

```
%Response of discrete system
T=0.01;
a=exp(-15*T);
num=[0 2*(1-a)];
den=[1 -(3*a-2)];
k=0:50;
[y]=dstep(num,den,k);
plot(k,y,'x',k,y,'-')
axis([0 50 0 1]);
grid;
title('Unit Step Response')
xlabel('k, sample time=0.01 sec')
ylabel('y(k)')
```

FIGURE 9.13   MATLAB program to calculate output sequence (sample time $T = 0.01$ s).

Then the closed-loop transfer function of

$$\frac{\omega(z)}{\omega_R(z)} = \frac{2(1-a)}{z-(3a-2)} \tag{9.62}$$

We would like to compute the system response for a step input, using the MATLAB command "dstep" assuming that α is 15 and the sampling time is 0.01 s. The code and the response are shown in Figures 9.13 and 9.14, respectively. Note that in the code, the number of samples selected is 50, which correspond to duration of 0.5 s.

The response is shown in Figure 9.14.

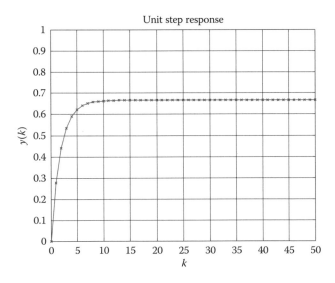

FIGURE 9.14   Unit step response using "dstep" command for 0.5 s.

EXAMPLE 9.14

In this example, we intend to get the response for Example 9.13 using simulation. We needed to compute the closed-loop transfer function in order to use the "dstep" command. Using the MATLAB Simulink modeling tool, the response can be plotted without having to compute the closed-loop transfer function. This is quite straightforward, using a Simulink model as shown in Figure 9.15.

A sample program is listed in Figure 9.16. The response obtained is shown in Figure 9.17, which is the same as Figure 9.14.

EXAMPLE 9.15

It is possible to simulate a sampled system directly using MATLAB. Consider the continuous system

$$G(s) = \frac{2}{s^2 + 7s + 10} \tag{9.63}$$

We have created a simulation model as shown in Figure 9.18. The MATLAB code to drive it is given in Figure 9.19, and the response is given in Figure 9.20.

9.7 TYPICAL CONTROLLER SOFTWARE IMPLEMENTATION

Controller transfer functions are written as difference equations to implement them in microprocessors. The program running on the processor performs the calculations in a sequence. We will illustrate how this is done by using a PID controller as an example. Such computer implementations of controllers are usually done using

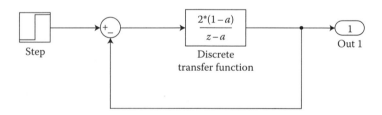

FIGURE 9.15   Simulink model for simulating a discrete system.

```
%Simulink based solution
T=0.01;
a=exp(-15*T);
[k,y,Out1]=sim('figure915mdl');
plot(k,Out1,'x',k,Out1,'-')
axis([0 50 0 1]);
grid;
title('Unit Step Response')
xlabel('k, sample time=0.01 sec')
ylabel('y(k)')
```

FIGURE 9.16   MATLAB code for closed-loop simulation of a discrete system.

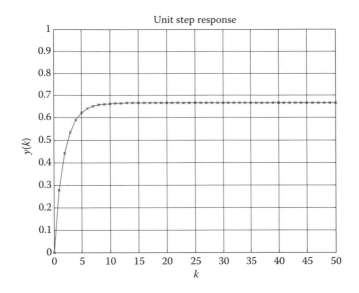

FIGURE 9.17   Response of discrete system simulation for 0.5 s.

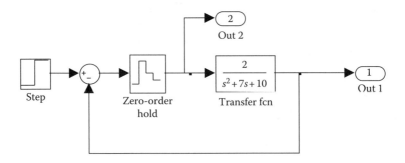

FIGURE 9.18   Simulation model of a system with ZOH.

```
T=0.25;%ZOH interval
%computation and & plot interval 0.01 sec
[t,x,Out1,Out2]=sim('figure918mdl');
figure(1);
subplot(2,1,1)
plot(t,Out1)
grid
title('sampled data simulation')
ylabel('output')
subplot(2,1,2)
plot(t,Out2)
grid;
ylabel('control variable')
xlabel('time, seconds (ZOH hold time T=0.25 seconds)')
```

FIGURE 9.19   MATLAB code for the simulation.

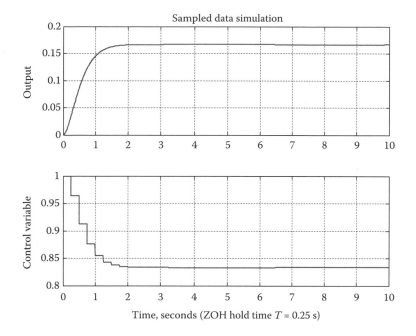

FIGURE 9.20 Response of the sampled data system.

approximations. A PID controller has three terms, namely, proportional, derivative, and integral. In time domain, it is written as

$$m(t) = K_p \left\{ e + \frac{1}{T_i} \int edt + T_d \frac{de}{dt} \right\}$$

or

$$m(t) = K_p e + K_i \int e \cdot dt + K_d \frac{de}{dt} \tag{9.64}$$

where K_p is the proportional gain, T_i is the integral time, and T_d is the derivative time.

It can be seen that the first term is proportional control, the second term is the integral part, and the third term is the derivative part. In general, the proportional part provides the basic control, the integral part tackles the steady-state error, and the derivative part speeds up the response.

K_i and K_d are defined appropriately as

$$K_i = \frac{K_p}{T_i} \tag{9.65}$$

$$K_d = K_p T_d \tag{9.66}$$

9.7.1 INTEGRAL CALCULATIONS

Let us assume that the values of sampled errors are denoted as $e(k)$, $e(k-1)$, $e(k-2)$, and so on and the manipulated variables calculated by the controller are denoted as $m(k)$, $m(k-1)$, $m(k-2)$, and so on. The k and $k-1$ indicate the current and the last sampling instant, respectively. Let the control interval be T seconds, which should be chosen after considering the closed-loop system time constants. At any sampling instant k, the integral of error can be approximated as

$$e_i(k) = e_i(k-1) + [e(k) + e(k-1)]\frac{T}{2} \tag{9.67}$$

where $e_i(k-1)$ is the last integral of system error.

9.7.2 DERIVATIVE CALCULATIONS

The current value of the derivative of error is approximated as

$$e_d(k) = \frac{e(k) - e(k-1)}{T} \tag{9.68}$$

Finally, the controller output can be computed as

$$m(k) = K_p \times e(k) + K_i \times e_i(k) + K_d \times e_d(k) \tag{9.69}$$

9.7.3 IMPLEMENTATION OF A DIGITAL CONTROLLER

A control system in which a digital processor controls a plant by sampling output data, calculating the manipulated variable, and outputting it to the plant in *real time* is called direct digital control (DDC). The term "real time" is used here because for every sample, the processor must respond before the next sample is taken. An important requirement of DDC is that the data acquisition, control computation, and outputting of control signal to the plant through ZOH should be performed at regular intervals. There are two ways of implementing constant sampling and control intervals. One method uses computation time to achieve the constant and control interval. A second method uses interrupts to achieve this. Let us consider both types of implementation in detail.

Loop time-based sampling and control: The controller performs calculations and it takes certain time duration. Most of the computations take fixed duration. After the computation of control output, it must be implemented and then the processor goes into a fixed time delay loop. By utilizing this method, we can create a pseudosampling period effect. It is desirable to select a processor which can perform all the calculations within a small fraction of sampling time, and the remaining time is decided by the delay loop.

Timer-based sampling and control: It is possible to use CPU interrupt as a time keeper to maintain constant sampling intervals. Within the timer interrupt service routine, an index is incremented and it can be used to decide when the next sampling and control is to be done. Furthermore, some data sampling/filtering may also be performed within the timer interrupt routines. After the program starts, the main loop performs instrumentation and control functions. As an example, let us assume that this process takes 2 ms and processor outputs the control signal. Let us also assume that the sampling interval has been fixed as 12 ms, and timer interrupts occur every 1 ms. As the main calculations are being executed at the main loop, two interrupts would have occurred within those 2 ms of computation, since timer interrupt occurs every 1 ms. This number is being counted at the interrupt service routine. After outputting the control signal, the processor waits in a loop, checking the counted interrupts. When the counted interrupts reach 12, the processor resets the interrupt count to zero and loops back. Then, in the next step, samples are taken, control outputs are computed and this cycle continues.

It is important to realize that in the above procedure, immediately after performing instrumentation and control computations, the control signal should be put to the output port. If the program waits for the end of the sampling time to do this, then we would be introducing one sampling time delay. This delay in fact was not part of the plant model, and hence the control may not work satisfactorily. Considering the above example, if the processor can take <1 or 2 ms, it is a small fraction of the total sampling time, and there is no delay introduced. The processor still can wait for the 12th interrupt before looping back. This is not a serious problem as modern processors are very fast.

9.8   DISCRETE STATE SPACE SYSTEMS

We presented some ideas on using state space techniques earlier in Chapter 8. We have stated that state space representation has many advantages over transfer function representation since the former is more suitable for multi-input/multi-output (MIMO) systems. Robots are typically MIMO systems. In modern controller implementations, the trend is not only to use digital controllers, but more and more such controllers are based on state feedback. Intuitively, this leads to the point that we ought to implement digital state space-based controllers. For this purpose, let us first learn a few techniques to derive discrete state space equations.

9.8.1   DISCRETE STATE SPACE SYSTEM FROM DISCRETE TRANSFER FUNCTIONS

There are a few ways of deriving discrete state space equations from discrete transfer functions, and they result in the following types of state models:

1. Controllable cannonical form
2. Observable canonical form
3. Diagonal canonical forms

They differ in the way the states are chosen and the specific purpose of the model. Since pulse transfer functions have one input and one output, in higher-order systems states need to be estimated. In the above methods, efforts needed to obtain state variables (for feedback) differ from each other. Regardless of the method used, the bottom line is that during the design process we need to be sure that the state estimation is possible or that states are readily available for measurement. Based on our experience, complicated calculation for state estimation is time consuming and therefore such calculation should be avoided wherever possible. Furthermore, in robotics, we are more interested in direct derivation of discrete state equations from continuous state models such that either states are directly measurable or they can be easily estimated. Here, MATLAB also can provide some useful tools in the design process.

9.8.2 DISCRETE STATE SPACE MODEL FROM CONTINUOUS STATE SPACE MODEL

Once we derive continuous time state space models, it is relatively straightforward to derive the discrete state space model from there. There are various methods for this, and we will discuss them in the following sections.

9.8.2.1 Analytical Method

Let us start with the continuous state space model

$$\dot{X}(t) = AX(t) + Bu(t)$$
$$y(t) = Cx(t) + Du(t) \tag{9.70}$$

If the model is sampled regularly at intervals of T seconds and $u(t)$ is fed through ZOH, we can write

$$u(t) = u(kT) \quad \text{for } kT < t < (k+1)T \tag{9.71}$$

This means that the input is taken in at time $t = kT$ and held constant by the latch (and converted to analog form by the DAC for analog systems), until $t = (k + 1)T$, when next value of manipulated variable u is supplied to the latch by the digital controller.

Usual discrete time state space representation is

$$\bar{X}(k + 1) = G\,\bar{X}(k) + Hu(k) \tag{9.72}$$

To make things clear, we will rewrite the required solution as

$$\bar{X}((k + 1)T) = G(T)\bar{X}(kT) + H(T)u(kT) \tag{9.73}$$

Hence, to obtain the discrete time state representation from the continuous time state representation, we need to compute the matrices $G(T)$ and $H(T)$ in Equation 9.73. The solutions are available in the control literature (see, for instance, Astrom and Wittenmark 1990; Nagrath and Gopal 1996; Ogata 1995). The results are given below as

$$G(T) = e^{AT} \text{ and } H(T) = \left\{ \int_0^T e^{At}\, dt \right\} B \qquad (9.74)$$

where

$$e^{At} = L^{-1}\left\{ [sI - A]^{-1} \right\} \qquad (9.75)$$

Furthermore, if A is nonsingular

$$G(T) = e^{AT}$$

$$H(T) = \left\{ \int_0^T e^{At}\, dt \right\} B = A^{-1}(e^{AT} - I)B = A^{-1}[G(T) - I]B \qquad (9.76)$$

Then, we can obtain the discrete time state representation as in Equation 9.73. The output can be written as

$$Y(kT) = C\bar{X}(kT) + Du(kT) \qquad (9.77)$$

From Equation 9.71, we can see that ZOH is built into this model. At the sampling instants, the model output is the same as the plant output. We have simply integrated the continuous time state model over a sampling period and gotten the discrete state model. Also, it is interesting to note that as T approaches zero, $G(T)$ approaches to become a unity matrix.

EXAMPLE 9.16

Let us derive the discrete time state model for the plant given by

$$G(s) = \frac{Y(s)}{U(s)} = \frac{1}{s^2 + 3s + 2} \qquad (9.78)$$

for a sample time of 1 s. Continuous time state model can be written as

$$\begin{bmatrix} \dot{x}_1 \\ \dot{x}_2 \end{bmatrix} = \begin{bmatrix} 0 & 1 \\ -2 & -3 \end{bmatrix} \begin{bmatrix} x_1 \\ x_2 \end{bmatrix} + \begin{bmatrix} 0 \\ 1 \end{bmatrix} u$$

$$y = \begin{bmatrix} 1 & 0 \end{bmatrix} \begin{bmatrix} x_1 \\ x_2 \end{bmatrix} \qquad (9.79)$$

Using Equation 9.74, we need to evaluate $G(T)$. Using the inverse Laplace transform method

$$e^{At} = L^{-1}\left\{[sI - A]^{-1}\right\} = L^{-1}\left\{\begin{bmatrix} s & -1 \\ +2 & s+3 \end{bmatrix}^{-1}\right\} = L^{-1}\begin{bmatrix} \dfrac{s+3}{s^2+3s+2} & \dfrac{1}{s^2+3s+2} \\ \dfrac{-2}{s^2+3s+2} & \dfrac{s}{s^2+3s+2} \end{bmatrix}$$

$$= L^{-1}\begin{bmatrix} \left(\dfrac{-1}{(s+2)}+\dfrac{2}{(s+1)}\right) & \left(\dfrac{-1}{(s+2)}+\dfrac{1}{(s+1)}\right) \\ \left(\dfrac{2}{(s+2)}+\dfrac{-2}{(s+1)}\right) & \left(\dfrac{2}{(s+2)}+\dfrac{-1}{(s+1)}\right) \end{bmatrix} \tag{9.80}$$

we obtain

$$e^{At} = \begin{bmatrix} \left(-e^{-2t}+2e^{-t}\right) & \left(-e^{-2t}+e^{-t}\right) \\ \left(+2e^{-2t}-2e^{-t}\right) & \left(+2e^{-2t}-e^{-t}\right) \end{bmatrix} \tag{9.81}$$

This leads to

$$G(T) = e^{AT} = \begin{bmatrix} \left(-e^{-2T}+2e^{-T}\right) & \left(-e^{-2T}+e^{-T}\right) \\ \left(+2e^{-2T}-2e^{-T}\right) & \left(+2e^{-2T}-e^{-T}\right) \end{bmatrix}_{T=1}$$

$$= \begin{bmatrix} 0.6004 & 0.2325 \\ -0.4651 & -0.0972 \end{bmatrix} \tag{9.82}$$

To compute $H(T)$, we use Equation 9.74

$$H(T) = \left\{\int_0^T e^{At}dt\right\}B = \left\{\int_0^T \begin{bmatrix} \left(-e^{-2t}+2e^{-t}\right) & \left(-e^{-2t}+e^{-t}\right) \\ \left(+2e^{-2t}-2e^{-t}\right) & \left(+2e^{-2t}-e^{-t}\right) \end{bmatrix} dt\right\}B \tag{9.83}$$

After integrating and using $T = 1$

$$H(T) = \begin{bmatrix} \left(\dfrac{e^{-2t}}{2}-\dfrac{2e^{-t}}{1}\right) & \left(\dfrac{e^{-2t}}{2}-\dfrac{e^{-t}}{1}\right) \\ \left(-\dfrac{e^{-2t}}{1}+\dfrac{2e^{-t}}{1}\right) & \left(-\dfrac{e^{-2t}}{1}+\dfrac{e^{-t}}{1}\right) \end{bmatrix}_0^1 B$$

$$= \begin{bmatrix} [(-0.6681)-(-1.5)] & [(-0.3002)-(-0.5)] \\ [(0.6004)-(1)] & [(0.2325)-(0)] \end{bmatrix}\begin{bmatrix} 0 \\ 1 \end{bmatrix}$$

$$= \begin{bmatrix} (0.8319) & (0.1998) \\ (-0.3996) & (0.2325) \end{bmatrix}\begin{bmatrix} 0 \\ 1 \end{bmatrix} = \begin{bmatrix} 0.1998 \\ 0.2325 \end{bmatrix} \tag{9.84}$$

Let us try out the other form given in Equation 9.76 here:

$$H(T) = A^{-1}[G(T) - I]B = \begin{bmatrix} 0 & 1 \\ -2 & -3 \end{bmatrix}^{-1} \left\{ \begin{bmatrix} 0.6004 & 0.2325 \\ -0.4651 & -0.0972 \end{bmatrix} - \begin{bmatrix} 1 & 0 \\ 0 & 1 \end{bmatrix} \right\} \begin{bmatrix} 0 \\ 1 \end{bmatrix}$$

$$= \begin{bmatrix} -1.5 & -0.5 \\ 1 & 0 \end{bmatrix} \begin{bmatrix} -0.3996 & 0.2325 \\ -0.4651 & -1.0972 \end{bmatrix} \begin{bmatrix} 0 \\ 1 \end{bmatrix} = \begin{bmatrix} 0.1998 \\ 0.2325 \end{bmatrix}$$

(9.85)

The result agrees with the earlier result. Hence, the model is

$$\begin{bmatrix} x_1(k+1) \\ x_2(k+1) \end{bmatrix} = \begin{bmatrix} 0.6004 & 0.2325 \\ -0.4651 & -0.0972 \end{bmatrix} \begin{bmatrix} x_1(k) \\ x_2(k) \end{bmatrix} + \begin{bmatrix} 0.1998 \\ 0.2325 \end{bmatrix} u(k)$$

$$y(k) = \begin{bmatrix} 1 & 0 \end{bmatrix} \begin{bmatrix} x_1(k) \\ x_2(k) \end{bmatrix}$$

(9.86)

9.8.2.2 MATLAB Approach

We can use MATLAB for obtaining discrete state space model from the continuous state space model. MATLAB provides the "c2d(A,B,T)" command where notations are standard. We will attempt the solution for Example 9.16 using MATLAB. A program segment for this purpose and the result obtained are shown in Figure 9.21. We can see that discrete time state equations can be obtained analytically from continuous time state equations, as well as by using software tools such as MATLAB.

9.8.3 TIME DOMAIN SOLUTION OF DISCRETE STATE SPACE SYSTEMS

To understand the nature of a system, we need to see the time domain response of that system. Furthermore, when we design controllers, we also need to see the response of the controlled system in order to assess the performance of the controller. For this, we need to calculate the time domain solutions. We will discuss two ways of obtaining the time domain solutions.

```
A=[0 1;-2 -3];
B=[0;1];
T=1;
[G,H]=c2d(A,B,T)

G =
        0.6004      0.2325
       -0.4651     -0.0972
H =
        0.1998
        0.2325
```

FIGURE 9.21   MATLAB commands and results obtained for getting digital state model.

9.8.3.1   Computer Calculations

Computer calculations are done from sample to sample. It starts from the knowledge of initial values of the states and the values of the input sequence. Calculating the states can be easily achieved by programming a computer for recursive calculations from sample to sample as long as the input $u(kT) < t < u((k + 1)T)$ is piecewise constant. That is, the input is held constant between controller outputs to the plant through the latch–DAC combination. Let us look at the case of time-invariant state equations only. Hence, the state equations become

$$X(k + 1) = GX(k) + Hu(k)$$
$$y(K) = CX(k) + Du(k) \tag{9.87}$$

With $k = 0$ in Equation 9.87, using $u(0)$ and $X(0)$ vector values, $X(1)$ vector can be calculated and hence $y(1)$ can also be calculated. The same calculation is repeated for $k = 1$. This leads to $X(2)$, which can be calculated from $u(1)$ and $X(1)$. This process is repeated for obtaining the complete solution. We have assumed that input u is a scalar. The procedure is similar for vector U as well.

9.8.3.2   Z-Transform Approach

Let us consider Equation 9.87 and write the Z-transform using the usual notation with initial conditions, keeping in mind that G and H are constant matrices:

$$zX(z) - zX(0) = GX(z) + Hu(z) \tag{9.88}$$

Then

$$(zI - G)X(z) = +zX(0) + Hu(z)$$
$$X(z) = (zI - G)^{-1} zX(0) + (zI - G)^{-1} Hu(z) \tag{9.89}$$

Taking the inverse Z-transform

$$X(k) = Z^{-1}\left[(zI - G)^{-1} z\right]X(0) + Z^{-1}\left[(zI - G)^{-1} Hu(z)\right] \tag{9.90}$$

The first part is the undriven part, and the second part is the driven part of the solution. This involves matrix inversion as well as the inverse Z-transform, and it may get complicated if the plant is more than second order. Let us see an example.

EXAMPLE 9.17

To describe the above concept, we repeat Example 9.16 here. For the system represented by discrete model given below, let us compute the total solution in terms of sample count k, assuming that the input is a unit step function.

$$\begin{bmatrix} x_1(k + 1) \\ x_2(k + 1) \end{bmatrix} = \begin{bmatrix} 0.6004 & 0.2325 \\ -0.4651 & -0.0972 \end{bmatrix} + \begin{bmatrix} 0.1998 \\ 0.2325 \end{bmatrix} u(k) \tag{9.91}$$

$$y(k) = [1 \quad 0] \begin{bmatrix} x_1(k) \\ x_2(k) \end{bmatrix} \quad \text{and} \quad \begin{bmatrix} x_1(0) \\ x_2(0) \end{bmatrix} = \begin{bmatrix} 1 \\ -1 \end{bmatrix} \tag{9.92}$$

Referring to Equation 9.89, the Z-transform of $X(k)$ is given by

$$Z[X(k)] = X(z) = (zI - G)^{-1}zX(0) + (zI - G)^{-1}Hu(z)$$
$$= (zI - G)^{-1}[zX(0) + Hu(z)] \tag{9.93}$$

Since, $u(k)$ is a unit step, we can write

$$zX(0) + Hu(z) = \begin{bmatrix} z \\ -z \end{bmatrix} + \begin{bmatrix} 0.1998 \\ 0.2325 \end{bmatrix}\left(\frac{z}{z-1}\right) = \begin{bmatrix} \dfrac{z^2 - 0.8002z}{z-1} \\ \dfrac{-z^2 + 1.2325z}{z-1} \end{bmatrix} \tag{9.94}$$

Hence, the output transform is

$$X(z) = (zI - G)^{-1}[zX(0) + HU(z)] = (zI - G)^{-1}\begin{bmatrix} \dfrac{z^2 - 0.8002z}{z-1} \\ \dfrac{-z^2 + 1.2325z}{z-1} \end{bmatrix} \tag{9.95}$$

Substituting for the first term on right-hand side, we get

$$X(z) = \begin{pmatrix} z - 0.6004 & -0.2325 \\ +0.4651 & z + 0.0972 \end{pmatrix}^{-1}\begin{bmatrix} (z^2 - 0.8002z/z - 1) \\ (-z^2 + 1.2325z/z - 1) \end{bmatrix} \tag{9.96}$$

After finding the inverse of the first term, we have

$$X(z) = \frac{1}{[(z - 0.6004)(z + 0.0972) + 0.2325 \times 0.4651]}$$
$$\times \begin{pmatrix} (z + 0.0972) & +0.2325 \\ -0.4651 & (z - 0.6004) \end{pmatrix}\begin{bmatrix} \left(\dfrac{z^2 - 0.8002z}{z-1}\right) \\ \left(\dfrac{-z^2 + 1.2325z}{z-1}\right) \end{bmatrix} \tag{9.97}$$

after simplifying, we obtain

$$X(z) = \frac{1}{[z^2 - 0.5032z + 0.0498](z - 1)}$$
$$\times \begin{pmatrix} (z + 0.0972)(z^2 - 0.8002z) + 0.2325(-z^2 + 1.2325z) \\ -0.4651(z^2 - 0.8002z) + (z - 0.6004)(-z^2 + 1.2325z) \end{pmatrix} \tag{9.98}$$

Further simplifying yields

$$X(z) = \frac{1}{\left(z^3 - 1.5032z^2 + 0.5530z - 0.0498\right)} \begin{pmatrix} (z^3 - 0.9355z^2 + 0.2088z) \\ (-z^3 + 1.3678z^2 - 0.3678z) \end{pmatrix}$$

$$= \begin{bmatrix} \dfrac{(z^3 - 0.9355z^2 + 0.2088z)}{\left(z^3 - 1.5032z^2 + 0.5530z - 0.0498\right)} \\ \dfrac{(-z^3 + 1.3678z^2 - 0.3678z)}{\left(z^3 - 1.5032z^2 + 0.5530z - 0.0498\right)} \end{bmatrix} \qquad (9.99)$$

We need to keep z in the denominator of the left-hand side of the equation for matching Z table entries.

$$\frac{X(z)}{z} = \begin{bmatrix} \dfrac{(z^2 - 0.9355z + 0.2088)}{\left(z^3 - 1.5032z^2 + 0.5530z - 0.0498\right)} \\ \dfrac{(-z^2 + 1.3678z - 0.3678)}{\left(z^3 - 1.5032z^2 + 0.5530z - 0.0498\right)} \end{bmatrix} \qquad (9.100)$$

Let us do some partial fraction expansions utilizing the "residue" command of MATLAB.

$$X(z) = \begin{bmatrix} 0.500\dfrac{z}{z-1} + 0.00\dfrac{z}{z-0.3678} + 0.5000\dfrac{z}{z-0.1354} \\ 0.000\dfrac{z}{z-1} + 0.000\dfrac{z}{z-0.3678} - 1.000\dfrac{z}{z-0.1354} \end{bmatrix} \qquad (9.101)$$

Note that we moved z back to the right-hand side of the equation after partial fraction expansion. We realize that the coefficient of pole term at 0.3678 became zero. If we check the roots of the first row of Equation 9.100, there is a pole-zero cancellation at 0.3678. Similarly, at the second row of Equation 9.100, there are two pairs of pole zero cancellation, one at 0.3678 and another one at 1 resulting in two coefficient terms in Equation 9.101, vanishing, although the original continuous model has no pole-zero cancellation. The results we have obtained here are accurate. Now, taking the inverse Z-transform, we get

$$\begin{bmatrix} x_1(k) \\ x_2(k) \end{bmatrix} = \begin{bmatrix} 0.5 - 0.0\,(0.3678)^k + 0.5\,(0.1354)^k \\ 0.0 + 0.0\,(0.3678)^k - 1.0\,(0.1354)^k \end{bmatrix}$$

$$= \begin{bmatrix} 0.5 + 0.5\,(0.1354)^k \\ -1.0\,(0.1354)^k \end{bmatrix} \qquad (9.102)$$

We can check the initial values from the above results, which is obviously

$$
\begin{bmatrix} x_1(0) \\ x_2(0) \end{bmatrix} = \begin{bmatrix} 1 \\ -1 \end{bmatrix}
\tag{9.103}
$$

To understand why some coefficients vanished, let us go back to Equation 9.93 and keep the undriven and driven solutions separate. We reevaluate Equation 9.93 as

$$
X(z) = \frac{1}{[(z - 0.6004)(z + 0.0972) + 0.2325 \times 0.4651]}
$$
$$
\times \begin{pmatrix} (z + 0.0972) & +0.2325 \\ -0.4651 & (z - 0.6004) \end{pmatrix} \left[\begin{bmatrix} z \\ -z \end{bmatrix} + \begin{bmatrix} 0.1998z \\ 0.2325z \end{bmatrix} \left(\frac{1}{z - 1} \right) \right]
\tag{9.104}
$$

then

$$
X(z) = \frac{1}{\left[z^2 - 0.5032z + 0.0498 \right]} \begin{bmatrix} z^2 + 0.0972z - 0.2325z \\ -0.4651z - z^2 + 0.6004z \end{bmatrix}
$$
$$
+ \frac{1}{\left[z^2 - 0.5032z + 0.0498 \right](z - 1)}
$$
$$
\times \begin{bmatrix} 0.1998z^2 + 0.1998 \times 0.0972z + 0.2325^2 z \\ -0.1998 \times 0.4651z + 0.2325z^2 - 0.2325 \times 0.6004z \end{bmatrix}
\tag{9.105}
$$

Collecting terms in the columns above, we get

$$
X(z) = \frac{1}{\left[z^2 - 0.5032z + 0.0498 \right]} \begin{bmatrix} z^2 - 0.1353z \\ -z^2 + 0.1353z \end{bmatrix}
$$
$$
+ \frac{1}{\left[z^2 - 0.5032z + 0.0498 \right](z - 1)} \begin{bmatrix} 0.1998z^2 + 0.0735z \\ 0.2325z^2 - 0.2325z \end{bmatrix}
\tag{9.106}
$$

Let us move one z term to the left as before and write

$$
\frac{X(z)}{z} = \begin{bmatrix} \dfrac{z - 0.1353}{[z^2 - 0.5032z + 0.0498]} \\ \dfrac{-(z - 0.1353)}{[z^2 - 0.5032z + 0.0498]} \end{bmatrix} + \begin{bmatrix} \dfrac{0.1998z + 0.0735}{(z^3 - 1.5032z^2 + 0.5530z - 0.0498)} \\ \dfrac{0.2325z - 0.2325}{(z^3 - 1.5032z^2 + 0.5530z - 0.0498)} \end{bmatrix}
\tag{9.107}
$$

After calculating the partial fractions, we have

$$
\frac{X(z)}{z} = \begin{bmatrix} \dfrac{1}{z - 0.3678} + \dfrac{0}{(z - 0.1354)} \\ \dfrac{-1}{z - 0.3678} + \dfrac{-0}{(z - 0.1354)} \end{bmatrix} + \begin{bmatrix} \dfrac{0.5}{z - 1} + \dfrac{-1}{z - 0.3678} + \dfrac{0.5}{z - 0.1354} \\ \dfrac{0}{z - 1} + \dfrac{1}{z - 0.3678} + \dfrac{-1}{z - 0.1354} \end{bmatrix}
\tag{9.108}
$$

In Equation 9.108, we still see some pole-zero cancellations. In the undriven term, if the initial conditions were different (say, $x_1(0) = 1$, $x_2(0) = 0$), the second term in partial fraction expansion would not have vanished. We can move z to the right-hand side and take the inverse Z-transform.

$$\begin{bmatrix} x_1(k) \\ x_2(k) \end{bmatrix} = \begin{bmatrix} (0.3678)^k \\ -(0.3678)^k \end{bmatrix} + \begin{bmatrix} 0.5 - (0.3678)^k + 0.5(0.1354)^k \\ +(0.3678)^k - (0.1354)^k \end{bmatrix} \qquad (9.109)$$

The first part of the right-hand side of Equation 9.109 represents the undriven solution and the second term represents the driven solution. Note that the driven solution vanishes for $k = 0$, and the undriven solution is the same, as the initial conditions as assumed. The total solution is written here for comparison with Equation 9.102.

$$\begin{bmatrix} x_1(k) \\ x_2(k) \end{bmatrix} = \begin{bmatrix} 0.5 + 0.5(0.1354)^k \\ -(0.1354)^k \end{bmatrix} \qquad (9.110)$$

9.9 DISCRETE STATE FEEDBACK CONTROLLERS

So far we have introduced the analytical solutions to the discrete state equations. Our next task is to look at the methods of designing stable controllers. We will focus only on two types of controllers, namely, pole placement controllers (PPC) and linear quadratic controllers (LQC). However, it is important to realize before starting to design controllers that the system is controllable and states are available for feedback. Hence, let us first see the concepts of controllability and state observability. In the following sections, we will utilize required results and formulas from control literature without going through their details and proofs, since our primary concern is to demonstrate application of these concepts in robotics.

9.9.1 CONCEPT OF STATE CONTROLLABILITY

The concepts of controllability were originally introduced by Kalman, and further work was done on it mainly by Gilbert (Nagrath and Gopal 1996). Kalman's work gives a solution to the problem based on system matrices. In literature, many definitions of controllability can be found (Astrom and Wittenmark 1990; Nagrath and Gopal 1996; Ogata 1995). A typical definition of controllability is that a control system is said to be completely controllable if the system can be transferred from any arbitrary initial states to any other arbitrary states within finite time using a control sequence, where the control magnitudes are unbounded. It implies that if any state is independent of control signal, then that renders the state uncontrollable and hence the system is not controllable. Consider the typical system equation given with usual notation. We assume that control variable is a scalar.

$$X((k + 1)T) = GX(kT) + Hu(kT)$$
$$Y(kT) = CX(kT) \qquad (9.111)$$

where $u(k)$ is the constant control signal from instant kT to $(k + 1)T$, $X(k)$ is an $n \times 1$ state vector, G is an $n \times n$ state matrix, H is an $n \times 1$ vector, $Y(k)$ is an $r \times 1$ vector, and C is an $r \times n$ matrix.

Here, we are concerned about discrete systems only. The necessary and sufficient condition for complete state controllability is that the rank of the controllability matrix $[CM]$ is n. Then the controllability condition can be stated as

$$\text{rank}[CM] = n \tag{9.112}$$

where the controllability matrix is defined as

$$CM = [H \; GH \; G^2H \cdots G^{n-2}H \; G^{n-1}H] \tag{9.113}$$

Gilbert suggested using Jordan's canonical form to derive a different condition for testing controllability.

Hence, if both conditions in Equations 9.112 and 9.113 are satisfied, we can conclude that it is possible to transfer any initial state to any final state in utmost n sampling periods, provided the control inputs $u(0)$ to $u((n - 1)T)$ are unbounded. The above conditions can also be interpreted, as that there exists a sequence of control inputs $u(0), u(T), \ldots, u((n - 1)T)$ to bring the initial state $X(0)$ to final state $X(t_f)$ within n sampling periods.

9.9.2 Concept of State Observability

In designing controllers for state space systems, we need to feed back the state variables, and these are called state feedback controllers. However, in complex systems it may not be possible to directly measure the states, simply because they may be hidden inside the systems or simply they are not physical quantities, such as voltage, current, or torque. In many cases, there are hidden modes of the systems. In that case, if we need to feed them back, we have to measure them or estimate them. This leads to the concept of observability. A simple definition of observability is that the system is said to be completely observable if the measurements of output samples taken over a finite duration are sufficient to compute all the initial states. There are a few versions of this definition in literature. The conditions of observability with the relevant proof have been well discussed in the literature (Astrom and WittenMark 1990; Nagrath and Gopal 1996; Ogata 1995). We will only present the results here. We can start by considering the system equations:

$$X((k + 1)T) = GX(T) + Hu(kT)$$
$$Y(kT) = CX(kT) + Du(kT) \tag{9.114}$$

where X is an $n \times 1$ vector, G is an $n \times n$ matrix, H is $n \times 1$, C is an $m \times n$ matrix, D is an $m \times 1$ matrix, and Y is an $m \times 1$ vector. We assume that only one control input exists for simplicity. The observability condition can be stated as that $nm \times n$ matrix given by

$$\begin{bmatrix} C \\ CG \\ CG^2 \\ \dots \\ CG^{n-1} \end{bmatrix} \tag{9.115}$$

should have a rank of n. The above equation can be transposed without changing the rank.

We went through the above two sections because any control designer needs to ascertain that the plant he is controlling is controllable in the first place. Once that is done, the problem of state feedback arises. The observability condition assures that states can be estimated for feedback, if not measurable directly.

9.9.3 Common Condition for Controllability and Observability of Sampled Data Systems

There is also an additional common condition for state controllability and observability for sampled data systems. It refers to the possible existence of complex roots, say, $\sigma \pm j\omega$. Supposing that the system has such complex roots with a natural frequency component of ω, then by selecting a wrong sampling time controllability can be jeopardized by sampling at wrong points periodically. To avoid such synchronous sampling at the wrong points, the condition can be stated as

$$T \neq \frac{i\pi}{\omega} \tag{9.116}$$

where ω is the natural angular frequency of the system, and i is an integer. Since the half period of natural oscillation will be π/ω, the sampling period T should not be a multiple of that (refer to Astrom and Wittenmark (1990) and Ogata (1995) for mathematical proof).

EXAMPLE 9.18

Let us consider a plant whose discrete model is given by

$$\frac{Y(z)}{U(z)} = \frac{z^2 - 1.7z + 0.6}{z^3 - 2.8z^2 + 2.5z - 0.728} \tag{9.117}$$

a. Check the controllability of the system.
b. Check the observability of the system.

We can get the state model by calculation or by using MATLAB. A useful MATLAB command for this conversion is "tf2ss." By entering command "[G,H,C,D] = tf2ss(numz,denz)" in the MATLAB command window, we get

$$G = \begin{bmatrix} 2.8 & -2.51 & 0.728 \\ 1.0 & 0 & 0 \\ 0 & 1.0 & 0 \end{bmatrix}$$

$$H = \begin{bmatrix} 1 \\ 0 \\ 0 \end{bmatrix}$$

$$C = \begin{bmatrix} 1 & -1.7 & 0.6 \end{bmatrix}$$

$$D = 0 \tag{9.118}$$

To check for controllability, we compute the controllability matrix $CM = [H \quad GH \quad G^2H]$, which can be evaluated as

$$CM = \begin{bmatrix} 1 & 2.8 & 5.33 \\ 0 & 1.0 & 2.8 \\ 0 & 0 & 1.0 \end{bmatrix} \tag{9.119}$$

Using the MATLAB command "rank(CM)," we obtain the rank of the matrix CM as 3. So the system is completely state controllable and hence the arbitrary pole placement is possible.

To check for observability, we compute the observability matrix in transposed form:

$$OM = [C^* \quad G^*C^* \quad (G^*)^2C^*] \tag{9.120}$$

Above, we have used the notation C^* and G^*, where the superfixes "*" indicate that they are the conjugate transpose of matrices C and G, respectively.

Matrix OM can be evaluated as

$$OM = \begin{bmatrix} 1 & 1.1 & 1.17 \\ -1.7 & -1.91 & -2.033 \\ 0.6 & 0.728 & 0.8008 \end{bmatrix} \tag{9.121}$$

and its rank is 3. Hence, the system is controllable as well as observable, and the state feedback control for arbitrary pole placement is possible.

9.9.4 Design of Pole Placement Regulators Using State Feedback

In this section, we present the controller design method called pole placement or pole assignment technique, which is well discussed in the literature (Ogata 1995; Astrom and Wittenmark 1990). We assume that all the state variables are measurable and are available for feedback. It can be shown that if all states are controllable, then the poles of the closed-loop system can be placed arbitrarily anywhere by means of state feedback. This is only true if the control signals are unbounded. If there is saturation, then the system turns nonlinear, and such design becomes invalid. Having said that, we have found in our experience that occasional saturation of u is usually well tolerated in many practical systems.

The first step in this procedure is to decide the "desirable locations" of the closed-loop poles based on the transient response or frequency response specifications such as speed, damping ratio, or bandwidth. Another consideration in designing sampled data controllers is the sampling period. Selecting a very small sampling time may result in large values of control signals, and this may lead to saturation. In what follows, we are considering a case where control signal is a scalar. Also, we assume that the necessary and sufficient condition for arbitrary pole placement is that the system is completely state controllable. We are going to present two methods for pole placement design, one by comparison of coefficients and the other one by "place" command from MATLAB.

9.9.4.1 Comparison of Coefficients Method

This is a straightforward method, and it does not require the use of any formula, which is applicable for systems of order 3 or less. We only consider the regulator problem from fundamental concepts, assuming that the system is completely controllable and hence arbitrary pole placement is possible. Let us consider the discrete system given by

$$X(k + 1) = GX(k) + Hu(k) \tag{9.122}$$

Let us formulate an admissible control law using a feedback gain vector K such that

$$u(k) = -KX(k) = -[k_1 \quad k_2 \cdots k_n]X(k) \tag{9.123}$$

where the vector $K = [k_1 \quad k_2 \cdots k_n]$.

Then substituting Equation 9.123 in Equation 9.122, we can write the closed-loop system equation as

$$X(k + 1) = GX(k) - HKX(k) = [G - HK]X(k) \tag{9.124}$$

The comparison of coefficients method involves determining the feedback gain values K, such that the characteristic equation of the closed-loop system represented by Equation 9.124 has desired roots. We will illustrate the method by an example.

EXAMPLE 9.19

Let us consider a plant whose discrete state model is given by

$$X(k + 1) = \begin{bmatrix} 2.8 & -2.51 & 0.728 \\ 1.0 & 0 & 0 \\ 0 & 1.0 & 0 \end{bmatrix} X(k) + \begin{bmatrix} 1 \\ 0 \\ 0 \end{bmatrix} u(k)$$

$$y(k) = \begin{bmatrix} 1 & -1.7 & 0.6 \end{bmatrix} X(k) \tag{9.125}$$

This is as the same system we saw in Example 9.18. Let us design a state feedback controller such that the closed-loop poles are at 0.5, 0.7, and 0.8.

Using the chosen roots, the closed-loop characteristic equation can be calculated as

$$(z - 0.5)(z - 0.7)(z - 0.8) = z^3 - 2.0z^2 + 1.31z - 0.28 = 0 \quad (9.126)$$

Furthermore, the closed-loop system is written as

$$X(k + 1) = [G - HK]X(k) \quad (9.127)$$

Taking the Z-transform of Equation 9.127, the closed-loop characteristic equation of the controlled system can be obtained as

$$\left| \begin{bmatrix} z & 0 & 0 \\ 0 & z & 0 \\ 0 & 0 & z \end{bmatrix} - \begin{bmatrix} 2.8 & -2.51 & 0.728 \\ 1 & 0 & 0 \\ 0 & 1 & 0 \end{bmatrix} + \begin{bmatrix} 1 \\ 0 \\ 0 \end{bmatrix} \begin{bmatrix} k_1 & k_2 & k_3 \end{bmatrix} \right| = 0 \quad (9.128)$$

The above determinant can be evaluated in polynomial form as

$$z^3 + (k_1 - 2.8)z^2 + (k_2 + 2.51)z + (k_3 - 0.728) = 0 \quad (9.129)$$

Equation 9.129 is the characteristic polynomial of the closed-loop system involving unknown gain values of k_1, k_2 and k_3, and Equation 9.126 is the characteristic polynomial having the desired poles as its roots. Then, if the closed-loop system is to have desired roots, the two polynomials have to be equated to evaluate suitable gain values, k_1, k_2, and k_3. When those gain values are used for state feedback, the closed-loop system will behave as though it has the desired closed-loop poles. Then, the left-hand sides of Equations 9.126 and 9.129 can be equated as

$$z^3 + (k_1 - 2.8)z^2 + (k_2 + 2.51)z + (k_3 - 0.728) = z^3 - 2.0z^2 + 1.31z - 0.28 \quad (9.130)$$

Comparing coefficients of z-terms with the same power from both sides, we can evaluate the feedback gain values.

Comparing the coefficients of z^2 on both sides

$$k_1 - 2.8 = -2.0$$
$$k_1 = 0.8 \quad (9.131)$$

Comparing the coefficients of z on both sides

$$k_2 + 2.51 = 1.31$$
$$k_2 = -1.2 \quad (9.132)$$

```
G=[2.8 -2.51 0.728;1.0 0 0;0 1.0 0];

H=[1;0;0];

C=[1 -1.7 0.6];

D=0;

poldes=[0.5 0.7 0.8]; % Define desired poles

K=place(G,H,poldes)

Result:

K =     0.8000    -1.2000      0.4480
```

FIGURE 9.22 Using "place" command.

Comparing the constant terms on both sides

$$k_3 - 0.728 = -0.28$$
$$k_3 = 0.448 \tag{9.133}$$

9.9.4.2 MATLAB Method of Pole Placement

What we have demonstrated above can be achieved by the "place" command in MATLAB. The dialog is given in Figure 9.22.

9.9.4.3 MATLAB Simulation of the Controller Performance

One possible Simulink model is shown in Figure 9.23. Let us study this simulation model. We have made some small changes to the C vector and call it C_x. By making

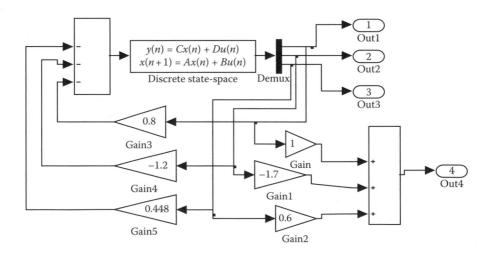

FIGURE 9.23 Simulation of state feedback control setup for pole placement control.

C_x a unity matrix of size n, we get all the n state variables as outputs. Also, we are using actual components of the C vector to weigh the state variables to obtain the system output y. Thus, we have state variables readily available for feedback and display as well as the system output synthesized. We have introduced an initial value of $x1(0) = 3$ to state 1, to see if the system stabilizes under the controller. Note that this appears in the MATLAB code given in Figure 9.24, which calls the Simulink model. Without this initial value, we will only see a flat response for all the states as well as the output.

The MATLAB program that drives the simulation model shown in Figure 9.23, is listed in Figure 9.24, and the simulation results are shown in Figure 9.25.

We note that the system in Example 9.19 was derived from the pulse transfer function given in Equation 9.117. By factorizing the numerator and denominator of that pulse transfer function, we can see that the open-loop system has zeros at 0.5, and 1.2 and poles at 1.3, 0.8, and 0.7. At the beginning, the plant is in the open-loop, unstable, and nonminimum phase. When we design the gains to place the poles at the desired locations, we see all the states, as well as the outputs, are quite stable upon the application of the pole placement controller.

```
%Discrete model simulation of PolePlacement gains
G=[2.8 -2.51 0.728;1 0 0;0 1 0];
H=[1;0;0];
% We need all states for feedback
Cx=[1 0 0;0 1 0;0 0 1];
Dx=[0;0;0];
x0=[3 0 0];
imax=50;
ymax=10;
[k,x,Out1,Out2,Out3,Out4]=sim('figure923mdl');
figure(1)
subplot(4,1,1)
plot(Out1);
v=[0 imax -ymax ymax];
axis(v);
grid;
title('Output and states in feedback control');
ylabel('state x1')
subplot(4,1,2)
plot(Out2);
v=[0 imax -ymax ymax];
axis(v);
grid;
ylabel('state x2')
subplot(4,1,3)
plot(Out3);
v=[0 imax -ymax +ymax];
axis(v);
grid;
ylabel('state x3')
subplot(4,1,4)
plot(Out4);
v=[0 imax -ymax ymax];
axis(v);
grid;
xlabel('k, sample count')
ylabel('Output y')
```

FIGURE 9.24 MATLAB listing for driving model in Figure 9.23.

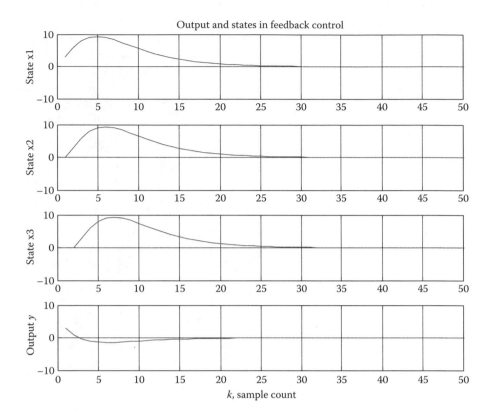

FIGURE 9.25 Results of pole placement regulator simulation.

9.9.5 STEADY-STATE QUADRATIC OPTIMAL CONTROL

It is possible to use a different criteria for deciding the feedback gain values. In optimal control, we define a cost function or index involving the system states and other relevant quantities such as control variables. After that we devise a strategy to minimize this cost function. This can be over a fixed duration of time or at steady state. We will provide the basic outline for steady state optimization design here.

We are considering the system described by

$$X(k + 1) = GX(k) + Hu(k)$$
$$y(k) = CX(k)$$

the feedback law is $u(k) = -KX(k)$. Then the performance index that we want to optimize is given below:

$$J = \frac{1}{2} \sum_{k=0}^{\infty} \left[X^*(k)QX(k) + U^*(k)RU(k) \right] \tag{9.134}$$

In the above equation, Q is a diagonal positive definite matrix of order n, comprising the weights we want to give to the respective state variables, and R is a diagonal positive definite matrix comprising the weights we want to use for the manipulated variables U. When we consider cases of single control variable, it is a single positive element.

For a system of order n with m control inputs, Equation 9.134 can be rewritten as

$$J = \frac{1}{2} \sum_{k=0}^{\infty} \left[\begin{array}{c} q_1 x_1^2(k) + q_2 x_2^2(k) + \cdots + q_n x_n^2(k) \\ + r_1 u_1^2(k) + r_2 u_2^2(k) + \cdots + r_m u_m^2(k) \end{array} \right] \tag{9.135}$$

where q_1, q_2, and so on are the diagonal elements of the Q matrix and r_1, r_2, and so on are the diagonal elements of the R matrix. The importance of state x_i in the cost function is decided by q_i.

Similarly, the importance of input u_i in the cost function is decided by r_i. This property can be used to influence the response of the individual states and control inputs. We will demonstrate this in the next chapter using a case study.

In a formal procedure for LQC design, we have to solve a form of matrix Riccati equation by iterative procedure and then solve for feedback gains. Since we are interested only in application, we prefer to use a MATLAB command for obtaining solutions for gain values. This is illustrated below.

9.9.5.1 Use of MATLAB in LQC Design

What we have presented above can be easily achieved by the "dlqr" command in MATLAB in the format [K,P,E] = dlqr(G,H,Q,R). This command yields gain vector K, P matrix, and closed-loop system eigenvalues E. Here, P matrix is a positive definite matrix, which is an intermediate result in design. However, we are only interested in the closed-loop system eigenvalues E, and the feedback gain vector K.

EXAMPLE 9.20

Consider a plant given in Example 9.19. It is represented by a state model with parameters

$$G = \begin{bmatrix} 2.8 & -2.51 & 0.728 \\ 1.0 & 0 & 0 \\ 0 & 1.0 & 0 \end{bmatrix} \quad H = \begin{bmatrix} 1 \\ 0 \\ 0 \end{bmatrix}$$

$$C = \begin{bmatrix} 1 & -1.7 & 0.6 \end{bmatrix} \quad D = 0 \tag{9.136}$$

We have stated earlier that the open-loop plant is an unstable and nonminimum phase system, but it is completely state controllable and state observable. Our objectives in this example are listed as follows:

a. Design a steady-state optimal regulator to minimize the cost function

$$J = +\frac{1}{2} \sum_{k=0}^{\infty} \left[X^*(k)QX(k) + U^*(k)RU(k) \right] \tag{9.137}$$

```
G=[2.8 -2.51 0.728;1.0 0 0;0 1.0 0];
H=[1;0;0];
C=[1 -1.7 0.6];
D=0;
% Define desired weights
Q=[10 0 0;0 5 0;0 0 5];
R=10;
[K,P,E]=dlqr(G,H,Q,R);

Result:

K =      1.9176    -2.0732     0.6466

E =
    0.2834 + 0.4214i
    0.2834 - 0.4214i
    0.3157
```

FIGURE 9.26   Using "dlqr" command.

where

$$Q = \begin{bmatrix} 10 & 0 & 0 \\ 0 & 5 & 0 \\ 0 & 0 & 5 \end{bmatrix} \quad \text{and} \quad R = 10. \tag{9.138}$$

    b. Check the closed-loop stability.
    c. Simulate the response of the controlled system by MATLAB.

First, we will use the MATLAB command "dlqr" to obtain the feedback gains, which will optimize the linear quadratic criteria and the closed-loop eigenvalues to check the stability.

Later, we will use a simulation diagram and a MATLAB program to check the performance of the controller. The MATLAB codes for controller design and simulation diagrams are shown in Figures 9.26 and 9.27, respectively. The code results in gain values and the closed-loop poles.

The MATLAB program to drive simulation is listed in Figure 9.28, and the results of the LQC-based state feedback are shown in Figure 9.29. We note that because of the optimization process, system response is faster than the pole placement controller performance shown in Figure 9.25. However, a better choice of the closed-loop poles may compare favorably with LQC response.

9.9.6   A SIMPLE SERVO CONTROLLER

In Sections 9.9.4 and 9.9.5, we saw cases where state variables are regulated against disturbances using state feedback. In robotics, there are instances where the output needs to track a reference input. Such cases are called servo control. A typical system with only one reference is shown in Figure 9.30.

The objective is to make $y(k)$ to track $y_r(k)$. We see that there is the usual state feedback for regulation and additional input $k_r y_r(k)$, which is called the feedforward input. Since the state feedback would have changed the gain of the system, we need

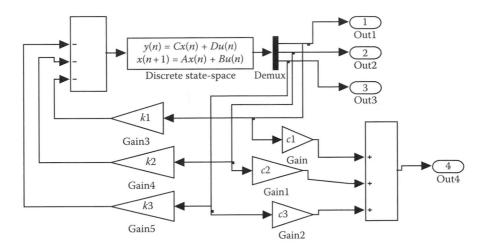

FIGURE 9.27   The Simulink diagram for LQC-based state feedback regulation.

to have an adjustable gain k_r to make the overall transmission gain from $y_r(k)$ to $y(k)$ unity so that accurate tracking can be achieved (Ogata 1995). The discrete state equation is given by

$$X(k + 1) = G\,X(k) + H\,u(k)$$
$$y(k) = C\,X(k)$$

(9.139)

Owing to the presence of the feedforward term, the manipulated variable is written with an additional term as

$$u(k) = -K\,X(k) + k_r\,y(k)$$
$$K = [k_1\ k_2 \cdots k_n]$$

(9.140)

Ignoring the feedforward term, it is straightforward to determine the feedback gain vector K, using one of the methods described earlier. Since K has been found, the only other unknown to be evaluated is the feedforward gain k_r. One popular method to find k_r is to find the pulse transfer function $(Y(z)/Y_r(z)) = T(z)$. By applying the final value theorem in Z-domain to $Y(z) = T(z)Y_r(z)$, an expression for $y(\infty)$ in terms of k_r can be written since $T(z)$ involves k_r. By equating the expression for final value $y(\infty)$ to y_r, k_r can be evaluated. However, the evaluation of $T(z)$ involves the inversion of nth-order matrix in terms of variable z. When the order is 3 or less, this is quite trivial.

Special case: We are interested in a class of systems where we need not use the above method, which involves the inversion of an n-dimensional matrix in terms of z. Let us assume that the plant has the following characteristics:

1. The output involves only one of the state variables, say, $x_1(k)$, which implies that $C = [1\ 0\cdots0]$, and hence

$$y(k) = x_1(k)$$

(9.141)

```
%Discrete simulation of LQC design and response
G=[2.8 -2.51 0.728;1 0 0;0 1 0];
H=[1;0;0];
%We want all states for feedback
Cx=[1 0 0;0 1 0;0 0 1];
Dx=[0;0;0];
C=[1 -1.7 0.6];
c1=C(1);
c2=C(2);
c3=C(3);
K=[1.9176    -2.0732    0.6466];
k1=K(1);
k2=K(2);
k3=K(3);
x0=[+3 0 0];
imax=50;
ymax=5;
[k,x,Out1,Out2,Out3,Out4]=sim('figure927mdl');

figure(1)
subplot(4,1,1)
plot(Out1);
v=[0 imax -ymax ymax];
axis(v);
grid;
title('Output and states in LQC feedback control');
ylabel('state x1')

subplot(4,1,2)
plot(Out2);
v=[0 imax -ymax ymax];
axis(v);
grid;
ylabel('state x2')

subplot(4,1,3)
plot(Out3);
v=[0 imax -ymax +ymax];
axis(v);
grid;
ylabel('state x3')

subplot(4,1,4)
plot(Out4);
v=[0 imax -ymax ymax];
axis(v);
grid;
xlabel('k, sample count')
ylabel('Output y')
```

FIGURE 9.28 The MATLAB program that drives the Simulink for LQC-based state feedback.

2. At steady state, all the state variables have to go to zero with the exception of $x_1(k)$. This implies that

$$x_2(\infty) = x_3(\infty) = \cdots = x_n(\infty) = 0 \qquad (9.142)$$

3. At steady state, the manipulated variable can be zero. Hence

$$u(\infty) = 0 \qquad (9.143)$$

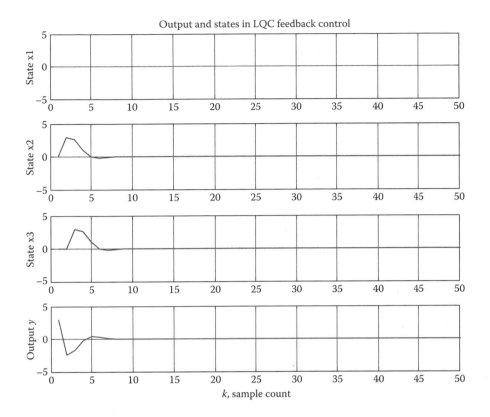

FIGURE 9.29 The output and state response of the LQC-based state feedback control.

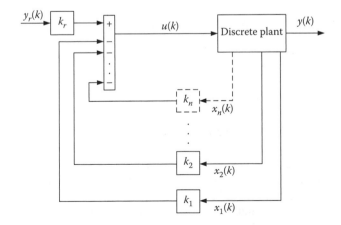

FIGURE 9.30 A servo system.

There are a few systems that satisfy the above assumptions. For example, a position control servo using a DC motor will satisfy the above conditions

Let us rewrite Equation 9.140 as

$$u(k) = k_r y_r(k) - k_1 x_1(k) - k_2 x_2(k) \cdots - k_n x_n(k) \qquad (9.144)$$

For $k = \infty$, using Equations 9.142 and 9.143 (assumptions 2 and 3), Equation 9.144 becomes

$$u(\infty) = k_r y_r(\infty) - k_1 x_1(\infty) = 0$$

From Equation 9.141 (assumption 1)

$$k_r y_r(\infty) - k_1 y(\infty) = 0$$

Since at steady state $y = y_r$

$$k_r = k_1$$

In general, if $y(k) = x_i(k)$

$$k_r = k_i \qquad (9.145)$$

We can redraw Figure 9.30 as shown in Figure 9.31. We will see such an example in the following chapter. However, this system will not work for all systems that do not satisfy the assumptions stated. Otherwise, we need to apply final value theorem to $Y(z) = T(z)Y_r(z)$ to find $y(\infty)$ and proceed as described earlier. Of course, there are many other sophisticated methods of servo control (Astrom and Wittenmark 1990; Ogata 1995).

9.10   TYPICAL HARDWARE IMPLEMENTATION OF CONTROLLERS

Since hardware changes fast, we will show only a generic setup in Figure 9.32, which is popular in the literature, avoiding hardware details. In the figure, the load can be the weight of a mobile robot or the weight that a robot arm is lifting.

Any manufacturer of a development platform provides a hardware setup consisting of an embedded system and provides a tool set for development, and the code is developed using a PC. The computational hardware used to be microcontrollers in the past, but the trend is rapidly changing. They were replaced by processors such as PIC systems. Recently, there have been many manufacturers that provide well-packed DSP processors, as embedded systems complete with digital and analog I/Os as well as good communication interface. The role of the PC ends at the completion of program development, and the system becomes standalone without the need for a PC. Once developed and downloaded, the control algorithm is executed by the processor. Sampling time is decided by the processor interrupts in software. The processor periodically takes in required feedback signals and computes a control signal

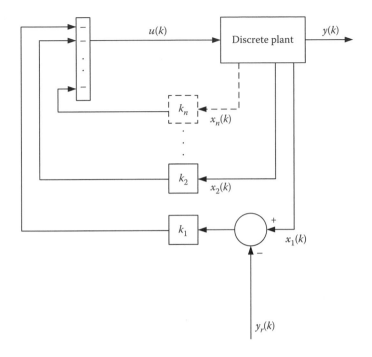

FIGURE 9.31 Servo control for a special case using reference as offset.

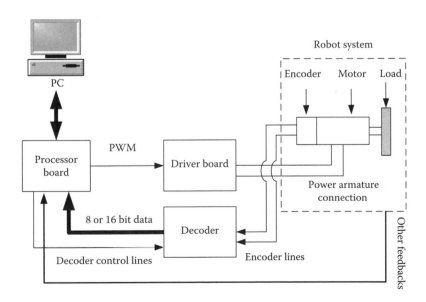

FIGURE 9.32 Development setup for a robot control system design.

based on them. In the block diagram above, the control signal manipulates PWM to the motor driver board. In the beginning of the program development, the robot may be operated along with the PC link for debugging purposes. Otherwise, for normal operation, especially for mobile robots, the PC link will not be there. The features of the block diagram shown in Figure 9.32 will be found in all the robots that will be presented in the following chapter on our case studies.

9.11 CONCLUSION

In this chapter, the basic ideas necessary to understand computer-controlled systems have been presented. This by no means is a complete treatise on digital control, but only a starting point. Most of the time, stabilizing a robotic structure needs some feedback. For example, a pole-balancing robot is unstable by its very nature. To stabilize it, we have provided some information on pole placement design. If we need to achieve speed and stability in response, optimal controllers will be very useful. If a robot arm has to be moved, then a servo control would be necessary. We have not gone much deeper into optimization or pole placement designs beyond what is necessary. However, we have provided enough material as starting point for a robot designer so that he/she can get started right way without spending too much time on learning the control aspects of the design. We took this approach because, as such, control theory is a deep and wide subject, and it takes a lifetime to master it. Furthermore, robots being a multidisciplinary subject, it requires knowledge in many other fields such as embedded systems, instrumentation, and actuation to name a few. To design and implement advanced controllers, further reading is recommended. The material presented in this chapter is satisfactory to start designing a robot controller. However, the references listed in this chapter are recommended for further reading.

REFERENCES

Astrom, K.J. and Kanniah, J. 1993. A fast adaptive controller for motion control. *Proceedings of IEEE Asia-Pacific Workshop on Advances in Motion Control*. Singapore, 63–68.

Astrom, K.J. and Wittenmark, B. 1990. *Computer Controlled Systems: Theory and Design*. 2nd Edition. Englewood Cliffs, NJ: Prentice-Hall.

Astrom, K.J., Carlsson, A., and Kanniah, J. 1994. A flexible system for adaptive motion control. *Journal of Mechatronics* 4: 99–112.

Cadzow, J.A. 1973. *Discrete-Time Systems: An Introduction with Interdisciplinary Applications*. Englewood Cliffs, NJ: Prentice-Hall.

Cavallo, A., Setola, R., and Vasca, F. 1996. *Using MATLAB, Simulink, and Control System Toolbox: A Practical Approach*. Hertfordshire, UK: Prentice Hall Europe.

Nagrath, I.J. and Gopal. M. 1996. *Control System Engineering*. 2nd Edition. New Delhi: New Age International (P) Limited Publishers.

Ogata, K. 1994. *Designing Linear Control Systems with MATLAB*. Englewood Cliffs, NJ: Prentice-Hall.

Ogata, K. 1995. *Discrete-Time Control Systems*. 2nd Edition. Englewood Cliffs, NJ: Prentice-Hall.

10 Case Study with Pole-Balancing and Wall-Climbing Robots

10.1 INTRODUCTION

In earlier chapters, we went through various components of knowledge needed to understand the basics of robot control. A study of robotics will not be complete unless we go through a design process and encounter real-world problems. In this chapter, we intend to give an overview of the steps involved in designing and implementing a workable robot for robotic games. However, there are other stipulations one has to take into account as listed below:

1. *Function of the robot*: This simply involves some knowledge of what the robot is expected to do. Of course, it can be obvious from the name itself. For example, the name "wall-climbing robot" (WCR) indicates what that robot is supposed to do. In some cases, the functions may not be so obvious. In summary, specific functions of that robot need to be defined.
2. *Specifications*: Beyond functions, a robot will have specifications for performance such as speed, load, and so on. Furthermore, in some robotic game events, the weight and dimensions of robots are limited by the game rules. The designer has to take note of these factors as well.
3. *Conditions of operation*: For example, it may be stipulated that the WCR should not use magnets, or it should not use nonelectrical drives. Or, in some cases, the power requirements may be limited, such as the robot cannot use internal combustion engines for locomotion.

It is impossible to go through the design process of every robotic game. However, we intend to describe the design steps of a few game robots so that the reader may be able to appreciate how theories are applied to achieve what is expected of the robots, along with component and material selection. We will study the design processes of two types of game robots, namely the pole-balancing robot (PBR) and the WCR, which compete in the Singapore Robotic Games (SRG 2012).

The PBR is expected to operate on a horizontal platform 3 m long and 1 m wide, moving from one end to other while balancing the free-falling pole. The first thing that the robot needs to do is to balance the pole for a given duration of time when started. After completing this task, it should move to the other side of the platform and then return to the starting point. It can repeat this as many times as possible within the permitted time. At the end, it returns to the starting point and stays there balancing the pole for another specific period of time.

The WCR game is held using a structure that consists of a horizontal surface, a vertical surface, and a ceiling, all of which are nonmagnetic. The WCR is expected to start on the horizontal part of the surface, climb the vertical surface, and move under the ceiling. The robot designed for this competition must be autonomous. Now, let us discuss design steps.

10.2 POLE-BALANCING ROBOT

As the name implies, the main challenge in this case is to design a motorized vehicle that supports an inverted pendulum using a pivot joint with one degree of freedom so that the pole is able to fall freely along the direction of motion of the robotic vehicle. The vehicle should keep the pole in the vertical position, without falling off and at the same time move along a straight line up and down. The basic system is shown with a block diagram in Figure 10.1.

In this particular game, we can see that the challenge is to design an appropriate control strategy to achieve the goal. We are leaving all the problems of instrumentation and power driving out of our discussion in this section to keep things simple. This game is particularly designed upon a well-known study problem in control theory. Hence, theoretical solutions for it have been discussed by many authors (Ogata 1990, 1995). The main thrust in these theoretical studies presented is force control of the vehicle using state variable feedback, where feedback gains are computed using either the closed-loop system pole placement technique or the linear quadratic control concept. Furthermore, it is cleverly assumed that the system is linear. If the

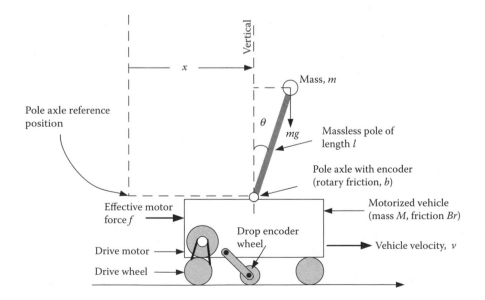

FIGURE 10.1   A block diagram of pole-balancing robot.

pole does not fall off too far from the vertical, say, not more than 7°, the linearity assumption holds good (since for small values of θ, sin $\theta \approx \theta$). Here, θ is the angle of deviation of the pole from the vertical axis. While the theoretical solutions for this problem are readily available, the practical implementation is not so straightforward. Apparently, the challenge put upon robot designers in this game is to put theory into practice. In this chapter, we would like to focus on implementation issues.

10.2.1 Mathematical Modeling

We believe that velocity control of the vehicle, as shown in Figure 10.1, is far easier to implement than force control because it involves torque control using motor back EMF and current measurements. Here, we develop our control strategy based on velocity control, even though force control models have been derived and given widely in control literature. Furthermore, those models ignore friction terms. We also need to include friction terms since friction plays a role in robot movement as well as in pole-swinging motion. We assume that mass, m, is attached to a rigid pole of length l. In reality, the pole weight may be uniformly distributed along a length of $2 \times l$. In that sense, the analysis is approximate and sufficient for a practical design. A more accurate analysis is quite involved, and it is not presented here.

A detailed analysis of inverted pendulum dynamics can be found in Ogata (1990). Here, we would like to include the friction terms in that analysis just by modifying the two proven equations. There are two kinds of friction: the friction in the motion of robotic vehicle and the rotary friction in the pole support system. The rotary friction force is depicted in Figure 10.2.

Let us define the following terms:

M = the mass of the vehicle in kg
m = the mass, attached to a weightless rigid of the pole in kg

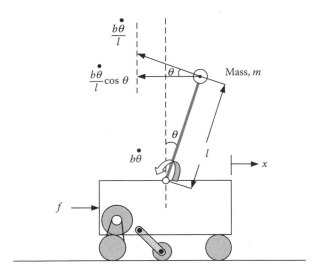

FIGURE 10.2 The rotary friction forces.

l = the length of rigid massless pole in meters
B_r = the linear friction coefficient of the vehicle motion in N/m/s
b = the rotational friction coefficient of the pole joint in Nm/rad/s
g = the acceleration due to gravity
x = the distance of the vehicle from reference in meters
f = the applied force in the horizontal direction in newtons

In the above definitions, we have used the terms B_r and b, which require some additional work to estimate. We have already explained how to estimate B_r in Chapter 6 on gear ratio design. We will see the method to estimate the value of rotational friction "b" in Section 10.2.9.

Referring to Figure 10.2, the frictional torque of the pole support joint is

$$t_f = b\frac{d\theta}{dt} \tag{10.1}$$

Then, the equivalent to linear friction force at the mass center of the pole can be written as

$$f_b = \frac{b}{l}\frac{d\theta}{dt} \tag{10.2}$$

Hence, the horizontal component will be

$$f_h = \frac{b}{l}\frac{d\theta}{dt}\cos\theta \tag{10.3}$$

The additional reaction due to linear friction of the robotic vehicle is given by

$$f_v = B_r\frac{dx}{dt} \tag{10.4}$$

Assuming that θ is small, and ignoring the additional friction terms, the linear force balance equation can be written as

$$(M + m)\frac{d^2x}{dt^2} + ml\frac{d^2\theta}{dt^2} = f \tag{10.5}$$

The additional friction terms are given in Equations 10.3 and 10.4. Since Equation 10.5 equates reaction forces on the left-hand side to the applied force on the right-hand side, we can add the additional frictional forces to the left-hand side and modify Equation 10.5 as given below:

$$(M + m)\frac{d^2x}{dt^2} + B_r\frac{dx}{dt} + ml\frac{d^2\theta}{dt^2} + \frac{b}{l}\frac{d\theta}{dt}\cos\theta = f \tag{10.6}$$

Again, since θ is small, we can write

$$(M + m)\frac{d^2x}{dt^2} + B_r\frac{dx}{dt} + ml\frac{d^2\theta}{dt^2} + \frac{b}{l}\frac{d\theta}{dt} = f \qquad (10.7)$$

This completes the motion dynamics in the horizontal direction. Next, we turn our attention to rotary motion dynamics of the pole. Ignoring the pole axle friction, the force balance equation for pole motion can be written as

$$m\frac{d^2x}{dt^2}\cos\theta + ml\frac{d^2\theta}{dt^2} = mg\sin\theta \qquad (10.8)$$

Multiplying Equation 10.8 by l, we get

$$m\frac{d^2x}{dt^2}l\cos\theta + ml^2\frac{d^2\theta}{dt^2} = mgl\sin\theta \qquad (10.9)$$

Now, we can identify the nature of the terms as follows. The moment of inertia of the pole with respect to axle is

$$I_m = ml^2 \qquad (10.10)$$

The torque due to pole mass m is

$$t_m = mgl\sin\theta \qquad (10.11)$$

The torque due to the pure linear motion of mass m is

$$t_x = m\frac{d^2x}{dt^2}l\cos\theta \qquad (10.12)$$

Since Equation 10.9 is a torque balance equation, we can add the axle friction torque given by Equation 10.1 to the left-hand side and write

$$m\frac{d^2x}{dt^2}l\cos\theta + ml^2\frac{d^2\theta}{dt^2} + b\frac{d\theta}{dt} = mgl\sin\theta$$

Using the assumption that θ is small, we can write

$$ml\frac{d^2x}{dt^2} + ml^2\frac{d^2\theta}{dt^2} + b\frac{d\theta}{dt} = mgl\theta$$

or

$$m\frac{d^2x}{dt^2} + ml\frac{d^2\theta}{dt^2} + \frac{b}{l}\frac{d\theta}{dt} = mg\theta \qquad (10.13)$$

Equations 10.7 and 10.13 include friction terms, and they are the two key equations that describe the dynamics of the PBR. Since robots have friction in wheel bearings, gears, motor, and pole axle, this approach better represents the actual robotic system.

Now, it is possible to form state equations and implement a control strategy to regulate the pole angle and distance using force f as the manipulated variable. However, at the beginning, we said that we wanted to eliminate force control, but we see that Equation 10.7 still has force, f, as an input. In the following sections, we show how we can avoid the force control and implement velocity control. Let us take the Laplace transform of both equations and obtain

$$[(M + m)s^2 + B_r s]X(s) + \left[(ml)s^2 + \left(\frac{b}{l}\right)s\right]\theta(s) = F(s) \qquad (10.14)$$

$$ms^2 X(s) + \left[(ml)s^2 + \left(\frac{b}{l}\right)s - mg\right]\theta(s) = 0 \qquad (10.15)$$

Let us peruse the second equation above and write a transfer ratio as

$$\frac{\theta(s)}{X(s)} = \frac{-ms^2}{mls^2 + (b/l)s - mg} \qquad (10.16)$$

We know that $v = (dx/dt)$; hence, $V(s) = sX(s)$. Equation 10.16 can be rewritten as

$$\frac{\theta(s)}{sX(s)} = \frac{\theta(s)}{V(s)} = \frac{-ms}{mls^2 + (b/l)s - mg} \qquad (10.17)$$

Let us now concentrate on Equation 10.14, which describes how the position depends on the force. We know that the effect of the second term consisting of $\theta(s)$ is very small since M is much larger than m and θ is quite small. Typically, robot mass M is a few kilograms, and the pole m is about 100 g. Hence, we want to make an approximation by ignoring this term and rewrite Equation 10.14 as

$$[(M + m)s^2 + B_r s]X(s) = F(s)$$
$$[(M + m)s + B_r]V(s) = F(s)$$

or

$$V(s) = \frac{1}{[(M + m)s + B_r]}F(s) \qquad (10.18)$$

where $F(s)$ represents the force. We can explore how we can represent force as a function of motor parameters and other electrical inputs. Before we proceed any further, let us define a few parameters and input quantities as follows:

E_s = the applied voltage to the motor in volts
E_b = the back EMF of the motor in volts
R_a = the armature resistance in ohms
K_b = the motor back EMF constant in volts/rad/s
K_t = the motor torque constant in Nm/A
N_g = the gear reduction ratio of the motor to the driving wheels
r_w = the radius of the drive wheel in meters
t_m = the torque of the motor in Nm
t_w = the torque of the drive wheel in Nm

The torque developed by the motor can be given as $t_m = K_t(E_s - E_b/R_a)$ by ignoring the effect of leakage inductance of the armature. This leakage is usually very small, and we can ignore it here. The torque on the drive wheel can be written as

$$t_w = N_g K_t \frac{E_s - E_b}{R_a} \tag{10.19}$$

The thrust on the wheel contact to push the vehicle is given by

$$f = \frac{t_w}{r_w} = N_g K_t \frac{E_s - E_b}{r_w R_a} = \frac{N_g K_t}{r_w R_a}(E_s - E_b) = T_f(E_s - E_b) \tag{10.20}$$

However, we know that

$$E_b = K_b \omega_m = K_b N_g \omega_w \tag{10.21}$$

since $\omega_m = N_g \omega_w$, where ω_m is the angular velocity of the motor, and ω_w is the angular velocity of the drive wheel.

Furthermore, the velocity

$$V = r_w \omega_w \tag{10.22}$$

Using Equation 10.22 in Equation 10.21, we get

$$E_b = \frac{K_b N_g V}{r_w} = \frac{K_b N_g}{r_w} V = H_b V \tag{10.23}$$

Using Equations 10.18, 10.20, and 10.23, a partial block diagram can be drawn as shown in Figure 10.3.

The closed-loop transfer function between applied voltage, E_s, and velocity, V, can be written as

$$\frac{V}{E_s} = \frac{1}{H_b + ((M + m)s + B_r)/T_f} = \frac{T_f}{H_b T_f + (M + m)s + B_r}$$

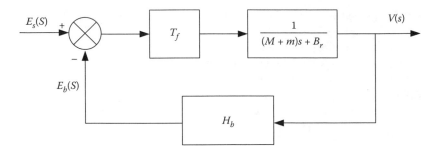

FIGURE 10.3   Partial block diagram from applied voltage to velocity.

Substituting for T_f and H_b

$$\frac{V}{E_s} = \frac{[K_t N_g r_w]}{[(M+m)R_a r_w^{\ 2}]s + [B_r R_a r_w^{\ 2} + K_t K_b N_g^{\ 2}]} \tag{10.24}$$

This describes the dynamics of the vehicle in response to the voltage applied to the armature of the drive motor. Obviously, this is a first-order system and not as complicated as it looks. We can set up a simple closed-loop control system to make the velocity respond to an applied reference velocity. The applied voltage E_s will be used as a manipulated variable. The reference velocity V_R will have to be compared to the actual velocity V and the error can be amplified using a gain G (in digital system it is just a multiplication instruction). We will use a pulse width modulation control (PWM) where amplified (multiplied) error signal will change the duty cycle of the PWM to handle the control of the motor speed. Such an implementation is shown in Figure 10.4.

A number of new variables are introduced in Figure 10.4, and they are

T_{pwm} = the PWM period in clock counts
δ = ON fraction
G = proportional gain

There are two kinds of connections possible to achieve the effect of the block diagram in Figure 10.4. In the type 1 connection shown in Figure 10.5, the direction-control

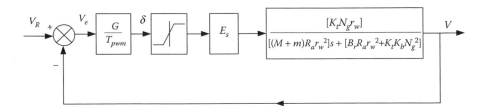

FIGURE 10.4   Velocity control implementation.

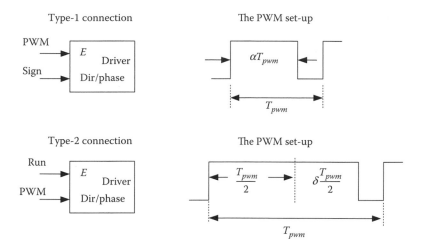

FIGURE 10.5 Two kinds of connection commonly used in H-bridge drivers.

pin of the H-bridge driver is made to respond to the sign of the manipulated variable computed by the control computer. The PWM signal goes to the enable pin of the H-bridge driver. Then, $\delta = 0$ duty cycle indicates zero push, and $\delta = 1$ indicates full push. However, the direction (or phase) signal input goes high for positive sign and low for negative sign. It implies that if the sign signal is positive (direction pin is high), PWM is 100% and motor thrust is maximum in the positive direction. On the other hand, if the sign is negative (direction pin is low) and PWM is 100%, the motor thrust is maximum in the negative direction. Obviously δ is limited within limits of ± 1. Note that the duty cycle cannot go negative. Hence, $\alpha = |\delta|$ is the fraction of the duty cycle, thereby providing an average voltage of αE_s to the motor. The direction pin decides how the voltage is connected to the motor, either forward or reverse. Thus, Figure 10.4 mathematically represents the effect of the type 1 connection shown in Figure 10.5 accurately.

In the type 2 connection shown in Figure 10.5, we connect the PWM signal to the direction input of the H-bridge driver and tie the enable pin high either by hardware or software. Then the PWM base is half of the PWM period and δ is within the limits of ± 1. When $\delta = 0$, the overall PWM is 50%. Since this goes to the direction pin, the motor current alternates equally between negative and positive. This happens at very high frequency in comparison to the time constant of the motor. Therefore, the motor does not move. When $\delta = 1$, the overall PWM is 100% and full forward voltage is applied to the motor resulting in full forward thrust. When $\delta = -1$, the overall PWM is 0% and full reverse voltage is applied to the motor resulting in full reverse thrust. These were discussed earlier in Chapter 6 on motors and drivers; however, for convenience, it is illustrated in Figure 10.5. Effectively, as δ changes from -1 to 0 to $+1$, motor power changes from negative maximum to zero and then toward positive maximum. Still, we see that Figure 10.4 mathematically represents the effect of the type 2 connection also (shown in Figure 10.5) accurately.

Usually, the PWM is implemented by introducing timer interrupts. These methods are very processor specific. Now, let us go back to the main aspects of our discussion, which is back to what is shown in Figure 10.4. Simplification will provide us with the overall closed-loop transfer function as below:

$$\frac{V}{V_R} = \frac{\left[GE_sK_tN_gr_w/T_{pwm}\right]}{[(M+m)R_ar_w^2]s + \left[B_rR_ar_w^2 + K_tK_bN_g^2 + (GE_sK_tN_gr_w/T_{pwm})\right]} \quad (10.25)$$

$$\frac{V}{V_R} = \frac{\left[GE_sK_tN_g/T_{pwm}R_ar_w\right]}{[M+m]s + \left[B_r + (K_tK_bN_g^2/R_ar_w^2) + (GE_sK_tN_g/T_{pwm}R_ar_w)\right]}$$

$$= \frac{\hat{G}}{(M+m)s + \hat{B}} \quad (10.26)$$

It may look complicated even though it is only a first-order system. Furthermore, it can also be written in time constant form as

$$\frac{V}{V_R} = \frac{(\hat{G}/\hat{B})}{[(M+m)/\hat{B}]s + 1} = \frac{\alpha}{Ts+1} \quad (10.27)$$

By comparing terms in Equations 10.26 and 10.27, we can compute the parameters of this system. Using Equations 10.17 and 10.27, the overall block diagram of this control system may be redrawn as shown in Figure 10.6.

In fact, an objective block diagram should also show the displacement, x, as an output as illustrated in Figure 10.7.

Sample calculations: Let us consider a sample case with the following given parameters:

$G = 1000$
$K_t = 0.033$ Nm/amp
$K_b = 0.033$ V/rad/s
$r_w = 0.03$ m
$R_a = 6$ ohms
$b = 0.01$ Nm/rad/s
$l = 0.5$ m
$T_{pwm} = 1000$ clock counts

FIGURE 10.6 Block diagram for pole angle control only.

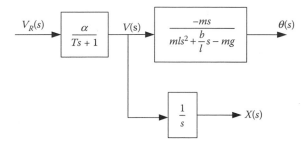

FIGURE 10.7   Overall block diagram for pole angle control and position control.

$N_g = 8$ (ratio)
$M = 2.5$ kg
$m = 0.13$ kg
$B_r = 2$ N/m/s
$E_s = 28$ V

Then, substituting the values in Equation 10.26, we obtain

$$\hat{G} = \frac{1000 \times 28 \times 0.033 \times 8}{1000 \times 6 \times 0.03} = 41.067 \tag{10.28}$$

$$\hat{B} = 2 + \frac{0.033 \times 0.033 \times 8^2}{6 \times 0.03^2} + \frac{1000 \times 28 \times 0.033 \times 8}{1000 \times 6 \times 0.03}$$
$$= 2 + 12.906666 + 41.06666 = 55.9733 \tag{10.29}$$

$$M + m = 2.5 + 0.13 = 2.63 \tag{10.30}$$

which yields

$$\alpha = \frac{41.0667}{55.9733} = 0.7337$$

$$T = \frac{2.63}{55.9733} = 0.047 \tag{10.31}$$

The above parameter calculations tacitly assume that there is no duty cycle saturation, even though saturation is provided in the original system, with a formulation shown in Figure 10.4. We imply that the ratio GV_e/T_{pwm} in Figure 10.4 is limited to −1 to +1.

Furthermore, we need to compute the pole angle dynamics to understand the system given in Figure 10.7. Then, we can write

$$\frac{\theta(s)}{V(s)} = \frac{-ms}{mls^2 + (b/l)s - mg} = \frac{-0.13s}{0.065s^2 + 0.02s - 1.2753} \tag{10.32}$$

Our primary objective is to control the position and the pole angle together. Let us downgrade our objective to control only pole angle θ while ignoring position x. Let us simply apply a proportional control and see how the closed-loop poles move as we change the gain.

10.2.2   Transfer Function for Pole Angle Control

To get some insight into the problem, we will use MATLAB® as a tool for further analysis. As a first step, we want to consider a system in which we intend to control θ only, instead of controlling x and θ at the same time, even though it was the original problem definition. Such a setup is shown in Figure 10.8. Note that the variable x is ignored.

Such a system can never achieve a steady-state value for θ. This exercise only serves the purpose of understanding the problem. We know that Equations 10.17 and 10.27 describe the cascaded system. We further take note that the values of α and T are dependent on G. These parameters are considered preset and are not disturbed, which are parts of the cascade controller that controls the vehicle velocity in response to reference velocity. This reference velocity plays the role of the manipulated variable. Let us try a simple program to find the open-loop poles and plot the root locus of the above system as the gain K changes. Even though we have done some sample calculations, we let MATLAB do all the calculations and do the pole trajectory plotting (Cavello et al. 1996). This gives us some insight into the nature of the problem. The code and the results are listed in Figure 10.9.

In the above code, variable "pden" shows that the overall open-loop system has one unstable root. The plot of the closed-loop system root locus is shown in Figure 10.10.

Apparently, as we close the loop, the system will not stabilize, irrespective of the magnitude or sign of K used in Figure 10.8. Here, we only make the point that in this particular case, simple feedback control will not work. We can try some other techniques, say, integral controller, PD controller, or PID controller. Even if we find a suitable controller that can regulate θ with $\theta_R = 0$, the robotic vehicle will drift. Our objective is to regulate θ and control x, with a single manipulated variable V_R. Such systems may be classified as single-input multi-output (SIMO) systems. We will pursue this objective in the following sections.

10.2.3   Pole-Balancing Robot State Model

Controller design for an inverted pendulum has been widely discussed and presented in the literature using force as the manipulated variable while ignoring friction terms

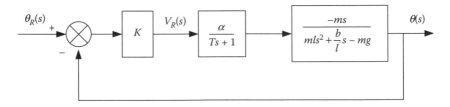

FIGURE 10.8   A simple θ control system with a proportional controller with gain K.

```
%This is the root-locus analysis for the pole angle control % only
ignoring the distance x controller parameters
G=1000;
Tpwm=1000;
%vehicle parameters
M=2.5;
Br=2;
m=0.13;
Ng=8;
rw=0.03;
%motor parameters
Kt=0.033;
Kb=0.033;
Ra=6.0;
Es=28;
Gm=G*Es*Kt*Ng/(Tpwm*Ra*rw)
Mm=M+m;
Bm=Br+Kt*Kb*Ng*Ng/(Ra*rw*rw)+Gm
Alpha=Gm/Bm;
T=Mm/Bm
Num_m=[0 Alpha]
Den_m=[T 1]
%
%pole angle dynamics parameters
%
l=0.99/2;
m=0.13;
b=0.01;
g=9.81;
Num_p=[-m 0]
Den_p=[m*l (b/l) -m*g]
%
%Overall transfer function velocity to theta
%
Num=conv(Num_m,Num_p)
Den=conv(Den_m,Den_p)
%
%Find the roots of numerator and denominator polynomials
%
znum=roots(Num)
pden=roots(Den)
%
%Plot the closed loop poles of a simple controller
%
rlocus(Num,Den)


Gm =   41.0667
Bm =   55.9733
Alpha=0.7337
T =      0.0470
Num_m =  [     0      0.7337 ]
Den_m =  [0.0470     1.0000]
Num_p =  [-0.1300            0]
Den_p =  [0.0644     0.0202    -1.2753]
Num =    [0    -0.0954            0]
Den =    [ 0.0030  0.0653   -0.0397    -1.2753]
znum =       0
pden =  -21.2826   -4.6115     4.2976
```

FIGURE 10.9 MATLAB code and result for computing the root locus of simple closed-loop system.

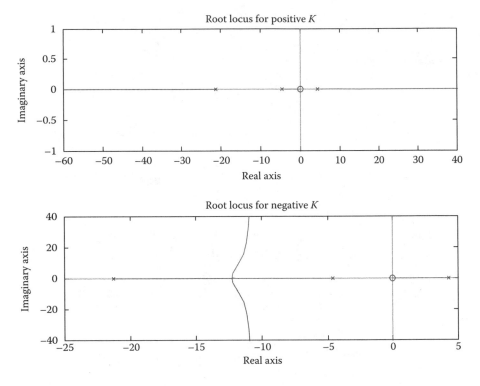

FIGURE 10.10 Closed-loop root locus plots for pure pole angle control for positive and negative gain values of K.

(Ogata 1990, 1995). The main difference in the presentation given here is that we include friction terms in the model, and we are using velocity reference as the manipulated variable. This applies to all controller designs described in the sections below. To proceed further, we need to write the state equations of the system. Let us define the states as follows:

$$x_1 = x$$
$$x_2 = \dot{x} = v = \dot{x}_1$$
$$x_3 = \theta$$
$$x_4 = \dot{\theta} = \dot{x}_3$$

(10.33)

We consider Equation 10.17 and do a trivial operation of cross-multiplying the terms

$$\frac{\theta(s)}{sX(s)} = \frac{\theta(s)}{V(s)} = \frac{-ms}{mls^2 + \dfrac{b}{l}s - mg}$$

$$mls^2\theta(s) + \frac{b}{l}s\theta(s) - mg\theta(s) = -msV(s) \tag{10.34}$$

By taking the inverse Laplace transform, we obtain

$$ml\frac{d^2\theta}{dt^2} + \frac{b}{l}\frac{d\theta}{dt} - mg\theta = -m\frac{dV}{dt} \tag{10.35}$$

Now, using Equation 10.33 in Equation 10.35, we get

$$ml\,\dot{x}_4 + \frac{b}{l}x_4 - mgx_3 = -m\,\dot{x}_2 \tag{10.36}$$

To form a state equation, we need to eliminate the dot term on the right-hand side of Equation 10.36. For this purpose, let us rewrite Equation 10.27 and process it as follows:

$$\frac{V}{V_R} = \frac{\alpha}{Ts + 1}$$

$$TsV(s) + V(s) = \alpha V_R(s)$$

By taking the inverse Laplace transform, we get

$$T\frac{dV}{dt} + V = \alpha V_R$$

$$T\,\dot{x}_2 = \alpha V_R - x_2$$

$$\dot{x}_2 = \alpha\frac{V_R}{T} - \frac{x_2}{T} \tag{10.37}$$

Using Equation 10.37 in Equation 10.36, we get

$$ml\,\dot{x}_4 + \frac{b}{l}x_4 - mgx_3 = -m\alpha\frac{V_R}{T} + m\frac{x_2}{T}$$

Now, we assemble the state equations:

$$\dot{x}_1 = x_2$$

$$\dot{x}_2 = -\frac{1}{T}x_2 + \frac{\alpha}{T}V_R$$

$$\dot{x}_3 = x_4$$

$$\dot{x}_4 = \frac{1}{Tl} x_2 + \frac{g}{l} x_3 - \frac{b}{ml^2} x_4 - \frac{\alpha}{Tl} V_R \tag{10.38}$$

The above equations can be brought together into a state equation as shown below:

$$\begin{bmatrix} \dot{x}_1 \\ \dot{x}_2 \\ \dot{x}_3 \\ \dot{x}_4 \end{bmatrix} = \begin{bmatrix} 0 & 1 & 0 & 0 \\ 0 & -\dfrac{1}{T} & 0 & 0 \\ 0 & 0 & 0 & 1 \\ 0 & \dfrac{1}{Tl} & \dfrac{g}{l} & -\dfrac{b}{ml^2} \end{bmatrix} \begin{bmatrix} x_1 \\ x_2 \\ x_3 \\ x_4 \end{bmatrix} + \begin{bmatrix} 0 \\ \dfrac{\alpha}{T} \\ 0 \\ -\dfrac{\alpha}{Tl} \end{bmatrix} V_R \tag{10.39}$$

This is of the general form

$$\dot{X} = A \cdot X + B \cdot u \tag{10.40}$$

where A is the state matrix and B is the input vector of dimensions 4×4 and 4×1, respectively. What we have done in this section can be summarized as follows:

1. We have derived a model where reference velocity V_R is the manipulated input to achieve the objectives of keeping the pole from falling and moving the vehicle according to command signal.
2. We have mentioned duty cycle-based chopper control as shown in Figure 10.4 in Section 10.2.1. It is important to note that the chopper duty cycle period would be far shorter than the time constants of the system and the sampling time to be used. This results in an equivalent cascade speed controller gain of α and time constant T, which can be considered as an analog system with the input V_R and the output V. Once proportional gain G is fixed, these two parameters above, α and T, remain fixed as it is evident from Equations 10.25 through 10.27.
3. We are just trying to understand the characteristics of the system by considering it as a continuous analog system. However, we have to bear in mind that no analog controller is implemented in such modern systems. At the end of this chapter, we will show how a digital controller can be implemented for this system.

10.2.4 STATE MODEL FOR THE POLE-BALANCING ROBOT FROM ROBOT AND MOTOR DATA

Let us derive the continuous state model from the above equations to process it further for controller design purposes. The following MATLAB code computes the continuous state model.

The resulting A and B matrices calculated with the code given in Figure 10.11 are important system matrices, which will be used hereafter.

```
%This is the code for system calculation and design of pole
%placement controller parameters
G=1000;
Tpwm=1000;
%vehicle parameters
M=2.5;
Br=2; %The friction coefficient
m=0.13;
Ng=8;
rw=0.03;
%motor parameters
Kt=0.033;
Kb=0.033;
Ra=6.0;
Es=28;
Gm=G*Es*Kt*Ng/(Tpwm*Ra*rw);
Mm=M+m;
Bm=Br+Kt*Kb*Ng*Ng/(Ra*rw*rw)+Gm;
Alpha=Gm/Bm
T=Mm/Bm
%parameters
l=0.99/2;
m=0.13;
b=0.01;
g=9.81;
A=[0 1 0 0;0 -1/T 0 0;0 0 0 1;0 1/(T*l) g/l -b/(m*l*l)]
Bt=[0 (Alpha/T) 0 (-Alpha/(l*T))];
B=Bt' % B implies state control matrix


System Results:

Alpha =
    0.7337
T =
    0.0470
A =

         0    1.0000         0         0
         0  -21.2826         0         0
         0         0         0    1.0000
         0   42.9952   19.8182   -0.3139


B =
         0
   15.6147
         0
  -31.5449
```

FIGURE 10.11   Plant model calculations.

10.2.5   POLE PLACEMENT CONTROLLER WITH SERVO INPUT USED AS OFFSET

First, let us look at the nature of the PBR. We have discussed in Chapter 9 (Section 9.9.6) about a class of systems where

  i. The output to be servo controlled involves only one state variable.
 ii. All other state variables go to zero at steady state.
iii. The system does not require a nonzero value of manipulated variable at steady state.

In the case of the PBR, the servo-controlled output is distance $x(k) = x_1(k)$ and all other states are only regulated. Hence, the first item is satisfied. At the beginning when there is no position command, and at the end of motion at steady state, the robot should stay still if there is no disturbance; hence, all states, except the output state, are zeros that satisfy the second item above. Since, at steady state, the robot is stationary, the velocity reference input has to go to zero, thus satisfying the third item. Thus, the PBR fits well into the description of the class of systems we discussed in Section 9.9.6. Then, we can use a control system as shown in Figure 9.31, which implies that we will try to move the robot just by introducing an offset to the state variable x, using a step function. We will not include any integrator in the system but just an offset. We also know that from the results we obtained with MATLAB code in Figure 10.9, there is one unstable open-loop pole. We want to achieve a system behavior such that the closed-loop system poles are where we want them, by using pole placement technique.

For the design, we normally start from the analog state model and from there obtain a discrete state model and finally proceed to design the controller. We convert the analog model to a digital state model with the sampling period of 9 ms. This is the actual sampling time used in our practical implementation. It was possible to implement such a sampling time using on-board DSP processor. We use MATLAB code to obtain the discrete state model matrices, G and H, from analog system matrices A and B. Matrices C and D do not change.

We discussed a few techniques for a discrete pole placement controller in an earlier chapter. However, we are not concerned with writing algorithms for such problems; we will simply use MATLAB command "$K = \text{place}(G,H,p)$," where G is the discrete system matrix, H is the control vector, and p is the vector consisting of the desired pole locations. We list the code assuming the values of A, B, C, and D matrices obtained in Section 10.2.4. The MATLAB code, and the results are listed in Figure 10.12. This code also calls a Simulink® model.

It is important to note that the closed-loop poles are just arbitrary for illustration purpose and are not thought through. The Simulink model of the closed-loop system that uses the newly designed controller is shown in Figure 10.13. The MATLAB command in the code that invokes the Simulink model is "`[k,x,Out1,Out2,Out3] =sim('figure1013mdl')`" where "`figure1013mdl`" is the file name of the Simulink model. This command takes the simulation model and runs it to get outputs Out1, Out2, and Out3 for plotting. When executed, simulation will plot the outputs of distance, pole angle, and velocity reference, which is the only manipulated variable.

The simulation result is shown in Figure 10.14.

At this point, by carefully examining the results, we can get some insights into the way this controller and the system functions as a whole. How is the robot made to move? It is interesting to see that at the starting instant, the pole falls forward and follows a certain pattern as shown in the "pole-angle" response in Figure 10.14. Because of the forward leaning of the pole, the robot is made to move forward to "catch" the pole. To understand this completely, we need to see how the controller achieves this action using the manipulated variable V_R. This can be seen with graph marked as V_{ref} manipulation in Figure 10.14. We see that the manipulated variable, speed reference, goes slightly negative. In response to this, the robot moves slightly backward; this causes the pole to fall forward and thus causes the chain of events.

```
A=[0 1 0 0;0 -21.286 0 0;0 0 0 1;0 42.9952 19.8182 -0.3139];
B=[0;15.6147;0;-31.5449];
C=[1 0 0 0];
D=[0;0;0;0];
st=0.009;
[G,H]=c2d(A,B,st);
%desired roots
droots=[0.96 0.97 0.98 0.99]
%Characteristic polynomial
K=place(G,H,droots)
k1=K(1);
k2=K(2);
k3=K(3);
k4=K(4);
Cx=[1 0 0 0;0 1 0 0; 0 0 1 0; 0 0 0 1];
Dx=[0;0;0;0];
x0=[0;0;0;0];
[k,x,Out1,Out2,Out3]=sim('figure1013mdl');
figure(1);
subplot(3,1,1);
plot(k,Out1)
title('Digital Servo just by inserting an offset into output)');
ylabel('position')
grid;
subplot(3,1,2);
plot(k,Out2)
ylabel('pole angle')
grid;
subplot(3,1,3);
plot(k,Out3)
xlabel('   k   ;            time = (k * 0.009) seconds')
ylabel('Vref manipulation')
grid;


Design Results

droots =

    0.9600    0.9700    0.9800    0.9900

K =

   -0.1301   -1.6066   -2.2751   -0.4931
```

FIGURE 10.12   MATLAB code and results for pole placement design.

We have seen the responses in Figure 10.14, and we would like to take a second look at the method we have used. In the above exercise, we did not make any effort to achieve servo control. We simply introduced the reference position as an offset in the distance measurement. This apparent "shortcut" will not always work for any arbitrary system. The response gives us further insight into why this method works here. The robot in its initial position as well as in its final position need not have any steady-state velocity and hence the steady-state value of the control input (velocity reference) is going to be zero as shown in Figure 10.14. At the end of the motion, the position minus its offset should become zero and none of the other states, such as dx/dt, θ, or $d\theta/dt$, can have a nonzero value, and this is confirmed by Figure 10.14. Since x and θ have stabilized, their derivatives cannot exist. Hence, at the steady state, all the feedback signals are zero, which makes the computed value of the manipulated variable zero, which is also confirmed in Figure 10.14. Since the steady-state velocity

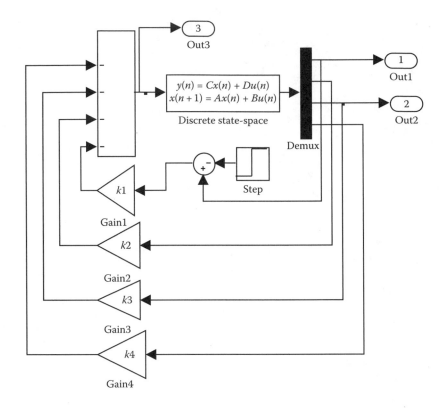

FIGURE 10.13 Simulation model of the PBR controlled by a pole placement controller.

reference can be zero, error integration is not needed in this case. Error integration is required only when a nonzero value of the manipulated variable is needed at steady state. Because of this, this method of just introducing an offset to position feedback works for this robot. In the above discussion, we have ignored disturbances, which are always taken care of.

We have just made use of the unique nature of the PBR and avoided the use of an integrator-based servo controller or any other sophisticated servo technique. Moreover, an integrator will, in general, slow down the response. In a competition environment, it is not desirable.

10.2.6 LQC Controller with Servo Input Used as Offset

The same model can also be controlled by the LQC-based controller. We have provided a MATLAB code to design and simulate the controller in Figure 10.15. The Simulink model used here is the same as shown in Figure 10.13. Note the diagonal matrices Q and R are chosen at first arbitrarily.

The response is given in Figure 10.16. Note that the response takes around 800 samples, and the maximum deviation of the pole angle is 0.07 radians and the velocity reference input is quite acceptable. The result can be influenced by the choice of

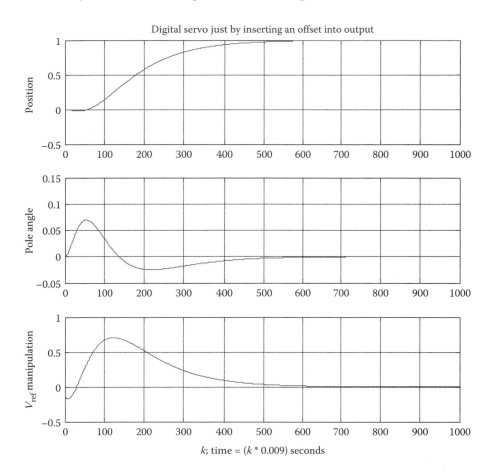

FIGURE 10.14   Responses for the simulation of robot for a step command (PPC design).

Q and R matrices. The weight given to position is 20, the weight given to pole angle is 200, and the value of R is 30.

10.2.6.1   Effect of a Change in Q Matrix

The response can be changed by choosing different weights for position and pole angle. Let us change the weight given to position as 100 and the weight assigned to pole angle as 10. Let us not change the value of R. The modified Q matrix in the MATLAB code becomes

```
Q = [100 0 0 0;0 10 0 0;0 0 10 0;0 0 0 10];
```

When this is used, replace Q in the MATLAB code in Figure 10.15, the design results obtained are shown in Figure 10.17.

The response obtained is shown in Figure 10.18. Notice that the position response has become faster. However, the pole deviation goes to 0.15 radians. The demand on reference velocity is actually higher than what we see in Figure 10.16.

```
A=[0 1 0 0;0 -21.286 0 0;0 0 0 1;0 42.9952 19.8182 -
0.3139];
B=[0;15.6147;0;-31.5449];
st=0.009;
[G,H]=c2d(A,B,st);
C=[1 0 0 0];
Cx=[1 0 0 0;0 1 0 0;0 0 1 0;0 0 0 1];
Dx=[0;0;0;0];
x0=[0;0;0;0];
Q=[20 0 0 0;0 10 0 0;0 0 200 0;0 0 0 10];
R=[30];
[KK,P,E]=dlqr(G,H,Q,R);
Gains=KK
EigenValues=E
k1=KK(1);
k2=KK(2);
k3=KK(3);
k4=KK(4);
%Get ready for state feedback by creating a fictitious Cx
%vector as 4th order unity matrix
[k,x,Out1,Out2,Out3]=sim('figure1013mdl');
figure(1);
subplot(3,1,1);
plot(k,Out1)
title('Servo control by adding offset to output(LQC)');
ylabel('position')
grid;
subplot(3,1,2);
plot(k,Out2)
ylabel('pole angle')
grid;
subplot(3,1,3);
plot(k,Out3)
ylabel('Vref manipulation')
xlabel(' k ;          time = (k * 0.009 ) seconds')
grid;

Design Results
Gains =

  -0.7584   -3.1588   -9.3780   -2.0379

EigenValues =

   0.7667

   0.9950

   0.9696

   0.9604
```

FIGURE 10.15   MATLAB code for LQC design and simulation with design results.

Competition environment: In game-playing robotics, speed of response is a consideration. In such a competition environment, to achieve the required speed of response, the weight given to the position has to be higher than the weight given to the pole angle. The price paid for this is that the pole angle deviation will be larger. Furthermore, there will be more demand on velocity reference, which is used as the manipulated input. This kind of explicit trade-off is possible in LQC design. Nevertheless, the weights must be chosen carefully. The designer needs to consider physical limits such as maximum thrust available and the slippage of the wheels

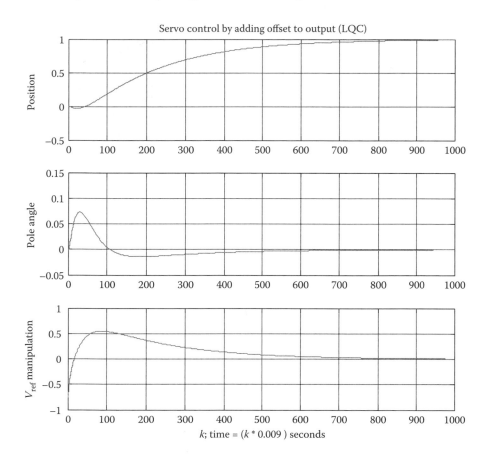

FIGURE 10.16   Response for LQC controller for the given Q and R values.

on the floor, and so on. Another consideration is that too large values of pole angle deviation will violate the linearity assumptions.

In pole placement design, the assigned closed-loop poles decide the response of the system modes. It is not possible to identify which pole controls which state variable. However, an experienced designer may still use pole placement design to achieve the desired results.

A practical constraint regarding step input: The step command for position is not a good idea in practice, even though it looks satisfactory in simulation. While responding to a step input, we have observed that real robots would rush forward and in the process the drive wheels would end up slipping on the floor. This will invariably cause instability. To avoid this problem, the position reference must be increased in small steps.

Servo control with integrator: What we have attempted here is to move the robot just by introducing an offset to the position feedback. We have not attempted servo control with an integrator. For more detailed information on servo implementation using an error integrator, refer to Ogata (1995).

```
Design Results

Gains =

   -1.6938    -3.6372    -10.2379    -2.2830

EigenValues =

   0.7659
   0.9671 + 0.0114i
   0.9671 - 0.0114i
   0.9890
```

FIGURE 10.17   Design results for the new Q matrix values.

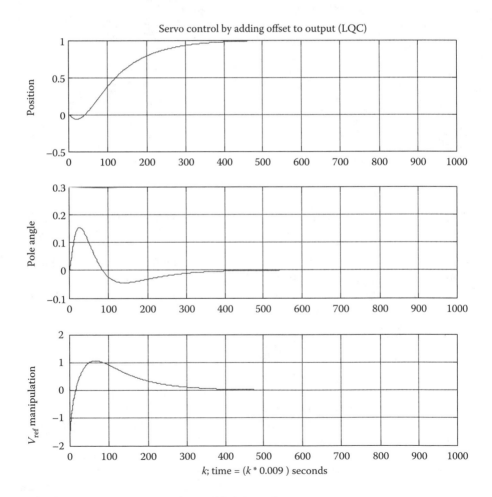

FIGURE 10.18   Response for the modified Q matrix.

10.2.7 IMPLEMENTATION OF THE POLE-BALANCING ROBOT CONTROLLER USING DSP PROCESSOR

It is possible to implement the controllers using many types of microcontrollers, microprocessors, or even from PCs. In this design, we employed a DSP processor. Many powerful DSP processors are available in the market, which have faster computational speeds than microcontrollers. Also, there are many vendors offering DSP-based motherboards with all the necessary accessories such as flash memory, RAM onboard, and communication means. We will discuss the general principles involved in such implementations using a DSP processor.

10.2.7.1 Hardware Setup

The system block diagram is shown in Figure 10.19. We have seen earlier that any digital controller will have the following hardware parts around the plant to be controlled:

1. The processor with program memory and data memory
2. The data acquisition system, which is used to collect data from the plant
3. The controller output connected to a power amplifier
4. The driver system, which provides the power to the motors

In many recent products, such hardware units may be integrated and the distinction between them may get fuzzy.

Figure 10.19 shows the robot and its control system. Note that the robot has a motor driving a wheel through a gear mechanism. The power for the drive motor

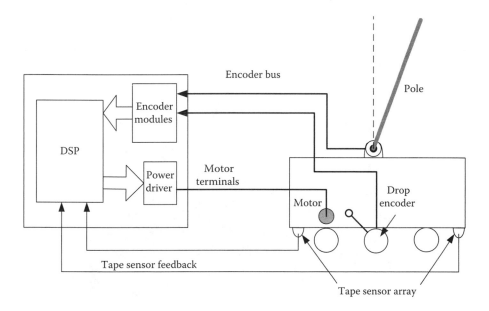

FIGURE 10.19 Simplified setup of a pole-balancing robot control system.

comes from the driver IC. It is necessary to make two kinds of measurement pole angles and distance measurements. Pole angle measurement is done by the incremental encoder. In fact, the pole is mounted on the shaft of the encoder so that, as the pole swings, the encoder provides the angle of swing. The distance measurement is done by another encoder attached to a drop wheel. It will be useless to just make an encoder to cling to one of the wheels, since due to strong torque the driven wheels will invariably slip on the floor. Then the distance measurements will be erroneous. A better practice is to have drop-encoders attached to the body such that an additional free encoder wheel is always in contact with the support surface due to gravity, or by spring loading it with just enough force to ensure contact. The figure also shows the side view of sensor arrays mounted at the bottom of the base board to give feedback on the cross-tapes on the platform.

Even though we show the processor and encoder driver assembly in one box, it is rarely possible to do that. DSP boards used in this design are pretty standard and are available off the shelf. Here, we use a Texas Instruments ezDSP 2047 board from Spectrum Digital Inc., which has a 32-bit fixed point DSP processor, with enough data as well as program memories. For practical reasons, such boards rarely come with power drivers. This board provides two channels of encoder inputs and many digital and analog I/Os.

Some systems may need additional encoders as well. There may be ground sensors to see the demarcation tapes stuck on the robot platform. All these support systems are usually provided on a sisterboard, and they are designed in such away to be attached to the motherboard consisting of the main DSP processor. While the motherboards are bought off the shelf, the sisterboards need to be designed and fabricated, unless a suitable board with drivers and encoders can be found. The safe practice in robotics is to avoid ribbon cables and multiconnectors as much as possible to link these boards. Typically, the boards are piggybacked on each other for firmer mechanical grip and reliable signal flow between them. These kinds of detail need to be worked out depending on the main platform the designer chooses. A system designed and used by the authors in an earlier version of their robot is shown in Figure 10.20. The figure shows

FIGURE 10.20   Photograph of the single-degree-freedom PBR with pole detached.

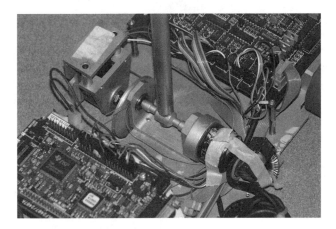

FIGURE 10.21   The drop-encoder and the pole support system.

the sensor arrays and the threaded pole supports bolt to which the pole can be fitted firmly, so that it can swing in the sagittal plane freely. Such a pole will have a single degree of freedom.

In this early design, note that there is a DSP motherboard, a driverboard linked by a ribbon cable contrary to our recommendation. Figure 10.21 shows the pole when fitted to the pole support system and a drop-encoder. The drop-encoder makes sure that the encoder is always in touch with the platform surface and reduces the error in distance measurement. There are ground LED sensors to locate the tapes, which are essential to synchronize and correct the distance measurements. Even though we use drop-encoders which are not "supposed" to slip, errors do occur due to high acceleration especially at the places where the robot changes direction. The encoders used were MTL MES 20-1000P. When fed to the 2016-quadrature decoders, they can give a resolution of 4000 counts per revolution. The driver IC used is the L6203 H-bridge driver, which has enough power capacity to drive the motor.

10.2.7.2   Software for the Robot

So far, we have seen the control methods to keep the pole close to the vertical position and move the robot at the same time. It is necessary for us to first enlist what we expect the robot to do, before we start discussing the program sequence. In some competitions, the first task of the robot is to balance the pole and stand in one region of the platform for a predefined period. This is called the static balancing part. Then, the robot is expected to move to the other side of the platform and then retrace the path back to the starting place. The robot may repeat this many times. Every time one up and down travel is completed, it is counted as one lap. A robot is expected to perform as many laps as possible within a given fixed period of time. To increase the challenge, if there is a curved path on the platform, then the robot should move along the curved path. All these functions should be fulfilled without any operator intervention. Hence, the robot should have the capability to keep track of the time. Usually, time keeping is performed by means of timer interrupts. These interrupts are set up at the beginning of the program, before the processor begins to start controlling the robot during initialization of the processor. We have mentioned the cascade control of the velocity with prefixed gain earlier, G. In our design, the basic interrupt period is set to 0.125 ms. The cascade control can be done inside the interrupt service routine or in the main loop. Because of this, the velocity control is transparent to the main loop of the program. We choose to change the manipulated variable every 9 ms in the main loop. We know that this is short enough to get smooth control of the robot motion and balancing. This is the robot sampling and control interval. Every 9 ms, the processor has to perform many operations including capturing data about the robot position, velocity, pole angle, and pole angular velocity. In addition, the robot calculates the feedback control value for the manipulated variable and outputs it as velocity reference. Every major sampling interval, which is 9 ms, the overall servo controller produces a velocity reference value. But, the cascade controller acts every basic sampling interval of 0.125 ms to obtain the velocity error and implements a duty cycle. This mimics the case of a continuous time cascade control. If cascade control cannot be done inside the interrupt service routine, it can be done even in the main loop every 9 ms without compromising the performance.

The first part of the general flowchart is shown in Figure 10.22. Let us summarize the events taking place at the initial part of the software. First, the processor is set up with necessary pins as inputs and output lines and the motors are disabled. We know that at this time that the motor power is switched off. In some cases, the enable line of the driver is directly controlled by the processor, and hence the software is used to disable the motor irrespective of the PWM value. Then, the timer interrupt is set up for creating interrupts every 0.125 ms. This value is decided empirically. The ezDSP 2047 is a complex processor with many timers. Some are dedicated for PWMs, and some are dedicated for encoder readings, and so on. We found that 0.125 ms is suitable, since we are using the same timer for generating interrupts as well as PWM signals. In any case, we have an instruction to keep the PWM neutral. What is neutral depends upon what connection we choose for the driver (refer to Figure 10.5). Then, we also have to set the reference velocity V_R to zero. Inside the interrupt routine, we need to maintain a count of these basic intervals (IntCount), and we set them to zero. After the completion of the initialization stage, we enable the interrupts. At this time, the robot operator would have placed the robot at the starting point and would be holding the pole vertically. The robot power switch has not been switched on yet.

Once the robot is powered on, the program enters the initial loop. At this point, the processor reads all the I/Os of pole angle and position encoders. It also sets the PWM neutral. During this part, the timer interrupts keep occurring and incrementing the "Intcount." Here, we wait for a count of 72 to occur since 72×0.125 ms $= 9$ ms, which is our control interval.

When this happens, the processor checks if the start button has been pressed. If it is not, then it moves on repeat the "idle" loop after resetting the "Intcount." If pressed, it implies that the motor has been switched on and the robot has to start the first static balancing part. Before we go any further, let us see what happens inside the interrupt service routine. This routine increments the interrupt count and also implements the cascade controller to make the robot move in accordance with the velocity reference provided by the controller. Even though this velocity reference changes every 9 ms, we may need to control the motor more often. This is shown

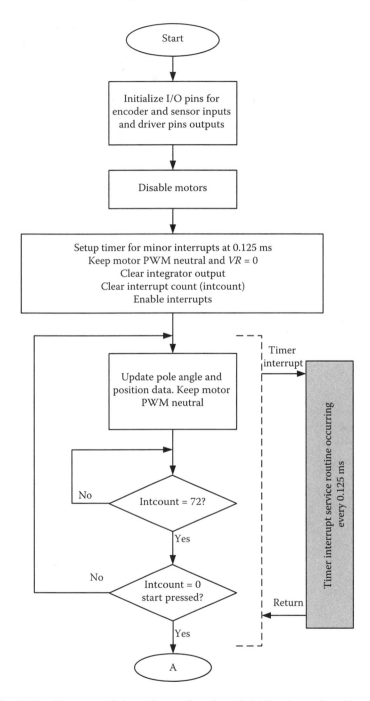

FIGURE 10.22 First part of the software flowchart (initialization and waiting for start button).

in Figure 10.23. However, we have given two versions of it. In version 1, the processors just increments the interrupt count and returns. It is a trivial action, but still it is needed to keep track of time. After all, the entire data processing cannot be completed in 0.125 ms. Since the system is implemented with one timer taking care of interrupts and PWM, the interrupt interval has to be small. With the addition of overheads of entering and leaving the service routine, all operations cannot be completed. If the processor is fast enough, some of the functions such as motor cascade control can be done inside the interrupt service routine. This is shown in version 2. In earlier models, we used the second version, but in complex cases which will be described latter, the version 1 was implemented. Then, one can conclude that in version 1, motor cascade control must be executed in the main loop and in version 2, that part can be left out of the main loop.

Now, we are ready to discuss the next part. We know that the robot has a few different functions. We may arbitrarily divide them as stages. We introduce a variable "stage" for convenience. Stage = 1 indicates that robot is doing static balancing keeping the pole balanced close to the vertical position while staying within the first boundary lines for the prescribed duration of time. During this time, the robot should not move beyond the demarcation of start position. Then, it becomes necessary to balance the pole by mildly moving the base and keep track of time to find when to start moving by switching the stage to 2. Stage = 2 indicates that robot is moving to the other side of the platform beyond the second boundary line. Stage = 3 indicates that the robot is returning to reach the starting region, and finally Stage = 4 indicates that

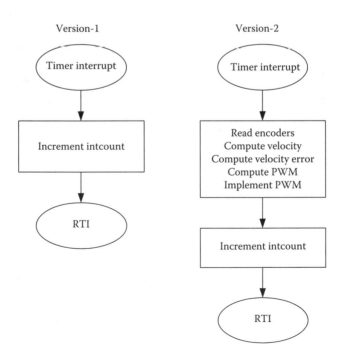

FIGURE 10.23   Interrupt service routine for cascade control.

the robot is doing final static balancing after repeating stage 2 and stage 3 as many times as possible, within the allocated time. We have terminated the flowchart at point A in Figure 10.22. We continue that in Figure 10.24. After we enter the loop at "A," we set stage = 1 and also we introduce a "LoopCnt" and set it to 0. The purpose is to keep a track of time. The next part of the software manages the robot through the stages while also changing stage numbers as the robot performance progresses.

The "Master Controller" box of Figure 10.24 is shown in detail in Figure 10.25.

The process of stage manipulation and motion management are done in an integrated manner starting from point "A" until the "Master Controller" box. Through this section, timer interrupts are active. Interrupts occur every 0.125 ms and IntCount is incremented in the interrupt service routine. In the main loop, after performing the Master Controller's action, the processor waits for "IntCount" to reach 72, which

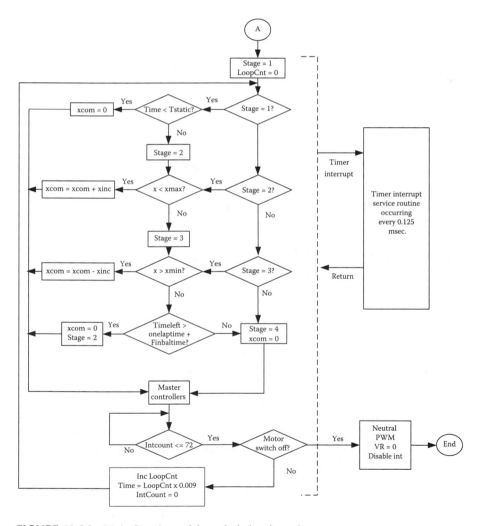

FIGURE 10.24 Main flowchart of the pole-balancing robot.

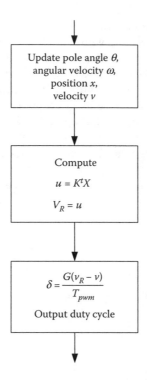

FIGURE 10.25 "Master Controllers" shown in Figure 10.24.

is equivalent to 9 ms and then major looping is executed. This guarantees that the major control interval is strictly 9 ms, and this is the basic requirement of DDC. After 9 ms elapses, if the motor switch is still ON, "LoopCnt" is incremented and the elapsed time is calculated. Here, the IntCount is also reset for the next loop, after which the processor loops back to the top.

Now, we look into how the stages are manipulated. "xcom" is introduced to indicate the position command. At Stage = 1, initially the processor loops through the static balancing part, keeping track of time. The robot does not move since xcom = 0. When the time goes beyond Tstatic, then the stage is set to 2. After this, the xcom is slowly incremented in steps of "xinc." Then, the processor keeps executing second horizontal segments of the flowchart until x goes beyond xmax, which is the upper demarcation distance on the platform. At this point, the stage is set to 3 and xcom is slowly decremented in steps. The processor keeps executing the third horizontal segment of the flowchart until x goes below xmin, which is the lower demarcation at the starting point of the platform. When this happens, the software checks if there is enough time for one more up and down travel and final balancing time (Timeleft > One lap time + Finbaltime?). If the answer is yes, the stage is set to 2 again and one up travel will be executed, followed by Stage = 3 for one down travel. This sequence will continue until there is not enough time to complete one more lap and final balancing time. Obviously, the programmer should have a good idea of how long it will take the robot complete one up and down motion. When there is no enough time, the stage is set to 4. This makes the robot keep balancing at the starting point with the "xcom" set to zero. When the operator switches the motor enable off, PWM is set to neutral and interrupts are disabled. This marks the end of the run. The robot we discussed in detail thus far was implemented and tested successfully. A video showing the robot in action can be viewed at (PBR-Single Degree 2012). In the video, the PBR travels on a platform which has flat and mildly sloping surfaces. It needs to be mentioned that when the terrain is inclined, pole angle calculations are adjusted to find the actual inclination of the pole from vertical. Since we are using a cascade velocity controller, this mild slope (5.7°) does not affect the performance.

10.2.8 Two-Degree-Freedom Pole-Balancing Robot

The robot described earlier has only one degree of freedom. The pole is free to move forward or backward, since it is fixed to horizontal shaft-supported bearings on both sides. In an advanced game event, it was stipulated that the pole should have two degrees of freedom. A picture of the support mechanism is shown in Figure 10.26a.

When the pole can fall in front–back (sagittal) as well left–right (transverse) directions, the robot base should be able to move in both X and Y directions. In other words, robot mobility requires "omni-wheels" as shown in Figure 10.26b.

10.2.8.1 Control Philosophy

If we write and analyze the state equations for controllers, we will have a state equation with twice the dimension and things will get mathematically out of hand. If we assume that the dynamics of X and Y directions are decoupled, then they can be treated separately. The situation is described below where As are system matrices, Bs are control matrices, and u_x and u_y are manipulated variables in the x and y directions, respectively.

(a)

(b)

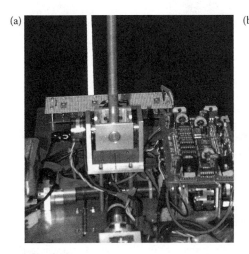
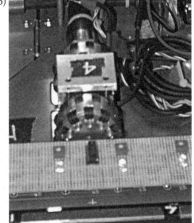

FIGURE 10.26 (a) The pole support for two degrees of freedom. (b) A typical omni-wheel.

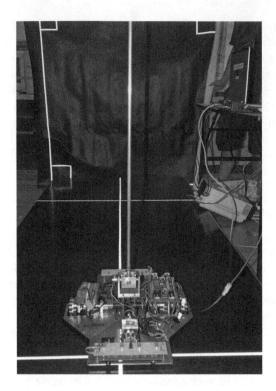

FIGURE 10.27   A two-degree-freedom pole-balancing robot.

$$\begin{bmatrix} \dot{X} \\ \dot{Y} \end{bmatrix} = \begin{bmatrix} A_{11} & A_{12} \\ A_{21} & A_{22} \end{bmatrix}\begin{bmatrix} X \\ Y \end{bmatrix} + \begin{bmatrix} B_{11} & B_{12} \\ B_{21} & B_{22} \end{bmatrix}\begin{bmatrix} u_x \\ u_y \end{bmatrix} \tag{10.41}$$

We can assume that the nondiagonal elements in A and B matrices are sparse or null and separate them as

$$\dot{X} = A_{11}\, X + B_{11}\, u_x$$
$$\dot{Y} = A_{22}\, Y + B_{22}\, u_y \tag{10.42}$$

Then, two controllers can be implemented independently. A picture of the PBR with two degrees of freedom is shown in Figure 10.27 and a video of it can be viewed at PBR-Two Degree (2012).

10.2.9   Estimation of Angular Friction Term b Used in PBR from Experiment

We have used friction coefficient, b, of the pole support system in Equation 10.17 and the robot friction coefficient, B_r, in Equation 10.18. Subsequently, the assumed values of those constants are used in many calculations, for example, in MATLAB programs given in Figures 10.9 and 10.11. We have illustrated the method to evaluate the robot friction coefficient B_r in Chapter 6. In this section, we illustrate a procedure to estimate b. The angular friction of a pole support system can be easily estimated by conducting a simple experiment and by treating it as a regular pendulum. Most analysis of pendulums does not take into account the friction term. For reasonable control accuracy, we need to have at least an idea of the order of magnitude of this friction. In robotic games, the limiting value of b is directly or indirectly specified.

First, let us undertake some modeling and analysis by referring to Figure 10.28. We reiterate our assumptions that the pole consists of a mass, m, attached to a rigid massless pole of length l. The unit of the angular friction term "b" is in Nm/rad/s.

Then, the frictional torque is given by

$$t_f = b \frac{d\theta}{dt} \tag{10.43}$$

If the force due to friction acting on the mass is F_f, then $lF_f = t_f = b(d\theta/dt)$. Obviously,

$$F_f = \frac{b}{l} \frac{d\theta}{dt} \tag{10.44}$$

Hence, the force balance equation can be written as

$$ml \frac{d^2\theta}{dt^2} + \frac{b}{l} \frac{d\theta}{dt} = -mg \sin\theta$$

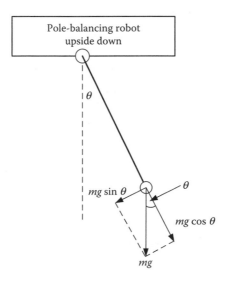

FIGURE 10.28 Pole-balancing robot kept upside down for pendulum experiment.

Since $\sin\theta \approx \theta$

$$ml\frac{d^2\theta}{dt^2} + \frac{b}{l}\frac{d\theta}{dt} + mg\theta = 0 \tag{10.45}$$

Taking the Laplace transform, we can write

$$ml\left[s^2\theta(s) - s\theta(0^-) - \frac{d\theta}{dt}(0^-)\right] + \frac{b}{l}\left[s\theta(s) - \theta(0^-)\right] + mg\theta(s) = 0$$

$$\left[mls^2 + \frac{b}{l}s + mg\right]\theta(s) = \left[mls + \frac{b}{l}\right]\theta(0^-) + ml\frac{d\theta}{dt}(0^-) \tag{10.46}$$

Let us assume that

$$\theta(0^-) = \theta_m \quad \text{and} \quad \frac{d\theta}{dt}(0^-) = 0 \tag{10.47}$$

Our analysis starting from the instant pendulum is let go at an angle of θ_m.

$$\theta(s) = \frac{mls + (b/l)}{mls^2 + (b/l)s + mg}\theta_m$$

$$\theta(s) = \frac{s + (b/ml^2)}{s^2 + (b/ml^2)s + (g/l)}\theta_m \tag{10.48}$$

Standard solutions for Equation 10.48 can be found in the control literature. Here, we derive the time domain solution from basics. For mathematical manipulation, we reorganize Equation 10.48 and obtain

$$\theta(s) = \left\{\frac{s + (b/2ml^2)}{\left(s + (b/2ml^2)\right)^2 + \left(\sqrt{(g/l) - (b^2/4m^2l^4)}\right)^2}\right.$$

$$\left. + \frac{(b/2ml^2)}{\left(s + (b/2ml^2)\right)^2 + \left(\sqrt{(g/l) - (b^2/4m^2l^4)}\right)^2}\right\}\theta_m \tag{10.49}$$

Let us define

$$a = \frac{b}{2ml^2}$$

$$\omega_n = \sqrt{\frac{g}{l}} \tag{10.50}$$

$$\omega = \sqrt{\frac{g}{l} - \frac{b^2}{4m^2l^4}}$$

Then, taking the inverse Laplace transform, we can write

$$\theta(t) = \left(e^{-at} \cos \omega t + \frac{a}{\omega} e^{-at} \sin \omega t \right) \theta_m \qquad (10.51)$$

Now, let us define an angle φ as in Figure 10.29.
Then, the solution can be rewritten as

$$\theta(t) = \theta_m \frac{\omega_n}{\omega} e^{-at} \sin(\omega t + \phi) \qquad (10.52)$$

$a = (b/2 \, ml^2)$ is the attenuation factor for the oscillation, which happens to be half of the coefficient of the s term in the denominator polynomial in Equation 10.48. In the control theory, this is a standard result. Obviously, at $t = 0$

$$\theta(t) = \theta(0) = \theta_m \qquad (10.53)$$

Now, let us consider Equation 10.52. While ignoring the actual wave form, we denote the amplitude by $A(t)$ and write

$$A(t) = \theta_m \frac{\omega_n}{\omega} e^{-at} \qquad (10.54)$$

It is very clear that the envelope of the sine wave decays according to an exponential law. At $t = t_1$, the amplitude of the envelope is A_1 and at $t = t_2$, the amplitude of the envelope is A_2. Then

$$\frac{A_1}{A_2} = \frac{\theta_m \frac{\omega_n}{\omega} e^{-at_1}}{\theta_m \frac{\omega_n}{\omega} e^{-at_2}} = e^{-a(t_1 - t_2)} = e^{a(t_2 - t_1)}$$

$$a = \frac{1}{(t_2 - t_1)} \ln \frac{A_1}{A_2} \qquad (10.55)$$

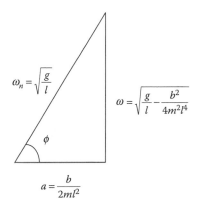

$$\omega_n = \sqrt{\frac{g}{l}}$$

$$\omega = \sqrt{\frac{g}{l} - \frac{b^2}{4m^2l^4}}$$

$$a = \frac{b}{2ml^2}$$

FIGURE 10.29 Definition of angle φ.

$$b = 2ml^2a = \frac{2ml^2}{(t_2 - t_1)}\ln\frac{A_1}{A_2} \tag{10.56}$$

EXAMPLE 10.1

Let us simulate a case with known b and verify.

Mass = 0.127 kg
Length = 0.99/2 m

From Equation 10.45

$$ml\frac{d^2\theta}{dt^2} + \frac{b}{l}\frac{d\theta}{dt} + mg\theta = 0$$

$$\frac{d^2\theta}{dt^2} = -\frac{b}{ml^2}\frac{d\theta}{dt} - \frac{g}{l}\theta$$

$$\frac{d^2\theta}{dt^2} = -C_1\frac{d\theta}{dt} - C_0\theta \tag{10.57}$$

Of course the angle θ can be solved by double integration, while taking care to provide appropriate initial conditions. A Simulink diagram for the friction estimation test is shown in Figure 10.30.

Let us assume that the friction coefficient b is 0.01 Nm/rad/s. Then, the initial angle, $\theta = 0.10$ rad and the $\omega = 0$. The MATLAB program for the simulation is listed in Figure 10.31.

$$b = 2ml^2a = \frac{2ml^2}{(t_2 - t_1)}\ln\frac{A_1}{A_2} \tag{10.58}$$

We notice from Figure 10.32 that at time

$$t_1 = 5s \quad A_1 = 0.045$$

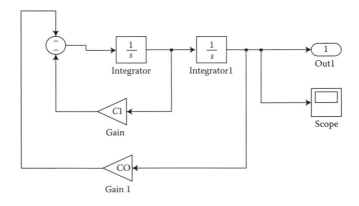

FIGURE 10.30   Simulink model for friction test example.

```
%The code for simulating friction test.
%parameters:
l=0.99/2;
m=0.13;
b=0.01;
g=9.81;
C1=b/(m*l*l);
C0=g/l;
theta0=0.10;
thetadot=0.0;
[t,x,Out1]=sim('figure1030mdl');
figure(1);
plot(t,Out1)
xlabel('Time, seconds')
ylabel('amplitude, radians')
grid;
```

FIGURE 10.31 MATLAB program for simulation of pole as a "normal" pendulum.

and

$$t_2 = 10s \quad A_2 = 0.021 \tag{10.59}$$

Hence

$$b = \frac{2 \times 0.13 \times 0.495^2}{(10 - 5)} \ln \frac{0.045}{0.021} = 0.00971 \tag{10.60}$$

This estimate is quite close to the assumed b value of 0.01 used in the simulation experiment.

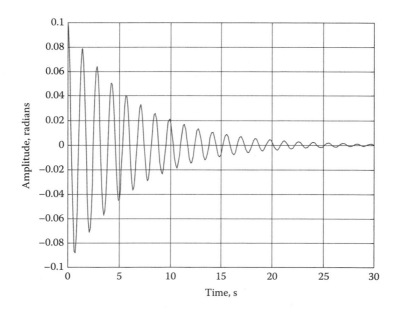

FIGURE 10.32 Simulation result.

10.3 WALL-CLIMBING ROBOTS

WCRs can be very useful in real world. The WCR competition stimulates techno-logical development in this field. While making a robot to climb a structured wall is quite straightforward, it may be difficult if the wall involved is unstructured. Furthermore, the recent trend in this competition is to restrict the clinging methods and in some games the walls are made of nonmagnetic material. Even in structured environments, nonmagnetic clinging may pose a problem. In this section, we will see two kinds of philosophies involved in making a WCR. Before we do that, let us see a sample structure used in wall-climbing competition concerned here. Figure 10.33 shows a diagram of the structure that is made of nonmagnetic material. The robot has to be placed in the start region. It is expected to move toward the wall, move up the wall, and travel under the ceiling beyond the finish line. Then, it is expected to travel back by retracing the path back to the start region. All along, a part of the robot should be in contact with the platform (that is flying or other means are not allowed). The robot should climb and travel by clinging to the surfaces. The travel-ing time from start to the finish and back to the start line is clocked, and the fastest robot is the winner.

10.3.1 FLIPPER WALL-CLIMBING ROBOT

The flipper WCR, shown in Figure 10.34, is similar to what we saw in the in the gear ratio design example discussed in Chapter 5, except that it has two main drive motors and one motor for cruising.

This flipper robot is much simpler to construct, program, and operate. It consists of two main arms, which can rotate through more than 180° using two driver motors. The robot also has four wheels placed on two axles, one in front and one in back at the one side of the arm linking the flipping arms, which is the main body on which the motherboard and driverboards are mounted. As shown in Figure 10.34, the robot has suction pads that are operated with a suction pump and valves. The suction pads provide the gripping needed for climbing. There can be two, four, or even six suction pads depending on the design. The robot has limit switches to indicate the limits of

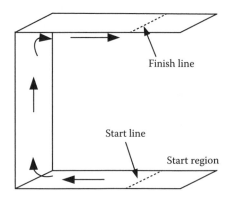

FIGURE 10.33 Wall-climbing robot competition platform.

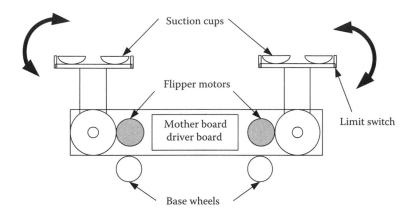

FIGURE 10.34 Diagram of the WCR-flipper robot.

motion and also marking the completion of the specific step. The wiring diagram of pneumatic system is shown in Figure 10.35.

Figure 10.35 shows a typical and simplest pneumatic wiring diagram used for the robot. The suction cups are named group A and group B. The pneumatic valves A and B are connected to the respective groups. These valves are electrically activated. If turned ON, then they do a straight connection from inlet to outlet (top to bottom), which is m to q and n to p as shown. If switched OFF, they cross-connect, by connecting m to p and n to q. The ON and OFF conditions vary from manufacturer to manufacturer. Off-the-shelf suction pumps are usually the diaphragm type and have

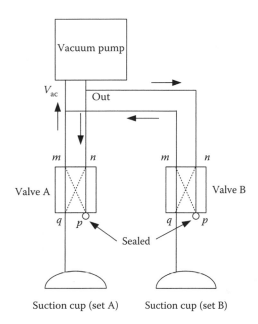

FIGURE 10.35 Pneumatic wiring diagram.

one inlet and one outlet. When switched on, they draw air through inlet and send air through the outlet. The above wiring is done in such a way that the pump is always ON. As can be seen in the wiring diagram, when valve A is ON, it is straight-connected and valve B is cross-connected. Also, note that the points marked p are sealed airtight on both valves. Obviously, the suction is applied to suction cups set A since path $m–p$ in valve B is sealed. Outgoing air from the pump has two paths. But, path $n–p$ of valve A is sealed. Hence, the outgoing air goes through path $n–q$ of valve B and exits via cups set B. So, while suction is applied to cups set A, the air is purged through suction cups set B. If switching is reversed, the suction cups set B will suck and get attached to the wall and suction cups set A will be purged away from the wall.

10.3.1.1 Overall System Configuration of Flipper WCR

Since the flipper WCR has three motors, sensors, vacuum pumps, and pneumatic valves, we would like to illustrate a simplified system diagram for the same in Figure 10.36.

The figure is self-explanatory. We can see that the three motors can be controlled using encoder feedback. The suction pumps can be turned on or off. Valves can be individually controlled to produce vacuum to stick the pad to the surface or purge the pad from surface. Pad sensors provide information regarding the respective pads sticking to the wall surface. The photograph of the entire robot is shown in Figure 10.37.

10.3.1.2 Control of Suction Pad Arms and Cruise Motor

The suction pads A and B are rotated using the flipper motors shown in Figure 10.34. Both joints are equipped with encoders. In most stages of climbing, typically, the motor controller implements a velocity control to rotate the joint of the suction

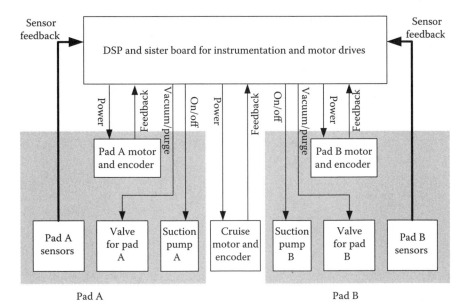

FIGURE 10.36 System configuration for flipper WCR.

FIGURE 10.37   Photograph of the flipping WCR.

pad, stuck to the wall, until the other suction pad touches the wall or ceiling, which is indicated by the sensors. When this happens, the controller switches to position control mode and the pad arms are held at the given positions briefly before the next move. Thus, depending on the need to hold position or rotate the suction pad arm with the given angular velocity, the software implements the either position control or velocity control, using a simple P-only controller. The software generates an appropriate velocity reference or a position reference to control the motors. For the cruise motor, only velocity control is implemented.

There are three motors in the robot, two for pad arm joints and one for cruise wheel. The controller for any particular motor is shown in Figure 10.38.

The processor can act as a position controller as well as a velocity controller. In the case of position control, a simple P-only control is computed as

$$u(k) = K_{p1} \times (\theta_r - \theta) \qquad (10.61)$$

where K_{p1} is the proportional gain for position control, θ_r is the reference position generated by the program, and θ is the position obtained from the encoder. In the case of velocity control, a P-only controller is computed as

$$u(k) = K_{p2} \times (\omega_r - \omega) \qquad (10.62)$$

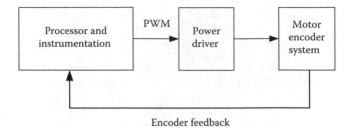

FIGURE 10.38 Control system for pad arms and cruise motors.

where K_{p2} is the proportional gain for velocity control, ω_r is the reference angular velocity generated by the program, and ω is the angular velocity calculated based on two consecutive encoder readings. There is no difference in wiring between position control and velocity control.

10.3.1.3   Operation Sequence of the Flipper WCR

We will explain the operation sequence of the robot using line diagrams given in Figure 10.39. Stages are indicated in circled numbers. In the beginning, the robot is placed on its base, so that it rests on the four wheels attached to the body. The arm A is turned horizontal in such a way that the pad holding the cups of set A is vertical as shown in stage 1. Then, power is applied to the cruise wheels fixed to the present bottom side of the frame, to make it go toward the wall. This motion continues until the limit switches in pad A sense the vertical wall as shown in stage 2. At this time, valve A is straight connected and the pump is also turned ON, and the set A cups get attached to the wall. Once the suction cups are attached to the wall, the motor of arm A is activated to turn the whole body counterclockwise so that the set B cups get to touch the vertical wall as shown in stage 3 of Figure 10.39. Subsequently, valve A is cross-connected and valve B is straight-connected. This causes the set B cups to get stuck to the wall, and the set A cups lose grip, ready to be purged and cranked away from wall. Now, joint B is activated so that the body turns again counterclockwise and at the same time joint A rotates to make pad A ready to face the wall, when it approaches the wall as shown in stage 4. At the end of stage 5, a complete flipping motion for climbing is accomplished. This process can go on until the upgoing arm cannot reach the vertical wall because the ceiling is obstructing. This indicates that the robot has reached stage 6. After how many upward steps this happens must be precalculated or found out by experiments.

At stage 6, both joints are activated, and they are all on the move. Here, we have not shown cruise wheels in the figure, since they can be on the either side of the body. Joints are not named, and we call them lower and upper joints. The speeds must be carefully programmed so that the upper suction pad's outer side touches the ceiling first. After sensing this, the upper joint is eased and the lower joint pushes further. This causes the upper pad to tilt until it completely touches the ceiling, activating all sensors. Appropriate time delays may need to be introduced. Now, suction is switched to the upper pads, and the lower pads will lose grip. Again, the upper joint

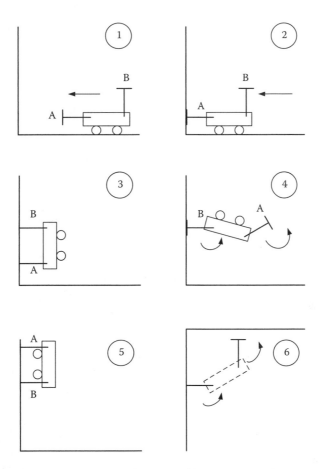

FIGURE 10.39   Stages of climbing action by the flipper wall-climbing robot.

is asserted and cranked counterclockwise, and the lower arm is flipped so that it is ready to touch the ceiling. Further motion is continued until the robot crosses the finish line. The steps are usually precalculated. The return journey is similar, except that on the way back, the ceiling to wall search has to be performed. At the end of downward motion, the situation will look as stage i shown in Figure 10.40. Now, the lower joint will be A, and pad A will be stuck to the wall. The upper joint is B and the set B cups will be purged. For a successful return cruising, this must be the situation. Once again, this is ensured by conducting experiments and by trial and error. The body length is adjusted so that at stage i, joint A is at the bottom. Also, as joint A is eased to let the cruise wheels touch the floor again by going through the motion shown in stage j, the clearance should be just sufficient and not too large for smooth landing. After landing, valve A is set to purge the set A cups. Now, the cruise wheels are powered to go back to the starting point. A video of the flipper WCR can be found in WCR-Flipper Type (2012). This video will also give a better understanding of the programming involved.

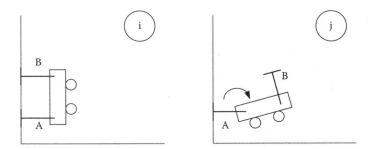

FIGURE 10.40   The transition from wall to base on return path.

10.3.2 DESIGN OF A WALL-CLIMBING ROBOT USING DYNAMIC SUCTION

The robot we describe here uses Bernoulli's equations to achieve the same objectives discussed for the WCR in the previous secton. Therefore, we will call it "Bernoulli's WCR." A line diagram of the robot is given in Figure 10.41.

Figure 10.41a shows the line diagram of the robot. It shows a box with two openings: one on the top and the other on the front side, fitted with cone-like structures extending inward. Both are fitted with very high power fans driven by brushless DC motors. These motors have been explained in Chapter 6 on actuators. At the bottom and top of the front side of the robot, we have driven wheels for moving on the horizontal part of the platform and climbing the front wall. On the horizontal part, the bottom wheels can do the job. But, while climbing up, both wheels are activated. While moving under the ceiling, only the top-driven wheel is sufficient and effective. On the front and top, we have four tiny caster wheels to keep the robot at a carefully chosen distance from the vertical and ceiling surfaces, respectively. The distances are chosen experimentally. The driven wheels are also placed to maintain the same clearance. Both driven wheels can be activated individually or together. They have encoders fitted to measure the distance traveled. Both front and top surfaces are fitted with limit switches as well. The system configuration is shown in Figure 10.42.

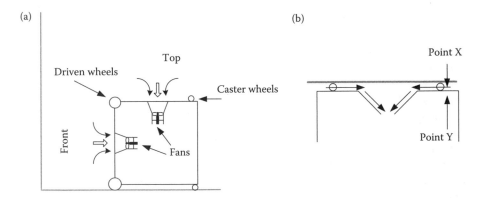

FIGURE 10.41   WCR using dynamic suction: (a) basic robot structure and (b) air flow.

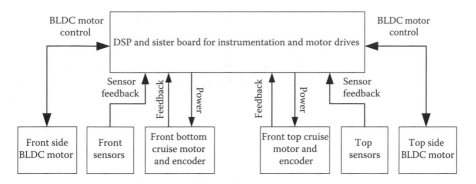

FIGURE 10.42 The system configuration for WCR using Bernoulli's principle.

10.3.2.1 Dynamic Suction Principle

Let us first describe the dynamic suction principle. For this purpose, we must study some illustrations before we embark on the concept of dynamic suction as applied to WCR.

The well-known Bernoulli's principle states that in the environment of steady flow of liquid, the increase in flow velocity causes a decrease in pressure and vice versa. This is shown in the case of a liquid flowing through a constriction in a pipe in Figure 10.43a and in the case of an airfoil in Figure 10.43b. These are the basic principles that we exploit in the design of a WCR design.

An experiment can be easily devised to demonstrate what happens when on one side of a foil or thin plate air flows fast and on the other side it is almost stationary. You will see that the foil experiences a force, which pushes it toward the side where there is high velocity airflow (Air-flow 2013).

The original equation was derived for the case of incompressible liquids by Bernoulli (Rajput 2011), and it is given below

$$\frac{p_y}{\rho} + z_y + \frac{v_y^2}{2g} = \frac{p_x}{\rho} + z_x + \frac{v_x^2}{2g} \tag{10.63}$$

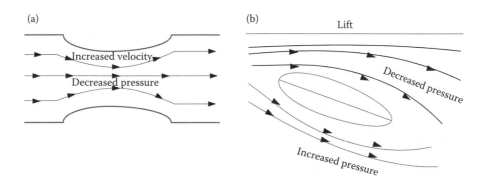

FIGURE 10.43 Bernoulli's principle in action.

where p_x and p_y are the static pressures at two points X and Y, respectively, and ρ is the density of the fluid at points X and Y. Since the fluid is incompressible, the density remains the same. z_x and z_y are the altitudes of points X and Y from an arbitrary reference and finally v_x and v_y are the velocities of air flow at points X and Y, respectively.

Bernoulli's equation is written in many forms. We have considered a type where dimensions of all terms in the equation are in meters of fluid head. For example, the unit of p_y/ρ is (kg/m²)(m³/kg) = m.

Similarly, the unit of $v_y^2/2g$ is (m²/s²)(s²/m) = m as well. Obviously, the unit of z_y is meter.

When we try to apply this theory to air, which is readily compressible, we face some discrepancies due to possible changes in densities. However, it is generally agreed that the density does not change considerably for speeds up to Mach 0.3. This translates to 100 m/s, which is hardly reached in our application so we can safely assume that density does not change.

Now coming back to our application, Figure 10.41b shows the air flow pattern between the ceiling and the top of the robot. We have marked two points X and Y. Let us apply Bernoulli's equation for these two points

$$p_y + \rho\, z_y + \rho \frac{v_y^2}{2g} = p_x + \rho\, z_x + \rho \frac{v_x^2}{2g} \tag{10.64}$$

where p_x and p_y are the static pressures at point X and Y, respectively. Let us rewrite Equation 10.64 as

$$p_y - p_x = \left(+\rho\, z_x - \rho\, z_y\right) + \rho\left(\frac{v_x^2}{2g} - \frac{v_y^2}{2g}\right) \tag{10.65}$$

Note that $v_x \gg v_y$ and in fact $v_y = 0$. Furthermore, we know that $z_x \approx z_y$. Hence, we can conclude that in the right-hand side of Equation 10.65, the second term dominates, and it is a large positive quantity, while the first term is insignificant. This indicates that there is a large difference in pressure from X to Y. That pushes the robot upward and keeps it stuck to the ceiling. The caster wheels maintain the appropriate gap. The same phenomenon applies to the front also when the front fan is activated in the vicinity of the front wall. The photograph of such a robot is shown in Figure 10.44.

10.3.2.2   Operation of the WCR Using Bernoulli's Principle

The operation of this robot is quite straightforward. First, the front cruise motor is activated. The robot moves forward from the base of the competition structure until front sensors indicate that the robot is pushing against the wall. At this point, the air flow fan motor for the front is activated, which makes the front side stick to the wall and both the drive motors are controlled to move the robot upward. When the top sensors indicate that the robot is pushing against the ceiling, the airflow fan motor

FIGURE 10.44   Photograph of the WCR using Bernoulli's principle.

for the top is activated, which makes the robot top stick to the ceiling and the fan motor for the front is deactivated. The robot moves under the ceiling until the destination line is crossed. Then, the cruise motor on top is reversed. A similar procedure is used to retrace the path. The operation of this particular robot during a competition can be viewed in WCR Using Bernoulli's Principle (2013).

10.4   CONCLUSION

In this chapter, we have considered a few cases of game robots. We have illustrated how the principles discussed in earlier chapters can be applied for the successful design of such robots. In addition to simulation studies, the videos cited provide a good idea of how the robots perform.

REFERENCES

Air-flow. 2013. http://www.youtube.com/watch?v=9GQk1rps6j8.
Cavello, A., Setola, R., and Vasca, F. 1996. *Using MATLAB, Simulink and Control System Tool Box: A Practical Approach*. Hertfordshire, UK: Prentice Hall Europe.
Ogata, K. 1990. *Modern Control Engineering*. 2nd Edition. Englewood Cliffs, NJ: Prentice-Hall.
Ogata, K. 1995. *Discrete-Time Control Systems*. 2nd Edition. Englewood Cliffs, NJ: Prentice-Hall.
PBR-Single Degree. 2012. http://guppy.mpe.nus.edu.sg/srg/srg03-media/pole-bal/index.htm.
PBR-Two Degree. 2012. http://www.youtube.com/watch?v=T_an16oQpsc.
Rajput, R.K. 2011. *A Text Book of Fluid Mechanics: In SI Units*. New Delhi: S. Chand and Company Ltd.
SRG—Singapore Robotics Games. 2012. http://guppy.mpe.nus.edu.sg/srg.
WallClimbingRobot-FlipperType. 2012. http://www.youtube.com/watch?v=CyQkoBEMEUM.
WCR Using Bernoulli's Principle. 2013. http://www.youtube.com/watch?v=v3X1WTiOd84.

11 Mapping, Navigation, and Path Planning

11.1 INTRODUCTION

An autonomous robot must be able to perceive its environment and act on it to achieve its goal. Sensors provide the robot with measurements about some physical phenomena. In some cases, these measurements are enough to make decisions, but at most of the time it is necessary to process these data to obtain more useful information about the environment. This process is called perception (Jones and Flynn 1993). The main information obtained by perception is the state, which is the representation of the environment and robot itself at some point of time. The robot's state would consist of what the robot is able to sense, process, and represent. For example, the state of a mobile robot will normally consist of a map of its environment, its position in the environment, its speed, its battery level, and so on.

The robot unit that makes decisions is called the controller. However, it is necessary to emphasize that the controller we are referring here is a higher-level control system that is responsible for the decision-making and planning of robot actions as a whole system. The robot control issues discussed earlier in Chapters 8 and 9 are referring to a lower-level control that handles the basic motions of the robot by acting upon its actuators. The remainder of this chapter, the term "controller" should be understood as high-level controller, unless otherwise stated. To achieve a high-level control, the robot utilizes data generated by the perception unit to make a plan, which is a set of actions that the robot must follow to achieve its goal. An important issue for a high-level controller is the rapid and timely decision-making capability. When a robot takes too long to deliberate about what actions it should take, those actions might be irrelevant and the plan may be invalid due to the changes in its environment. Normally, the environment for autonomous robot games is dynamic and unpredictable.

In the following sections, we will discuss the roles of perception and decision-making when developing autonomous robots. Different strategies will be discussed to provide a better understanding of the alternatives for perception and decision-making, but the final decision of what to use indeed depends on the environment, the robot structure, and the task.

11.2 PERCEPTION

Robots understand their environment through the limited information gathered by their sensors. As explained earlier in Chapter 3, there is a wide variety of sensors to measure different types of physical phenomena. We presented some of the sensors that are suitable for designing robots for games. Here, we will focus on the way that the sensor data are processed to become meaningful and to make the robot aware of its environment. Perception is more than merely reading the measurements from the sensors, but it is a process of understanding the environment by organizing and interpreting the information collected from the sensors.

11.2.1   FROM SENSOR MEASUREMENTS TO KNOWLEDGE MODELS

The data collected from sensors are inexact due to sensor noise and limitations of the hardware. To overcome the uncertainty, in the state of a robot or its environment, the perception system employs models that minimize the effects of uncertainty.

There are mainly three strategies when processing sensor measurements to deal with uncertainty. These three strategies that will be presented here are similar to those strategies in the following chapter about decision-making. The fundamental reason for this is that the way sensor information is used to deal with uncertainty is intrinsically related to the decision-making. Processing sensorial data and reducing its uncertainty is a fundamental part in decision-making that tries to minimize the possible failures in the system (Siegwart and Nourbakhsh 2004). Let us target these strategies here from a data-processing angle.

The first strategy is to use the raw sensorial information of each individual sensor to control or influence directly the robot behavior. In this strategy, the information about previous states of the sensors is not relevant, only the current measurement is considered, regardless of its accuracy. This strategy will yield to fast robot actions since it is looking for a particular stimulus in the sensors to respond accordingly. Inaccuracy of sensors may induce uncertainties that produce false reactions. These types of systems are continually assessing their sensor values; hence, it is expected that the uncertainty will reduce overtime. For example, a robot designed for micromouse competition moves forward, but it must turn left or right once a wall has been detected in front of it. If we consider a robot with three sensors, one on the left, one on the right, and one in front, this robot will make a turn as soon as the front sensor detects the wall; besides, the current information of the side sensors will be used to determine the direction of the turning. In this example, the robot only uses the current information of the sensors; it is not important to consider the previous information captured from sensors to avoid the wall in front.

The second strategy is to generate a higher level of representation of the environment through the information from one or more sensors. This high-level representation can then be used to trigger the appropriate robot behavior. This strategy requires extracting features that are relevant in the robot's task, but these features may not be obtained directly from the raw sensorial data or from one single sensor. Needless to say, this strategy is slow when compared to the first strategy of using the raw sensor data, but it handles uncertainty by using previous information recorded from sensors. However, if the process of feature extraction takes too long, there is the risk that the processed information would be irrelevant due to the dynamic nature of the world. For example, if we consider robots competing in the Humanoid League of the RoboCupSoccer competition, the position of robots and ball are changing constantly. Furthermore, the state of a robot might be changing due to external unforeseen circumstances such as being pushed by another robot or being moved to a new position by a robot handler in the field, and so on. In this type of task, the robot should not take too long to process sensor data; otherwise, the resulting information will be irrelevant to make adequate decisions. On the other hand, it is necessary to process current and past data to obtain essential information like robot position, an estimation of whereabouts of other robots and objects in the field, and so on.

The third strategy is a hybrid process that combines the two methods already discussed. Some of the raw information is utilized to control "survival" behaviors, while the feature extraction is used to identify relevant information to complete a given task. This strategy takes advantage of the quick response of the first strategy as well as the strength of feature extraction of the second strategy, while trying to minimize the response time to the changes in the environment. For example, in humanoid RoboCupSoccer, the robots should use current and past data for determining the position of the robots and the objects in the field. This process is slow and it might take a long time to compute. Nevertheless, robots should also be able to react fast if the ball is detected nearby, and the robot needs to kick the ball. Obtaining the position of objects in the field or at least a good estimation of them is useful so that a humanoid robot can make a decision to kick the ball to score a goal (Acosta-Calderon et al. 2010).

Let us discuss more about the feature-extraction process that is used in these strategies. The process consists of first collecting raw measurements from one or more sensors, second, filtering the raw measurements to remove redundant data, and finally extract distinct features from the filtered data. Feature extraction is a powerful technique to enable robots with high-level information. Collecting sensorial data requires a huge amount of memory space and heavy computation, which takes a long period of time to make sense of the data. One of the benefits of feature extraction is that it reduces the volume of data to represent a feature by simply producing an abstract representation of it from the raw data. These features could then be combined to produce a set of high-level abstracted features found about the environment. Subsequently, accessing and using high-level information speeds up the computation process, which is also another advantage of feature extraction.

Figure 11.1 illustrates the process of feature extraction with a mobile robot. Let us consider a mobile robot that uses laser ranging information from a sensor that provides 100 measurements per sample, 10 samples per second; the information can be recorded in polar coordinates (a distance and an angle), along with the odometry measurement of the robot (X, Y positions and a heading angle of the robot). After 20 s, the robot will collect 20,000 laser samples and 200 odometry measurements. From this large set of data, information about the two walls can be extracted. The question is now to represent the walls. One possibility is to store the entire raw data from the laser and the odometry readings, but this will require too much memory space just for a wall. The wall can also be represented as a straight line with only two points representing the beginning and the end of it. This way, all the raw data recorded previously can be discarded. Using representation saves considerable memory space, but with a risk of losing some important information. So, the way that information and features are stored in the memory will impact the control of the robot.

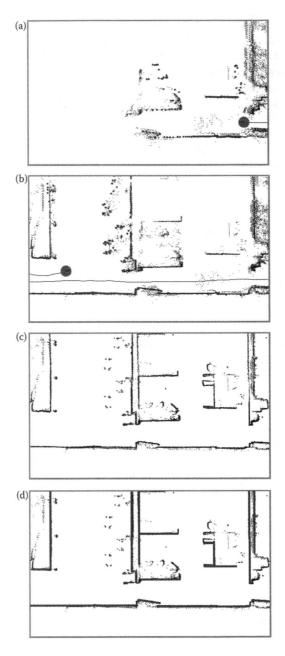

FIGURE 11.1 A mobile robot collects information from its laser ranger; the distance information is combined with its position and orientation to generate a map of its environment. (a) The
robot is only able to generate points in the map space that represents space that might be occupied
by an obstacle. (b) As the robot collects more data, the points cluster in different locations, increasing the likelihood that those locations are occupied by obstacles. (c) A postprocessing of the data
collected by the robot helps to connect points to identify wall, and to remove erroneous data in the
map. (d) The final step is the extraction of higher-level features such as wall, rooms, and corridors.

The type of features that the robot would be able to extract should be considered in the design of the entire system. The physical phenomena to be measured and the level of uncertainty of the sensors should be considered to see if it is suitable for the desired environment where the robot would operate. In this regard, the sensors and the environment need to be considered wisely in the robot design as they will influence the feature extraction process, and consequently the performance of the robot. For example, a robot navigating in an office environment will benefit by using range sensors to detect walls, corridors, rooms, and objects. In contrast, the same robot will not be so useful in a rescue scenario where the environment is a disaster area without structured walls, corridors, rooms, or doors. The rescue robot then may benefit from being equipped with sensors that would enable the robot to detect human victims, fire, gas, and so on. Earlier in Chapter 4, we have discussed the methods for detecting features from images obtained by a camera. A combination of what is described in this chapter and the features obtained with the camera is likely to result in a more robust perception tool.

11.2.2 MAP REPRESENTATION

Autonomous robots should be able to represent their environment in an efficient way. The representation of the environment is of particular importance for mobile robots, since they need to plan their paths to reach target locations, as well as to know which locations they have visited. This representation of the environment is better known as a map. The way that the environment is represented in a map affects the choices that the robot has to plan its path, as well as the representation of the robot's position on the map. There are many map representations, and the decision of choosing a particular representation should be based on the following points:

- The features represented in the map and its precision of must be based on the information extracted by the robot's sensors. For example, a robot equipped with a laser ranger would be able to detect the distance to the object in the same plane of the laser; this means that this robot can only represent 2D features in its map. In contrast, a robot equipped with a stereo camera would be able to extract the depth information for every single pixel of the image, as well as color information of the objects; these additional data will enable the robot to add a 3D representation of the features into its map.
- The size of the information used to represent the features in the map also determines the computational complexity required to handle the map for adding new features, searching, and planning. We consider two robots, one equipped with a laser ranger and the other with a stereo camera. The 2D map generated by the laser will require less storage space, and thus it will be faster to process compared to the 3D map generated by a stereo camera system.

Let us consider two most common map representations: the metric map and the topological map. The advantages and disadvantages of both representations will be discussed in the following sections.

11.2.3 Metric Map

The metric map is perhaps the most common representation technique used in mobile robotics. It is a two-dimensional representation of the space and the objects in the environment at the level of sensor point of view. In other words, objects are not represented with their volume, but the space they occupy at the sensor level. For instance, a table will appear with four blobs in the map since the robot sensor's plane intersects the legs of the table.

The space and shape of the objects are accurately represented in the map, which implies that a large memory space is required to store the map. There are a few types of metric maps, the most common one is the occupancy grid.

The occupancy grid map is based on the principle of fixed decomposition that transforms the environment into a discrete approximation. That is, the environment is represented by a discrete, coarse, fixed-size cell grid instead of having millimeter range accuracy for each feature in the map. The map is represented by a grid and the accuracy of the map will depend on the size of the cells in the grid. As described in the previous section, the size of the cells for this map depends on the task as well as the sensor accuracy. For example, the cells will be 10 cm^2, assuming that sensors have an accuracy of ±5 cm. Another approach for choosing the size of the grid cell is to use the size of the robot. For instance, CoSpace robot, shown in Figure 12.5b, is 16.5 cm length by 17 cm width. For this robot, the grid cells could be of a size between 17 and 20 cm^2.

The values of the cells in the map represent obstacles, free space, or unknown space. A robot equipped with a range-based sensor (ultrasonic, laser ranger, etc.) combines the sensorial data with the robot position to update the odds that grid cells are occupied. The way that the cell values are updated may differ depending on the method employed. Let us consider a robot equipped with an ultrasonic sensor mounted in front of the robot. Assuming that the ultrasonic sensor has a cone beam of 30°, as depicted in Figure 11.2, the probability that a cell within the cone is occupied or free is determined by the probabilities as shown in the figure. The values of each cell in the map are then updated with Bayes' rule. A simplified version of this process can be described as follows.

Each cell in the map is initialized as "−1," which will represent the unknown space. As the sonar cone passes by the cells, they will be set to "0," which represents free space. If the cell is within the range of the sensor, then this cell value will be increased. When the value of a cell passes a threshold value, then the cell is considered an obstacle. Values of the cells could also decrease when the sonar beam travels through the entire range without an encounter. Cells beyond the range of the sensor are not updated (see Figure 11.3).

Obtaining the map of an environment is useful and an autonomous robot should be able to store the maps to use them in future tasks. The occupancy grid maps are easy to generate, but instead of just storing the data of the grid, further processing can be applied to this map if we have some knowledge about the environment. Most of the man-made environments such as hallways, corridors, doorways, room, and so on can be modeled by connected straight lines. Thus, extracting line segments can improve the representation of objects and save space in the map. There are online

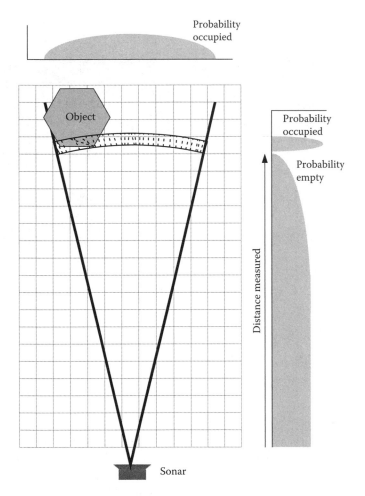

FIGURE 11.2   The probability model for an ultrasonic sonar sensor. The signal cone hits an object on one side, but it is impossible to predict if the object is located on the center or any side of the cone with certainty. The probability information of the sonar model can be then combined to update the value of the cells in the map with Bayes' rule.

and offline methods to extract this information. A typical method is the Hough transform that was discussed earlier in Chapter 4 for image processing. In this case, the Hough transform is applied to the collected data by range sensors instead of the pixels of an image. This is an offline method, and it is performed after the robot completes the data collection.

Let us discuss how to apply the Hough transform for an occupancy map. The occupancy map generated by the robot produces a matrix or grid with cells that represents the free space, unknown, and obstacles. We are interested in cells that represent the obstacles. Let us assume that a set of cells corresponds to a wall; unfortunately, they are not fully aligned and there will be a few missing connections. These errors will happen mainly due to the inaccuracy of the sensors, drifting position of

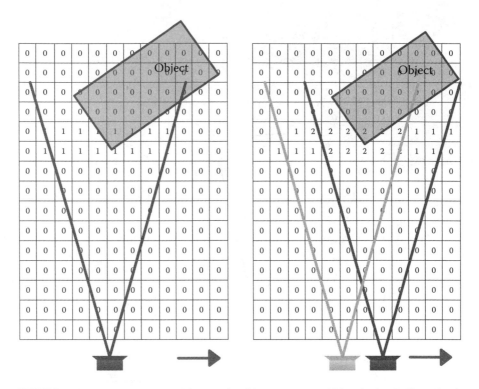

FIGURE 11.3 An occupancy grid example of how a map could be obtained. The robot in this figure consists of only one ultrasonic sensor.

the robot, and even the surface of the wall. The Hough transform is used to identify straight lines that fit to these points. The occupancy grid map is also a two-dimensional matrix, and it can be treated as an image assuming that each cell in the grid is like a pixel in the image where a black pixel represents obstacles and white pixels free space or vice versa.

The Hough transform considers the points of a line to be represented in polar coordinates (r,θ) rather than in Cartesian coordinates (x,y) as discussed earlier in Chapter 4. The parameter r represents the distance between the point and the origin of the image, while θ is the angle of the vector from the origin to this point. So, each cell in the map would be treated as a point and be represented with the parameters (r,θ) and be projected to the Hough space. As discussed before, a point in the Hough space corresponds to a sinusoidal curve, which is unique to that point. When the curves corresponding to the two points are superimposed in the Hough space, the point these two curves intersect corresponds to a line in the metric map that passes through both points. Finding the parameters with more number of intersections will result in finding the more salient lines in the map.

The resulting Hough space is examined and the maximum intersection points are interpreted as potential line segments. However, this does not guarantee that straight lines in the map are obtained accurately. Usually, a postprocess is applied to the information returned by the inverse Hough transform. This is mainly due to

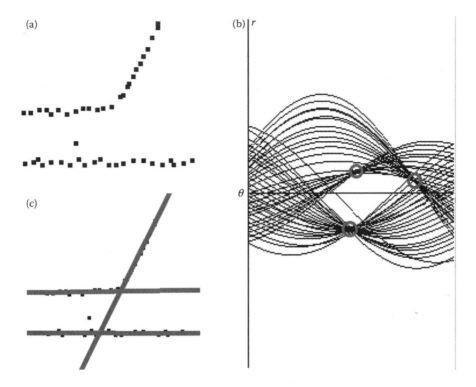

FIGURE 11.4 (a) The metric representation of the robot's environment generated by the ultrasonic readings. (b) The straight lines obtained from the Hough transform. (c) The Hough space for the points in the metric map.

two factors: First, the inverse Hough transform is able to find the line, but it is not able to say where the line is starting or ending in the image (see the output of the inverse Hough transform in Figure 11.4c). Second, false lines may also be found due to scattered points in the map that may look like forming a possible line. A common post-processing method is to overlap the lines detected by inverse Hough transform on to the map and compare them with actual points to validate.

The computational cost of the Hough transform depends mainly on the dimension of the map to process, as well as the accuracy of the samplings involved for the θ in the Hough space. These two parameters determine the additional memory required for the Hough space, as well as the number of iterations for each point in the Hough space.

11.2.3.1 Case Study

The RoboCup@Home league is part of the RoboCup competition, which aims to develop service and assistive robot technology with high relevance for future personal domestic applications (RoboCup@Home 2012). All the challenges are conducted in a real-world living room scenario. The RoboCup@Home league limits the mode of interaction between human and robot to natural ways like speech and gestures. In addition, to be able to perform all the challenges, the robots should be able to safely navigate in the environment without colliding with humans or obstacles in the room. Most of the challenges include some type of navigation abilities from the robot; this means that the robot should be able to recognize the environment and be able to localize itself in the environment. A map helps the robot with these two issues.

Figure 11.5 shows the Ariel robot from Singapore Polytechnic. This social robot has been developed to take part in the RoboCup@Home competition. The robot uses a Pioneer 3-AT robot, a four-wheel drive robotic platform, for navigation. The robot is also equipped with a laser rangefinder SICK LMS100; using the odometry system of the mobile robot and the information gathered by the laser, it is possible to generate a metric map.

Since the laser rangefinder has a higher accuracy, the odometry of the robot is also rather accurate and consequently so is the metric map generated with it. The resolution of the map is 20 mm; needless to say, this will produce a large volume of data and demand a large memory space to store it. Moreover, high accuracy is a key factor for this robot, so that it is capable of identifying not only walls and corridors, but also objects in the environment.

Let us assume that the laser rangefinder only returns one point instead of an array of points to simplify the explanation. At each iteration, the controller system

FIGURE 11.5   The Ariel robot is a social robot interacting with public in an exhibition.

will have the robot position as $[X,Y,\theta]$, where θ is the orientation or heading of the robot, with respect to its origin, and the laser rangefinder readings that consist of the distance for each point and its orientation $[D,\alpha]$. Since the laser rangefinder is not located at the center of the axis of the robot (the point considered as reference for the odometry), it is important to consider the displacement (m) of the sensor in our calculations.

Figure 11.6 shows the frame reference for the robot and how coordinates of a point, detected by laser rangefinder, can be calculated. The figure also shows the relation between the laser rangefinder frame and the mobile robot frame. In this example, there is only one displacement of the frames, which is on the X-axis. There is no displacement in the Y-axis or a distinct orientation of the laser range finder on the robot body. Also note that in this scenario all the information is in 2D; thus, the Z-axis is ignored.

We can calculate the position of the point provided by the laser rangefinder by considering all the previously discussed features. First, we obtain the components for the point read by the laser rangefinder as

$$x_{\text{laser}} = D\cos(\alpha) + m \tag{11.1}$$

$$y_{\text{laser}} = D\sin(\alpha) \tag{11.2}$$

After that, we can rotate and translate them to the global coordinate system as follows:

$$P_x = (x_{\text{laser}}\cos(\theta) - y_{\text{laser}}\sin(\theta)) + X \tag{11.3}$$

$$P_y = (y_{\text{laser}}\cos(\theta) + x_{\text{laser}}\sin(\theta)) + Y \tag{11.4}$$

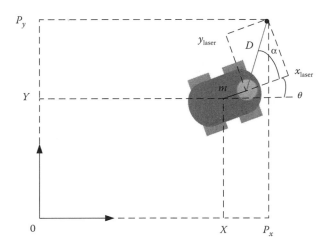

FIGURE 11.6 The frame reference for the Ariel robot base.

Once the Cartesian position of the point is calculated, it is possible to mark this point in the map by fitting it to the nearest cell in the map. It might be necessary to do further processing for the cells between the robot and the detected point in the map to indicate the free space or increase the probability of cells being empty or occupied. This could be done in a similar manner as described previously in this section for the ultrasonic sensor. The entire process will repeat for all the points in the laser rangefinder. This is a computationally heavy task, as it is repeated in every iteration. Once all these points are collected, it is possible to find fitting lines by applying the Hough transform as before.

11.2.4 TOPOLOGICAL MAP

Metric maps consider every feature in the environment, which translates into a huge amount of memory to store the map as well as a slower search or analysis of the map. An alternative to this representation is the topological map. Topological maps consider certain distinctive features in the environment and the relationship between these features without representing them in the map. The features used for these maps are called landmarks. A landmark must be a distinctive object or place of interest that the robot is able to perceive. Landmarks could be artificial. The features can be embedded in objects or locations to ease the recognition of the landmark. For example, colored markers or signs that are put on doors can indicate locations for robots. Landmarks can also be natural elements such as gateways or junctions.

Topological maps are built on top of grid-based maps. The free space of a grid-based map is partitioned into a small number of regions, and the regions are separated by critical lines. The critical lines correspond to passages such as doorways or other landmarks. As shown in Figure 11.7, the nodes are the critical points and the lines connecting each critical point are the critical lines. The critical lines partition the free space into disjoint regions. The lines provide information that the robot uses for planning and navigation; some of the information stored in this representation is the orientation and the distance between the nodes.

Gateways are commonly used as landmarks in robotics. They also provide an opportunity to change direction for a robot. The most common representation of the topological map is a relational graph as shown with the example in Figure 11.7. Note that the information of how these landmarks are related is embedded in the graph; however, it is not as explicit or as accurate as in the metric map. For example, for node 6, the exact orientation of the room (node 7) is not clear. However, it is clear that the room is located somewhere on the right-hand side of the robot coming from the direction of nodes 4–6.

11.2.4.1 Case Study of Topological Map

In the micromouse robotics competition a small autonomous robot equipped with few infrared sensors and differential motors must find its way to the destination point in a maze (see Figure 11.8). The robot is allowed to do multiple runs within 5 min. The initial runs are usually referred to as "searching runs," and they are meant for the robot to build the map of the maze and find the shortest path to the destination. Once the search has been completed, the robot will move as fast as it can, following the shortest path it has computed. The fastest robot wins the competition (Singapore

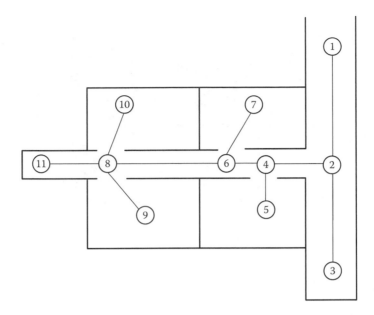

FIGURE 11.7 A topological map. The gateways are used as the main landmarks in this representation. The information of how these landmarks are related is stored as a relational graph.

Robotic Games 2012). Since the maze is made of square and rectangular shapes, all the turns should be 90° to the left or the right; this means that a corridor a gateway can be easily determined since it is either on the left or on the right of the robot. Thus, when a gateway is found, it is possible to follow the corridors that connect to the gateway. The micromouse robot tries to explore the maze as much as possible while mapping the gateways and its connecting corridors. It is also possible to record

FIGURE 11.8 Micromouse developed at the Singapore Polytechnic.

(a)   (b)

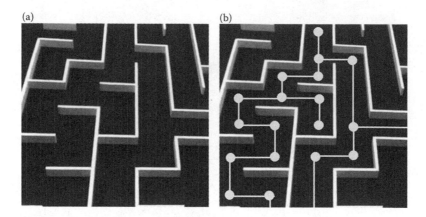

FIGURE 11.9   (a) A section of the maze that the micromouse has to explore. (b) Topological map built to represent this section of the maze.

certain corridors that have not been mapped yet. So, the robot can map those corridors once it completes working on its current corridor.

Figure 11.8 illustrates a micromouse robot that consists of six infrared sensors (three on each side), two motor wheels for differential control, and a caster wheel. The latest generation of micromouse robots also incorporates encoders on the wheel motors. This helps the robots to add distance to the segments that connect nodes as well as to discover paths that loop back to the main corridor. It is also possible to incorporate a digital compass to provide further information about the orientation of the robot, although this is not so common.

As the micromouse moves through the maze, the infrared sensors placed in front and on the sides of the robot assist identifying the wall. The robot will also maintain its position in the center of the corridor without crashing into the walls. If no obstacles are detected in front of the robot, then it will keep moving forward. When the sensors detect an open space in a direction other than the current direction of the robot, the location will be considered as a gateway, and the open space will be marked as one possible direction. According to the strategy of the team, the robot then follows any one of the newly discovered directions or it can just continue its exploring moving forward in the current corridor. Figures 11.9a and 11.9b show a part of the maze that the micromouse has to explore and its correspondent topological map. After the map has been built, a planner helps to choose the shortest path to the destination point. In the next section, we will discuss more on planners and navigation using topological maps.

11.3   NAVIGATION

A desirable ability for an autonomous mobile robot is to be able to go from one place to another. The term "navigation" refers to the way that a robot moves in its environment to reach its destination. The locomotion system of the robot tells the actuators of the robot how to move, whereas the navigation system tells the robot about its destination.

There are several behaviors that are involved in navigation and locomotion; some of these behaviors are simple and some others are more complicated. What behaviors should go into a robot depend directly on the type of task and the design of the robot. The final performance of the robot may vary according to the type and number of sensors as well as the mobile configuration of the robot. Nevertheless, these behaviors can be used to produce more complicated behaviors with a subsumption architecture, or a hybrid architecture as will be described in Section 12.3.3.

As discussed in the previous section, the uncertainty in the perception of the robot's sensors makes any kind of decision making a difficult task. As the robot moves, its odometry system will have errors, and as the robot keeps moving, the error will accumulate. This causes the uncertainty that the position of the robot is not really known. There are different methods that can help to reduce the uncertainty of the robot's whereabouts during motion. In the majority of the methods discussed below, it is necessary to have a map built about the environment. The following sections discuss some of these methods and their application to robotic competitions.

11.3.1 WALL FOLLOWING

Wall following is a behavior that makes a robot move smoothly and follow the contour of the wall. To produce a smooth motion, the alignment of the robot is the key component; thus, it is essential that the robot maintains a parallel heading to the wall during its motion. If robot has a distance sensor facing toward the wall side, then wall-following behavior will maintain a constant distance from the wall. This means that if the distance increases, the robot must turn toward the wall, and when the distance decreases, the robot must turn away from the wall (see Figure 11.10). The rotation of the robot toward the wall or away from the wall should stop when the robot is within the range of the constant distance required. In the figure, the dotted line shows the distance D that the robot should maintain from the wall. It is better to use a tolerance range X, in centimeters, so that distance from wall is within $D \pm X$. This will minimize the swinging of the robot while trying to follow the wall.

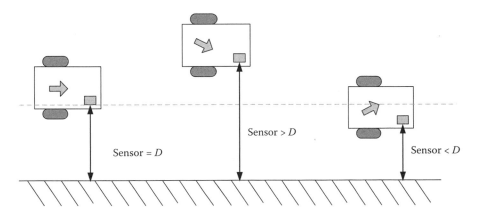

FIGURE 11.10   Wall-following behavior for a mobile robot with a distance sensor.

```
%
% The class DiffRobot the members of the class robot are declared as:
%
% robot.ultrasonic_sensors(3)   - 1 - Left, 2 - Right, 3 - Front
% robot.motor_vel(2);           - 1 - Left, 2 - Right
%

classdef DiffRobot < handle
    properties
        motor_vel = [ 0.0 0.0 ];
        ultrasonic_sensors = [ 0.0 0.0 0.0 ];
    end
    methods
        function obj = DiffRobot() % constructor
        end
    end
end


% This is the main function of the mobile robot. It will
% loop reading the sensors, Processing, and sending the motor command.
%

function MainRobot()

% Initialized robot
robot = DiffRobot();

while 1
    % Read the New Values of the Sensors
    ReadSensors(robot);

    % Follow the wall
    WallFollowing(robot, Desire_Dist);

    % Write to Motors
    SetMotorCmd(robot);
end
end

function ReadSensors(robot)
% Obtain the new readings of the sensors
end

function WallFollowing(robot, Desire_Dist)

% Difference between the wall and our desired distance from it
% it only uses the right sensor, to follow from left to right the wall.
diff = Desire_Dist - robot.ultrasonic_sensors(2);

% New Translational and Rotational Veloci
trans = Const_Vel;
rot = gain * diff;

% Convert velocities to Left and Right Velocities
robot.motor_vel(1) = ( 2*trans - rot ) / 2;    % Left Velocity
robot.motor_vel(2) = rot + robot.motor_vel(1); % Right Velocity

end

function SetMotorCmd(robot)
% Send the motor velocities to the robot
End
```

FIGURE 11.11 MATLAB code for the wall-following behavior presented in Figure 11.10.

Figure 11.11 presents the MATLAB® code of the wall-following behavior. It is important to mention that the rotation will be proportional to the difference of the distance of the robot from the wall. This means that if the robot deviates from the desired distance severely, then robot angular velocity will change rapidly to compensate the error. Similarly, if the deviation is small, then the compensation of the angular velocity will be minor. The value of the gain needs to be adjusted to specify how fast or slow the robot reacts to the difference of distance. In practice, we know that no sensor is entirely accurate, so it is important to consider the noise of the sensor to minimize the possible effect of the noise in the motion of the robot.

In the code presented in Figure 11.11, the constant value assigned for the translational velocity can be adjusted to a suitable value for the hardware used in robot design. The rotational and translational velocities need to be converted to left and right velocities since the robot used in this example is a differential drive robot. Furthermore, the code presented here only follows the wall from left to right. If the robot must follow from the opposite direction, the left ultrasonic sensor can be used, and both left and right ultrasonic sensors can be used to follow a wall on either side, or when moving in a corridor.

11.3.2 Obstacle Avoidance with Vector Force Histogram

Vector field histogram (VFH) was first presented by Johann Borenstein and Yoram Koren. It was further improved by Iwan Ulrich and Johann Borenstein and it was renamed VFH+. It is lately renamed VFH*. VFH is a real-time obstacle-avoidance method for mobile robots. The VFH is a computationally efficient method for navigation that has proven to be fast and reliable, especially when roaming in environments with many obstacles. In this approach, a robot does not need to know the environment prior to its navigation. This is advantageous when navigating in a highly dynamic environment, like the one where rescue robots operate. VFH enables the robot to detect obstacles and steer while avoiding collisions along its desired route.

The VFH method uses a two-dimensional Cartesian grid, which is updated by its environment continuously with the information of the robot's sensors, such as sonar or laser rangefinder. This grid is similar in principle to the metric map generated by an occupancy grid (presented in Section 11.2.2); however, this grid is smaller and it only represents the surroundings of the robot, much more like egocentric view because the robot will always remain in the center of this grid (see Figure 11.12a). This egocentric grid provides information about obstacles that are near. Let us imagine a vector coming from the center of the robot toward the obstacle or to the end of the grid, and we now repeat these vector calculations for 360° around the robot. Thus, all these vectors are further transformed into a polar histogram, where all the vectors are placed side by side and compared by their magnitude.

Three valleys are obtained from the histogram. Valley A is located in front of the robot in Figure 11.12a, valley B is located in the adjacent corridor besides the box, and it is also because of this obstacle that the valley is quite small. Finally, valley C is located behind the robot. Figure 11.13 shows the polar histogram. The horizontal dotted line in the polar histogram represents a threshold that indicates the robot is too close to obstacles.

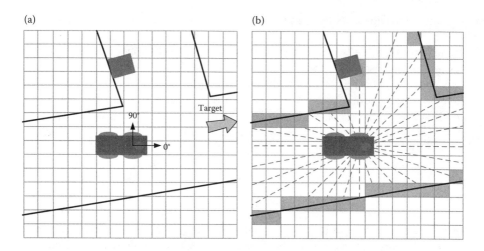

FIGURE 11.12 The egocentric grid of a robot moving in a corridor (a) Robot, which is always at the center of egocentric view, corridors and target direction for the robot. (b) A representation of obstacles in grid cells and vectors from the robot to the obstacles.

From this polar histogram, it is possible to compute a candidate valley, which is an area where sectors with polar obstacle density are below a threshold. The candidate valley is also selected based on the proximity toward the robot-desired route. In Figure 11.13, the vectors calculated from Figure 11.12b are represented in the polar histogram. Three valleys are obtained in the histograms; these are marked with the letters A, B, and C. According to the target direction presented in Figure 11.12a, valley A is chosen as the candidate valley for the navigation of the robot.

VFH+ improves the smoothness of the mobile robot's navigation by taking into account the size of the robot, thus allowing the candidate valley to be more precise. Obstacle looking-ahead is a feature of VFH+ that will discard candidate sectors that appear to be unobstructed in egocentric view, but they are obstructed outside the egocentric view. Finally, the last improvement of VFH+ is adding the cost function to better characterize the performance of the algorithm.

There are also disadvantages and limitations of the VFH. For example, it does not concurrently search for the optimal path toward the destination because it only uses local information instead of global information. The robot will also face difficulties at narrow paths due to the tolerances used in the VFH algorithm. The VFH cannot guarantee to reach the desired location since it is only using local information.

The VFH algorithm is still widely used in robotics for real-time obstacle avoidance to maintain the robot's path towards its destination despite all the above shortcomings. It is for this reason that VFH is used in combination with the path planners that can produce waypoints that the robot must reach before it can reach its final destination. Thus, these waypoints are input to the VFH method to reach the waypoint while avoiding any obstacles in the way.

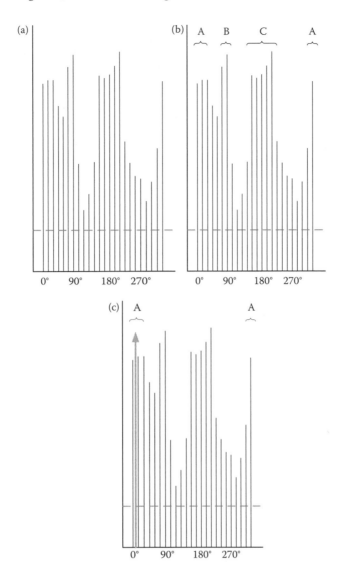

FIGURE 11.13   Polar histogram obtained from the vectors calculated in Figure 11.12b.

11.3.2.1   Case Study of Obstacle Avoidance with Vector Force Histogram

Most mobile robots implement some form of obstacle avoidance. Robots taking part in competitions are of special interest since they must achieve their task without hitting obstacles or without deviating from a desired path. RoboCup@Home, RoboCup Rescue, Robot Colony, and Intelligent Robot are some of the robot competition events where the obstacle-avoidance algorithm is needed. However, any competition that involves mobile robots will require robots to avoid collisions.

The code shown in Figure 11.14 is for a mobile robot to go to a target while avoiding obstacles employing the VFH method. In the example code, the function

```
% This is the main function of the mobile robot. It will
% loop reading the sensors, processing, and sending the motor command.
%
function MainRobot()
% Initialized robot
robot = DiffRobot();

while 1
    % Read the New Values of the Sensors
    ReadSensors(robot);

    % Obtain the new ec_grid
    UpdateGrid(robot, ec_grid);
    % Vector Force Histogram
    [angle dist] = vfh(robot, targetPos, ec_grid);
    % Drive the robot by the angle and distance
    GoTo(robot, angle, dist);

    % Write to Motors
    SetMotorCmd(robot);
end
end

function ReadSensors(robot)
% Obtain the new readings of the sensors
end

function SetMotorCmd(robot)
% Send the motor velocities to the robot
end

%
% Obstacle Avoidance with Vector Force Histogram
% [angle dist] = vfh(robot, targetPos, ec_grid)
%
% targetPos: the goal point [x,y]
% ec_grid: the ego-centric grid around the robot the grid values are
%          0 - free space, and 1 - obstacles.
%
% Returns:
% angle: the angle of the best direction of travel in radians
% dist: the distance to the target or to the boundary of the grid
%
```

FIGURE 11.14   MATLAB code of the VFH algorithm.

```
function [angle dist] = vfh(robot, targetPos, ec_grid)
% 72 sectors at 5 degres each sector
alpha = deg2rad(5);
nSectors = 72;
% Initialize each sector as if there is no obstable in this sector
sectors=ones(1,nSectors)*10;


nObstacles = 0;
obstacles = [];


nValleys = 0;
valleys = [];
gValleys = [];


% Minimum number of sector in a valley to be consider wide valley
minSectorsValley = 3;
% Threshold for a sector to be consider part of a valley
threshold=7;


% Corresponding sector to the target position
targetSector = Sector([robot.X robot.Y], targetPos, nSectors, alpha);


% Obtain the information of each obstacle
[M N] = size(ec_grid);
for (i = 1 : N)
    for (j = 1:M)
        if (ec_grid(i,j) == 1)
            nObstacles = nObstacles + 1;
            obstacles(nObstacles,:) = [robot.X + j-round(M/2) robot.Y +
round(N/2)-i];
            % update the sectors with the information of the obstacles
            % distance between robot and obstacle
            d = norm(obstacles(nObstacles,:) - [robot.X robot.Y]);
            k = Sector([robot.X robot.Y], obstacles(nObstacles,:),
nSectors, alpha);
                sectors(k) = min(sectors(k), d);
            end
        end
    end
```

FIGURE 11.14   (continued) MATLAB code of the VFH algorithm.

```
% calculate the candidate valleys
k=1;
while(k<=nSectors)
    if(sectors(k) > threshold)
        kr = k; % right border of a valley
        while(k <= nSectors && sectors(k) > threshold)
            kl = k; % left border of a valley
            k=k+1;
        end
        % calculate candidate valley direction
        if(kl-kr >= minSectorsValley)
            % Center of the valley
            nValleys = nValleys + 1;
            valleys(nValleys) = round((kr+kl)/2);
            % is the target Sector is within the valley?
            if(targetSector >= kr && targetSector <= kl)
                nValleys = nValleys + 1;
                valleys(nValleys) = targetSector;
            end
        end
    else
        k=k+1;
    end
end


% Determine the nearest valley to the target direction
% Calculate cost of taking each valley by considering:
%  - Distance between the valley and the target
%  - Distance between the valley and the heading of the robot
for(i=1:nValleys)
    gValleys(i) = 0.5 * DiffSectors(valleys(i), targetSector, nSectors) +
0.2*DiffSectors(valleys(i), robot.Theta/alpha, nSectors);
end
% Select the valley nerest to target direction
[value imin] = min(gValleys);
angle = valleys(imin)*alpha;


% Calculate the distance to the target
dist = min(norm(targetPos - [robot.X robot.Y]), sectors(valleys(imin)));
end
```

FIGURE 11.14   (continued) MATLAB code of the VFH algorithm.

```
% Calculates the difference between two sectors
function diff = DiffSectors(sec1, sec2, n)
diff =abs(sec1-sec2);
if ( diff > abs(sec1-sec2-n) )
    diff = abs(sec1-sec2-n);
elseif ( diff > abs(sec1-sec2+n) )
    diff = abs(sec1-sec2+n);
end
end


% Calculates the corresponding sector for a point from the centre
function sector=Sector(centrePoint, targetPoint, n, alpha)
dy = targetPoint(2) - centrePoint(2);
dx = targetPoint(1) - centrePoint(1);
angle = atan(dy/dx);
if(dx < 0)
    if(dy > 0)
      angle = pi - abs(angle);
    else
      angle = pi + abs(angle);
    end
else
    if(dy < 0)
      angle = 2*pi- abs(angle);
    end
end
sector = round(angle/alpha);
if(sector == 0)
    sector = n;
end
end
```

FIGURE 11.14 (continued) MATLAB code of the VFH algorithm.

VFH() takes three arguments. The first argument is a structure that contains the current position and heading of the robot. The second argument is a target position for the robot in the format of [X, Y]; this is used to specify the target direction that the robot should pursue. The final argument is an egocentric grid of the robot; this can be obtained from the metric map by specifying a square grid around the robot. The dimension of the grid can be defined by the task or about the maximum range of the distance sensor used. However, to simplify the further processing, the egocentric grid data should be in the format of zeros and ones to represent free and occupied space, respectively. In the code, the VFH() function returns the angle and the distance. The distance returned by VFH is not the distance to the target position. It indicates that the free space within the egocentric map that robot can move without encountering an obstacle. This distance can be used for speed control. Let us consider the following values for the arguments of the VFH() function:

```
robot.X = 0;
robot.Y = 0;
robot.Theta = 1.5;
targetPos = [20 20];
ec_grid =[0  0  1  1  1  0  0  0  0  1  1  1;
          0  0  1  1  1  0  0  0  0  1  1  1;
          0  0  0  0  0  0  0  0  0  1  1  1;
          0  0  0  0  0  0  0  0  0  0  0  0;
          1  1  0  0  0  0  0  0  0  0  0  0;
          1  1  0  0  0  0  0  0  0  0  0  0;
          1  1  0  0  0  0  0  0  0  0  0  0;
          1  1  0  0  0  0  0  0  0  0  0  0;
          1  1  0  0  0  0  0  0  0  0  0  0;
          0  0  0  0  0  0  0  0  0  0  0  1;
          0  0  0  0  0  0  0  0  1  1  1  1;
          0  0  0  0  0  0  0  1  1  1  1  1;]];
```

This will produce an output as shown in Figure 11.15. The output from the VFH() function corresponds to the angle 75° represented by a bold line in the figure. It is also possible that the robot navigates toward the target by taking an angle of 0°. The VFH() function takes into consideration the current heading of the robot, which is around 85°. Thus, the robot heading that is closer to the valley is represented by 75°.

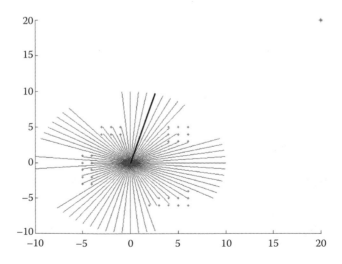

FIGURE 11.15 Representation of the egocentric grid with the vectors. The vector in bold represents the output from the VFH() function.

11.4 PATH PLANNING

Path planning is the process of finding a path from the robot's current location to its next destination. Assuming that both positions are known and represented in the robot's map, the path planner will look for the optimal path between these points. An optimal path is a path that satisfies the criteria defined by the robot task. This means that in most cases, checking all the possible paths will result into a heavy computation time.

The autonomous robot navigation problem consists of the calculation of a path between two points, a starting and a target point. The local navigation approach should produce an optimal (usually shortest) path, avoiding the obstacles present in the working environment.

However, not all path planners work in the same way. Some search only for part of a path and others look at the complete path. Both types of planners have advantages and disadvantages. For instance, while partial planners are fast they do not always warrant a shortest or optimal path. In contrast, full-path planners are able to compute the shortest or the optimal path that satisfy a given criterion, but the time required for this computation depends on the size of the map used. In addition, some methods only work with specific map representations. As discussed in the previous section, building a map representation is mainly determined by the task and the robot perception system. The following sections will discuss the use of two planners, the first one is for metric maps, and the second method is for topological maps.

11.4.1 WAVEFRONT PLANNER

Wavefront is a common algorithm used to determine the shortest path between two points that works on occupancy grid maps. In this method, a full-path planner assumes that each cell in the map is able to fit the robot.

Let us assume that a robot generated two-dimensional occupancy grid map (as described in Section 11.2.2) that represents its environment, and both the current position of the robot and the desired position are known in the map.

A second map is used for the wavefront method to update the values of each cell. The information in the original map is ported to the new map during the first step of the algorithm. The values in the cells will be updated by using neighborhood connectivity. The connectivity between the cells can be expressed as four-point or eight-point connectivity. The choice between four- or eight-point connectivity mainly depends on the size that each cell represents in the map as well as the locomotion capabilities of the robot to move between cell units. It is important to determine the connectivity, since this will not only affect the way that the values in the map are updated, but also the way that the planner will search for the shortest path. Once the new map is ready and the neighborhood connectivity has been selected, we can start with the algorithm, which employs the following steps:

1. All the free space in the map is set to the value of zero, the cells with obstacles are set to the value of one, the current position of the robot is labeled as "start," and the desired position of the robot is labeled as "destination," and it is set to the value of two.

2. Starting from the destination cell, increment the value of all the adjacent cells of the free space by one.
3. Repeat the previous step, but this time starting from only recently modified cells.
4. Repeat the previous step until completing the map. Cells with zero values should only exist in unreachable regions.

Figure 11.16 shows an example map and its update with wavefront planner. The method only deals with occupied and free space; however, it is possible to mark an unknown space as occupied space instead of free space to avoid the uncertainty that the unknown space represents for the planner. After filling the map with values as

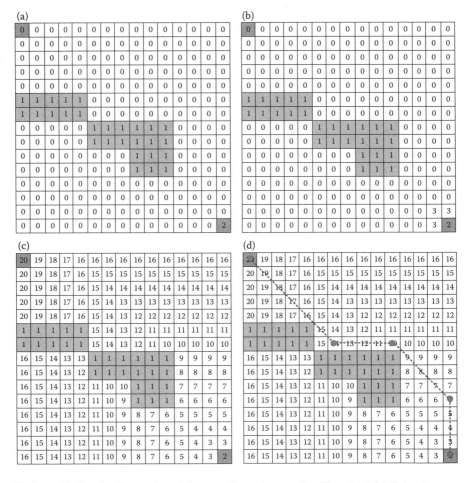

FIGURE 11.16 An illustration of the wavefront planner algorithm. (a) Initializing the map with zeros for free space and ones for the occupied areas. (b) Expansion from the destination to fill the map. (c) The resulting map after updating all the cells. (d) The shortest path to reach the destination.

shown in Figure 11.16c, it is now possible to search for the shortest path using the following steps:

1. From the starting position, follow the adjacent cell with the lowest value toward the direction of the starting position.
2. Move to the cell and repeat the previous step from the current cell.
3. Repeat the previous steps until the destination is reached.

The wavefront planner may discover more than one path. In all these paths, cells are connected by an uninterrupted sequence of decreasing numbers that leads to the destination. Figure 11.16d shows a path found for the given map. The path can be represented as a list of waypoints. A waypoint is a point in the map that indicates direction change for the robot on its path. In Figure 11.16d, the waypoints are indicated with circles along the path.

Once a path has been found, the robot must navigate from one waypoint to the next to reach its destination. Usually, VFH is employed to navigate the robot along waypoints since the VFH method ensures navigation without collisions.

11.4.2 PATH PLANNING USING POTENTIAL FIELDS

The main idea behind the path finding of the potential field method is to generate attraction and repulsion forces within the working environment of the robot to guide it to the target. The approach used is to generate the artificial potential fields to have obstacles exert repulsive forces onto the mobile robot, while the target applies an attractive force to the mobile robot.

Potential field is an array of force vectors represented in space. This will produce a force field analogous to a magnetic or gravitational field. Thus, the robot will be affected by this field and the force of attraction or repulsion of each object in the environment will contribute to this field. The vectors in each part of the field will be translated to direction and speed of the robot during its motion. Figure 11.17 shows an example of a two-dimensional map consists of attractive and repulsive forces.

Considering a metric map with two-dimensional space, and each cell in the map is defined by (x,y), it is possible to calculate the vector field F for a single element in the map; this is given as a vector sum of the two forces:

$$F(x, y) = \sum F_G(x, y) + \sum F_O(x, y) \tag{11.5}$$

where F_G is the attraction force toward the goal, and F_O is the repulsive force from the obstacles. Potential fields represent the description of the environment, which can be obtained completely *a priori* at the start of the motion process. Vector fields represent a map of actuator values, the orientation and magnitude.

The first step is to define what type of potential field primitive would be used for each element in the map. Figure 11.18 shows the different primitives for the potential

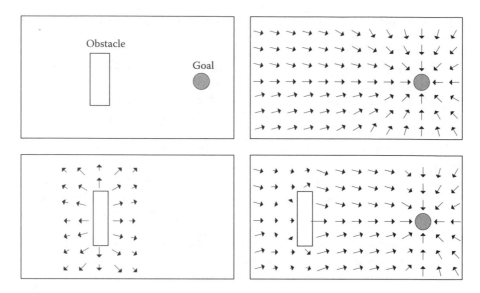

FIGURE 11.17 Obstacle on the left-hand side has repulsive forces for the robot to avoid. The target point on the right-hand side has attraction forces for the robot to move toward it.

fields. After choosing the primitives, it is essential to associate a magnitude profile with the objects for their potential field primitive. This magnitude of the vectors as mentioned before will be used to control the velocity of the robot. Thus, the profiles could indicate a constant velocity, a linear decrease, or even an exponential increase of the velocity. The calculation of the magnitude of the forces according to its profile will also need to specify a distance where the field is acting from the object. For example, the attraction field of the goal target should act in the entire map, in contrast the repulsion field of the objects should only act within the vicinity of the objects (see Figure 11.17).

11.4.2.1 Case Study of Path Planning Using Potential Fields

Let us consider the RoboCup@Home robot that needs to navigate from the entrance of a room to a table located at the other end of the room. The robot is required to

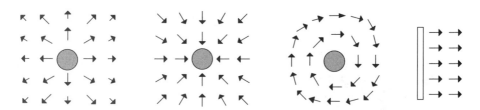

FIGURE 11.18 Potential field primitives: (a) repulsion, (b) attraction, (c) tangential, and (d) perpendicular.

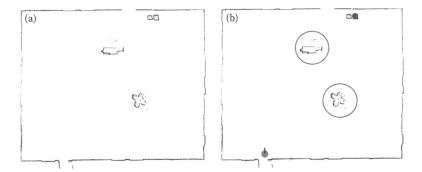

FIGURE 11.19 (a) The metric map of the room where the robot will perform the manipulation task. (b) The center and the sofa marked with a circle to ensure that the repulsion force will be strong enough to repel the robot if it gets closer to this circle. The goal position "table" is at the top right side of the map, and the robot is at the bottom left.

perform a manipulation task once it reaches the table, but first it needs to plan its way to reach the table without colliding with obstacles.

The potential field method will be employed for this purpose. An existent map of the room is shown in Figure 11.19; the map is a metric map. The potential field method will calculate the force in each robot location to produce the robot path to reach the table.

The tables and sofa in the room have a repulsive primitive in all directions; however, we do not want this repulsion to be felt everywhere. Thus, we specified a distance of 3 m radius in any direction for the force of repulsion to be felt. Furthermore, the repulsion force will be linearly increased as the robot approaches the objects in the map (see Figure 11.19b). For the goal, a constant attraction in all directions has been chosen, and this force should be felt everywhere in the room. Figure 11.20 shows the MATLAB functions used to calculate the force exerted by each obstacle and the goal to the robot.

There are two approaches for path planning. The first one is to generate a full path that the robot should follow to the target. In the second method at each location that robot visits, a local calculation of the next position will be performed. The second approach is widely used since it allows robot to cope with the changes in a dynamic environment. In Figure 11.20, the function "MainRobot()" initializes the robot and the position of the objects and goal. In the actual competition, after the robot has generated the metric map, a human user marks the location of the obstacles and goal position in the map and specifies the radius of the circles enclosing the objects. The function, "MainRobot()," will continuously read the sensor values, calculate the next vector in the path according to its current position, and consequently navigate the robot according to the vector obtained. The process continues until the robot reaches the goal position.

The function "PotentialField()" in Figure 11.20 calculates the force that the obstacles and the goal are exerting to the current position of the robot. The arguments

```
%-------------------------------------------------------------------------
% The class DiffRobot the members of the class robot are declared as:
%
% robot.X, robot.Y              - Position of the robot x,y
%-------------------------------------------------------------------------
classdef DiffRobot < handle
    properties
        X = 0;
        Y = 0;
    end
    methods
        function obj = DiffRobot() % constructor
        end
    end
end
%-------------------------------------------------------------------------
% The main function of the mobile robot. It will loop reading the sensors,
% Processing, and sending the motor command.
%-------------------------------------------------------------------------
function MainRobot()


% Initialized robot
robot = DiffRobot();


% Initial values
% Force Primitive of the obstacle
obstacles(1).type = 'Repulsion';
% Position of the obstacle
obstacles(1).x = 5000;
obstacles(1).y = 6000;
% Radius of the obstacle
obstacles(1).rad = 200;
obstacles(2).type = 'Repulsion';
obstacles(2).x = 6400;
obstacles(2).y = 3000;
obstacles(2).rad = 200;
% Goal Position
goal = [7300 7600];
```

FIGURE 11.20 MATLAB code for the potential fields of the obstacles and the target as illustrated in Figure 11.19.

```
% For Simulation
% Robot Position
robot.X = 3500;
robot.Y = 2000;

figure(1);
hold on;
plot(goal(1), goal(2), 'b>');
for (i = 1 : 2)
    plot(obstacles(i).x(1), obstacles(i).y(1), 'r>');
end
hold off;

%while 1                  % For Real robot
while (robot.Y <= goal(2)) % For Simulation
    % Read the New Values of the Sensors
    ReadSensors(robot);

    % Path Planning
    vector = PotentialField(robot, obstacles, goal);
    % Set the motor commands according to the new vector
    TurnNForward(robot, vector);

    % Write commands to Motors
    SetMotorCmd(robot);
end
end
%-----------------------------------------------------------------------
function ReadSensors(robot)
% Obtain the new readings of the sensors
end
%-----------------------------------------------------------------------
function SetMotorCmd(robot)
% Send the motor velocities and command to the robot
end
%-----------------------------------------------------------------------
% Path Planning by Potential Field
function vector = PotentialField(robot, obstacles, goal)
nObstacles = length(obstacles);
```

FIGURE 11.20   (continued) MATLAB code for the potential fields of the obstacles and the target as illustrated in Figure 11.19.

```
% Calculate the force that each obstacle exerts on the robot position
for (i = 1 : nObstacles)
    [a d] = Vector([robot.X robot.Y], [obstacles(i).x(1)
obstacles(i).y(1)]);
    switch(obstacles(i).type)
        case 'Repulsion'
            if (d > 3000)
                % 3mts from the object the force is zero
                fObstacles(i).mag = 0;
                fObstacles(i).dir = 0;
            else
                % within 3mts from the object the force increases linearly
                % when getting closer to the obstacle
                fObstacles(i).mag = (3000.0 + obstacles(i).rad - d)/
3000.0;
                % and with opposite orientation from the robot
                if ( a >= 0 )
                    fObstacles(i).dir = a - pi;
                else
                    fObstacles(i).dir = a + pi;
                end
            end
%           if need it add other types of Force Primitives
%           case 'Perpendicular'
%               ...
    end
end
% Attraction force to the goal point
[fGoal.dir d] = Vector([robot.X robot.Y], [goal(1) goal(2)]);
fGoal.mag = 1;
% Add all Forces
vector = SumForces(fObstacles, fGoal);
end
%-------------------------------------------------------------------------
function TurnNForward(robot, vector)
% Make the robot turns to the given direction and move forward with
% (magnitude * MAXSPEED) velocity.
% For simulation
figure(1);
hold on;
[x,y] = pol2cart(vector.dir, vector.mag*500);
```

FIGURE 11.20   (continued) MATLAB code for the potential fields of the obstacles and the target as illustrated in Figure 11.19.

```
plot([robot.X robot.X+x], [robot.Y robot.Y+y], 'g-');
plot([robot.X+x], [robot.Y+y], 'bo');
hold off;
% For Simulation - Update robot position
robot.X = robot.X + x;
robot.Y = robot.Y + y;
end
%-----------------------------------------------------------------------
function [direction magnitude] = Vector(p1, p2)
% Calculate a vector from two points
dp = bsxfun(@minus, p2, p1);
direction = atan2(dp(:,2), dp(:,1));
magnitude = hypot(dp(:,2), dp(:,1));
end
%-----------------------------------------------------------------------
function vR = SumForces(v1, v2)
nV1 = length(v1);
i = 1;
% vR = v1(1) + v2
ax = v1(i).mag * cos(v1(i).dir);
ay = v1(i).mag * sin(v1(i).dir);
bx = v2.mag * cos(v2.dir);
by = v2.mag * sin(v2.dir);
rx = ax + bx;
ry = ay + by;

bx = rx;
by = ry;
% vR = vR + v1(2) + v1(3) + ...
i = i + 1;
while (i <= nV1)
    ax = v1(i).mag * cos(v1(i).dir);
    ay = v1(i).mag * sin(v1(i).dir);
    rx = ax + bx;
    ry = ay + by;

    bx = rx;
    by = ry;
    i = i + 1;
end
vR.mag = hypot(rx, ry);
vR.dir = atan2(ry, rx);
end
```

FIGURE 11.20 (continued) MATLAB code for the potential fields of the obstacles and the target as illustrated in Figure 11.19.

for the function are the robot structure that contains the current (x,y) position of the robot, a list of the obstacles (each obstacle has (x,y) position, radius, and type of primitive), and finally the goal position for the robot in (x,y) coordinates. For the different types of force primitives, the function will calculate the force to the current position of the robot. In our sample code, only the repulsion force has been implemented. As described before, the repulsion force can only be felt when the robot is within 3 m from the object, and it linearly increases when the robot gets closer to the obstacle. After each obstacle's force has been calculated, the attraction force of the goal is computed. This force is always constant and it will be felt everywhere. The last part of the function "PotentialField()" is to add all these forces to produce a vector that will tell the robot the new orientation and magnitude of the maximum speed to use in order to navigate. Figure 11.21b shows the path generated by the resulting vector forces, assuming that the robot is able to navigate accurately to the position described by each calculated vector.

11.4.3 PATH PLANNING USING TOPOLOGICAL MAPS

Topological maps contain only essential information (landmarks); unnecessary details are eliminated. Topological maps are graphs that use nodes to represent different fixed objects, for example, rooms and doors. Using topological maps saves processing time and memory space, as the robot does not have to process so many details. Earlier in Section 11.2.3, topological maps were discussed, and building those maps was described.

The path planning for topological maps takes an initial node and a goal or target node as input. Using an algorithm such as Dijkstra's or Floyd–Warshall's shortest path, it is easy to find the shortest path plans between the two nodes in a topological map. Dijkstra's shortest path algorithm (Sniedovich 2011) is illustrated in Figure 11.22. From the initial node, the algorithm will calculate the distance between the node and its direct neighbors. Then, it will choose the shortest distance to all of its adjacent nodes and mark them according to the calculated distance. Once all the neighbors of the node have been covered, the algorithm will proceed to the next node with the shortest distance. This will continue until the algorithm reaches to the goal node and then it will terminate. The robot will then follow the path pointing toward the shortest line.

11.4.3.1 Case Study of Path Planning Using Topological Maps

In the micromouse competition, the objective is to complete the maze as fast as possible. Apparently, the micromouse robot should find the shortest path in the maze to achieve a fast trial. As discussed in Section 11.2.3, it is possible to generate a topological map from the micromouse maze. When the micromouse detects a gateway, distance and orientation from the previous gateway will be used to connect the new gateway with the previous one. Then, the robot will explore a new direction in the gateway and repeat the previous method when a new gateway is found. The change in direction is also recorded to help identify the gateways. The following matrix provides an example of a topological map:

(a)

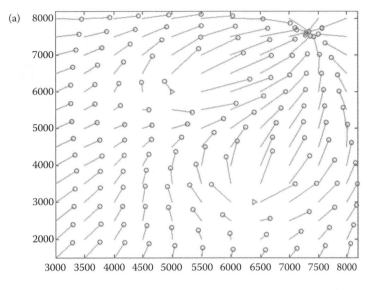

(b)

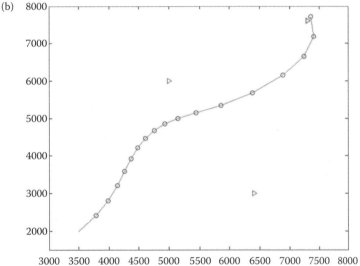

FIGURE 11.21 (a) The resulting force from the objects and the goal in every part for the map. The magnitude of the force is represented by the length of the line and the orientation is shown by the arrow side of the vector. (b) The calculated path.

```
Map = [-1  5 -1 -1 -1 -1;
        5 -1  2 -1 -1 -1;
       -1  2 -1  6  7 -1;
       -1 -1  6 -1 -1 -1;
       -1 -1  7 -1 -1  8;
       -1 -1 -1 -1  8 -1;];
```

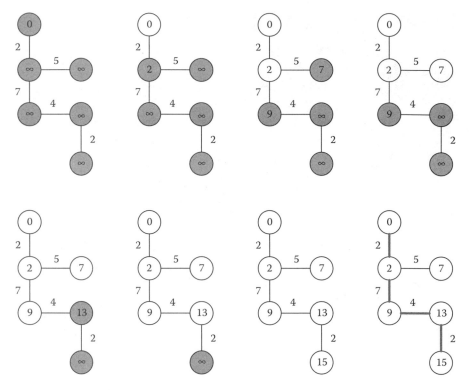

FIGURE 11.22 Dijkstra's shortest path algorithm for finding the shortest path between the initial node and the goal node.

In this map, each row and column represents a node (gateway) in the map, the number in the matrix represents the distance between the node in the row position and the node in the column position. If the distance is −1, it means that there is no direct connection between the nodes. In this map, only the relationship between the nodes and distances is presented, but not the orientation (which can be stored in another matrix).

In Figure 11.23, the MATLAB code for the path planning in a topological map is presented. The Dijkstra algorithm is used to find the shortest path. The function "Dijkstra()" receives three arguments, the map as a relational matrix with the distance between the nodes, the initial node, and the goal or target node. The function "Dijkstra()" will return a path from the initial node to the goal node. If there is a path, it will be a list of waypoints that robot must follow to reach the goal node. Note that in the function "MainRobot()" the path is calculated first before the robot starts moving; hence, this is a full path planner. This approach is different than the partial path planner presented for the force field in Section 11.4.2.1.

```
%
% The class DiffRobot the members of the class robot are declared as:
%
% robot.ultrasonic_sensors(3)    - 1 - Left, 2 - Right, 3 - Front
% robot.motor_vel(2);            - 1 - Left, 2 - Right
%


classdef DiffRobot < handle
    properties
        motor_vel = [ 0.0 0.0 ];
        ultrasonic_sensors = [ 0.0 0.0 0.0 ];
    end
    methods
        function obj = DiffRobot() % constructor
        end
    end
end


% This is the main function of the mobile robot. It will
% loop reading the sensors, Processing, and sending the motor command.
%


function MainRobot()

% Initialized robot
robot = DiffRobot();

% For Simulation
map = [ -1   5 -1 -1 -1 -1;
         5  -1  2 -1 -1 -1;
        -1   2 -1  6  7 -1;
        -1  -1  6 -1 -1 -1;
        -1  -1  7 -1 -1  8;
        -1  -1 -1 -1  8 -1;];
src = 1;
dst = 5;

% Find the shortest path from node src to node dst
```

FIGURE 11.23 MATLAB code for the micromouse during a race run. It will first calculate the path and visit each node until it reaches the target node. The path is found by using the Dijkstra algorithm.

```
path = Dijkstra(map, src, dst);
nNodes = length(path);
currentNode = 1;


while (currentNode <= nNodes)
    % Read the New Values of the Sensors
    ReadSensors(robot);


    % Navigate
    GoTo(path(currentNode));
    if (Reached(path(currentNode)))
        robot.motor_cmd = STOP;
        currentNode = currentNode + 1;
    end


    % Write to Motors
    SetMotorCmd(robot);
end
end


function ReadSensors(robot)
% Obtain the new readings of the sensors
end


function SetMotorCmd(robot)
% Send the motor velocities and command to the robot
end


% Shortest path from node src to node dst using Dijkstra algorithm.
function path = Dijkstra(map, src, dst)


nNodes = size(map,1);
% Assign to all nodes a tentative distance value of infinity, and the
% initial node a value of zero. Set all the previous nodes to undefined.
Distance = Inf * ones(nNodes, 1);
Distance(src) = 0;
PreviousNode = -1 * ones(nNodes,1);


% Mark all nodes unvisited. Set the initial node as current.
```

FIGURE 11.23 (continued) MATLAB code for the micromouse during a race run. It will first calculate the path and visit each node until it reaches the target node. The path is found by using the Dijkstra algorithm.

```
Visited = zeros(nNodes, 1);   % 0 - unvisited, 1 - visited
CurrentNode = src;


% If the destination node has been marked visited or if the smallest
% tentative distance for the current nodes is infinity, then stop.
while (~Visited(dst) && Distance(CurrentNode) ~= Inf)


    % For the current node, consider all of its neighbors and calculate
    % their tentative distances. If this distance is less than the
    % previously recorded tentative distance of the neighbor, then overwrite
    % that distance.
    for (i = 1 : nNodes)
        if (map(CurrentNode,i)>0)
            alt = map(CurrentNode,i) + Distance(CurrentNode);
            if (alt < Distance(i))
                Distance(i) = alt;
                PreviousNode(i) = CurrentNode;
            end
        end
    end


    % When we are done considering all of the neighbors of the current node,
    % mark the current node as visited. its distance recorded now is final.
    Visited(CurrentNode) = 1;


    % Set the unvisited node marked with the smallest tentative distance as
    % the next "current node".
    smallestDist = 0;
    for (i = 1 : nNodes)
        if (~Visited(i))
            if (smallestDist == 0)
                smallestDist = i;
            elseif (Distance(i) < Distance(smallestDist))
                smallestDist = i;
            end
        end
    end
    CurrentNode = smallestDist;


end
```

FIGURE 11.23 (continued) MATLAB code for the micromouse during a race run. It will first calculate the path and visit each node until it reaches the target node. The path is found by using the Dijkstra algorithm.

```
% Obtain the path from the src to the dst
i = 1;
path(i) = dst;
while (path(i) ~= -1)
    i = i + 1;
    path(i) = PreviousNode(path(i-1));
end
path(i) = [];
path = fliplr(path);
end
```

FIGURE 11.23 (continued) MATLAB code for the micromouse during a race run. It will first calculate the path and visit each node until it reaches the target node. The path is found by using the Dijkstra algorithm.

REFERENCES

Acosta-Calderon, C.A., Mohan, R.E. and Zhou, C. 2010. Distributed architecture for dynamic role behaviour in humanoid soccer robots. *Robot Soccer*, Ed. Kordic, V. Vienna, Austria: IN-TECH, 121–138.

Jones, J.L. and Flynn, A.M. 1993. *Mobile Robots: Inspiration to Implementation*. Wellesley, MA: A. K. Peters.

RoboErectus@Home, 2012, RoboCup//www.robo-erectus.org/HomeLeague.php.

Siegwart, R. and Nourbakhsh, I.R. 2004. *Introduction to Autonomous Mobile Robots*. Cambridge. MA: The MIT Press.

Singapore Robotic Games website. *Micromouse Competition* 2010, http://guppy.mpe.nus.edu.sg/srg/srg10/mm.pdf

Sniedovich, M. 2011. *Dynamic Programming: Foundations and Principles*. Boca Raton, FL: CRC Press.

12 Robot Autonomy, Decision-Making, and Learning

12.1 INTRODUCTION

In most robotic games, the robots are expected to display some level of autonomy. In robotics, autonomy is understood as the ability to perceive the environment and take decisions about the actions that would help to accomplish a given task. Furthermore, autonomy can be expressed at different levels, from a fully autonomous robot to a teleoperated robot. A fully autonomous robot is robust to the changes in the dynamic environment and requires no human intervention to accomplish a given task (Jones and Flynn 1993). A teleoperated robot, on the other hand, needs every decision from the user to perform its actions to achieve its task. To achieve autonomy, a robot should have a well-structured mechanism to link the sensor information to the action with a purpose. This mechanism uses the sensor information to make the best decisions, as well as to deal with situations when the actions fail to achieve the desired states in the environment. The robot architecture defines how sensor information is used as an input to make decisions and how actions are monitored until the desired state has been achieved. The selection of the robot architecture will depend on the task at hand, but most importantly on the level of autonomy we would like to achieve. Some of the behaviors that we would like a robot to demonstrate are very hard to program. In those situations, learning methods can help us to achieve the desired behavior. In this chapter, we will discuss robot autonomy and the different types of robot architectures and how they help the decision-making process. This chapter also shows some learning algorithms to achieve certain robot behaviors.

12.2 ROBOT AUTONOMY

Autonomous robots are able to carry out useful tasks without human supervision. As mentioned, robots exhibit different levels of autonomy. Fully autonomous robots are able to make their own decisions and execute actions without any human intervention. Semiautonomous robots are partially controlled by a human operator; these robots might make some decisions and execute some actions on their own, but the human operator might control the robot or overwrite its decisions. A teleoperated robot is a remotely controlled system that receives commands from the human operator. What a robot can perceive and what decisions it can make depend on its degree of autonomy. Consequently, the design of what a robot should accomplish is directly

(a)   (b)

FIGURE 12.1   (a) RC Sumo robot game. Humans observe the state of the game and decide what actions should be executed. These actions are transmitted to the robot via a remote control. (b) Autonomous Sumo robot competition. Humans are only allowed to place the robots in the ring and press the start button. The robot then senses the opponent and tries to push it out of the ring.

related to what type of sensors it has and the processing and decision-making capability based on this sensorial data.

The RC Sumo robot competition is a good example of teleoperated robots in robotic games (RC Sumo 2013). The objective in this competition is to push the opponent's robot out of the ring. In this game, the human operator observes the ring and both the robots. The operator makes the decisions to win the game by simply estimating the behavior of the opponent. In this game, the robots are not equipped with any sensors since the human operator is able to do the perception, decision-making, and controlling of the robot (see Figure 12.1a).

On the other hand, the Autonomous Sumo Robot competition has the same objective as the RC Sumo robot, which is simply to push the opponent out of the ring (Autonomous Sumo Robot 2013). However, in this case, the robots are not controlled by human operators. Robots are equipped with sensors and controller boards to collect the sensor data, make sense of what is detected, and take action to win the game (see Figure 12.1b). As discussed in the earlier chapters, a robot needs to perceive its environment and extract relevant information to perform its task. For example, if a robot is equipped with ultrasonic sensors, it is able to detect the position of the opponent by using the data provided by the sensors.

12.3   DECISION-MAKING

Decision-making is a crucial part of robot autonomy. Autonomous robots form models of the environment and the objects around them using sensorial data, and eventually take actions to complete their goals. Using environment models gives an advantage in delivering a plan. However, executing the plan will not be so easy when the environment is constantly changing. The real world is unpredictable and the robot should work under that assumption. What goes between the sensing and the acting determines the success or failure of the robot significantly. For example, a robot that connects sensing to the acting directly may not demonstrate a great

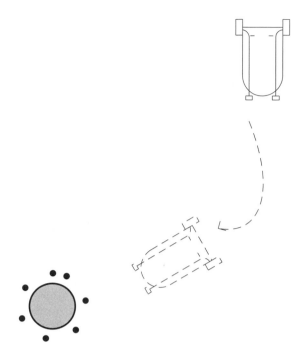

FIGURE 12.2   A simple Braitenberg vehicle displaying "love" for the light source.

intelligence capability. However, its actions will be rapid since it does not make lengthy computations and deliberate on them.

Simple behaviors can be achieved by connecting sensors to actuators, as is proven by the work of Valentino Braitenberg (Braitenberg 1984). Braitenberg developed a model of simple vehicles with sensors and actuators and simply provided interconnections between these two components. These connections produce behavior that is not as simple as the vehicle itself. Figure 12.2 shows a Braitenberg vehicle with two light sensors and two motors; the behavior depicted in the figure is often interpreted as love because the vehicle is attracted to the light.

In the following sections, we will discuss three different approaches for decision-making: the classical approach, the reactive approach, and the hybrid approach. These approaches will sound much like the three strategies to process the sensor measurements discussed earlier in Chapter 11. The reason for this is that in robotics it is almost impossible to separate completely the perception, decision, and actuation.

12.3.1   CLASSICAL DECISION-MAKING

The classical architecture to control a robot is defined as "sense–plan–act." This architecture performs each of these three processes in sequence; the robot will collect data from the sensors, then these data are used to obtain a plan, and finally the plan is executed. However, sense–plan–act has serious complications for most real-world applications. To generate a plan for the real dynamic world, the robot may need to collect a lot of information, which results in a huge amount of data and slow computation of a proper plan. The dynamic nature of the real world also implies that the robot needs to update its sensorial data constantly. Consequently, the world model needs to be updated. This may invalidate the current plan that the robot is executing and a new plan must be computed again.

12.3.2 REACTIVE DECISION-MAKING

Reactive control for robots is based on the direct connection between the sensors and the actuators with minimal information for representing the state of the world. A reactive robot presents instinctive responses to particular situations, like a reflex to dangerous situations. Reactive rules or behaviors can be seen as independent modules (as shown in Figure 12.3) that are triggered by stimuli in the sensors to produce a specific action. For example, a robot makes a turn to avoid an obstacle sensed in front, or the robot may follow a wall that is sensed on its right-hand side. Reactive rules should be designed as unique situations, where only a sensor triggers a particular action. However, this is not always possible. In those cases, it is necessary to find a way to solve the conflicts between the reactive rules. For example, consider that the above examples of reactive rules are implemented in a robot. If the robot faces an obstacle in front and a wall on its right, which one of the two behaviors should be triggered? There are different techniques of solving conflicts between the reactive rules or behaviors. One method is by arbitration, which means an action will be chosen from multiple candidates (see Figure 12.3a). Another alternative is fusion, where the commands to the actuators from the actions are fused into a single signal for the actuators (see Figure 12.3b). Using fusion will produce more different types of behaviors than just arbitration, and it will increase the level of complexity of the system. It is important to note that in some situations the robot is required to perform more sophisticated behaviors to achieve its goal.

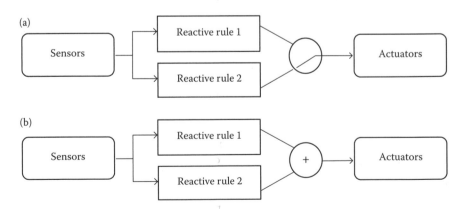

FIGURE 12.3   Two ways of solving conflicts between the reactive rules: (a) arbitration and (b) fusion.

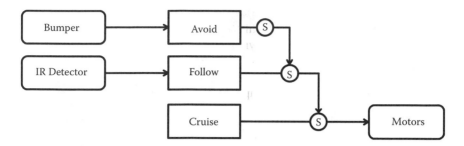

FIGURE 12.4 A subsumption controller consisting of three reactive rules (or behaviors). The lower-level behaviors are located at the bottom, and their output can be supressed by the higher-level behaviors on top.

Another approach used in resolving conflict is the subsumption method. The subsumption architecture helps to solve conflicts between rules and enables to create complex reactive systems that consist of simple parts that can be added or removed to change the functionality of a system. This architecture uses a prioritized arbitration scheme to resolve any possible conflict among the reactive rules (Brooks 1985). This means that reactive rules or behaviors will have priorities that should be considered in the design of the system. A higher-level behavior can temporarily supress lower-level behaviors as shown in Figure 12.4. The implementation of the subsumption controller always starts from the lower-level behaviors. The higher-level behaviors are implemented only after the lower-level behaviors are debugged and tested. This way the complexity of the system will be reduced during implementation (not for the execution). The complexity is also reduced because the connection among the behaviors is reduced to a simple suppression or inhibition of the sensor or motor signals.

12.3.2.1   Case Study on Reactive Decision-Making

The theme of the RoboCupJunior Rescue competition is disaster scenarios where a robot must follow a line and deal with obstacles and victims on its path. During the competition, the robot faces different challenges. For example, the robot needs to follow a line, and these lines may be broken. The robot may need to avoid obstacles, climb up and down slopes, and identify and rescue victims. Of course, the robot tries to perform all this as fast as it can. Figure 12.5a presents a scenario of RoboCupJunior Rescue, where a robot follows the line and detects an obstacle on its path. The robot should avoid colliding with the obstacle and go around it, find the line again, and continue its course by following the line.

For this study case, we use the RoboCup CoSpace robot, shown in Figure 12.5b, which is equipped with two infrared sensors located at the front of the robot facing the floor; three ultrasonic sensors, one at the front and one on each side of the robot; and two independent motor wheels and one caster wheel (RoboCup CoSpace 2012). Let us also use the scenario presented in Figure 12.5a for this study case. There are two situations the robot needs to address: the line following and the going around the obstacle. The CoSpace robot achieves line following by using two infrared sensors that are facing down toward the floor and located in front of the robot. Obstacle detection is performed by the three ultrasonic sensors located in front of the robot

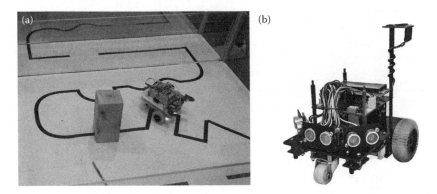

FIGURE 12.5 (a) A robot trying the rescue field during the RoboCup Singapore Open 2012. (b) The RoboCup CoSpace robot is a simple robot used for RoboCupJunior and educational purposes. The robot was developed at the Advanced Robotics and Intelligent Control Centre of the Singapore Polytechnic.

collectively covering a wide range. Although it is possible to use other types of sensors to detect obstacles, in this case study, the robot employs ultrasonic sensors to detect them from a distance.

The robot controller should consider two situations: follow the line and avoid the obstacles. Since the robot has to follow the line most of the time, this should be the simplest of the behaviors and thus Follow-Line will be the lower-level behavior. The obstacle avoidance should take control of the system to avoid a collision with the obstacles; however, since the obstacles are located on the line, the robot should be able to go around the obstacle and find the line again; thus, Round-Obstacle will be a higher-level behavior. With these design considerations, the subsumption controller for the rescue robot is presented in Figure 12.6.

As discussed before, the implementation will start from the lower-level behaviors. The Follow-Line behavior will use the data from the infrared sensors; the value returned by the sensors will be different for an area white in color (the arena space) and an area black in color (the line). The idea is to implement these behaviors in parallel; however, if the microprocessor or the processing unit that is used does not support this, then the behaviors can be implemented as functions. The code shown in Figure 12.7

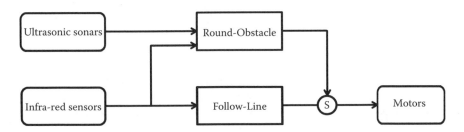

FIGURE 12.6 The proposed controller for the rescue challenge. It consists of two behaviors: Follow-Line and Round-Obstacle. Follow-Line is the lower-level behavior that can be supressed by the higher-level behavior Round-Obstacle.

```
%
% The CoSpace function uses an structure as argument the member
% of the structure robot are declared as:
%
% robot.ir_sensors(2)            - 0 - Left, 1 - Right
% robot.ultrasonic_sensors(3)    - 0 - Left, 1 - Right, 2 - Front
% robot.motor_command;           - 0 - stop, 1 - Forward,
%                                  2 - Turn Left, 3 - Turn Right
% robot.state                    - 0 - NO_STATE, 1 - OBSTACLE_FRONT_LINE,
%                                  2 - OBSTACLE_LEFT, 3 - OBSTACLE_LEFT_LINE
%

function ret = CoSpace(robot)
% This is the main function of the CoSpace. It will loop reading the
% sensors, following the line, and sending the motor command.

% Initialized robot
ret = Init(robot);

while ret == 1
    % Read the New Values of the Sensors
    ReadSensors(robot);

    % Follow the Line
    FollowLine(robot);

    % Write to Motors
    SetMotorCmd(robot);
end

function ret = Init(robot)
% Any necessary setup for the robot...
ret = 1;

function ReadSensors(robot)
% Obtain the new readings of the Infra-Red and Ultrasonic

function FollowLine(robot)

delta = robot.ir_sensors(RIGHT) - robot.ir_sensors(LEFT);
if (abs(delta) > IR_DEAD_ZONE)
    if (delta > 0)
        % Line on left, turn left
        robot.motor_command = TURN_LEFT;
    else
        % Otherwise turn right
        robot.motor_command = TURN_RIGHT;
    end
else
    % No difference, deactivate
    robot.motor_command = FORWARD;
end

function SetMotorCmd(robot)
% Send the motor_command to the robot
```

FIGURE 12.7   MATLAB code of the implementation for the Follow-Line behavior for the CoSpace robot.

presents the Follow-Line behavior and subsequent behavior considering a sequential processing instead of parallel processing.

The main function in this code is CoSpace, and it has two key components, the first one is to initialize the robot by calling the function "Init()" and the second one is to execute the controller of the robot in an endless loop. The initialization of the robot is left blank in this example since it is different from robot to robot. However, everything necessary to set up the robot and the communication with the hardware must be placed inside this function. When all the initialization is completed, the function must return the value of "one" that indicates that everything is ready to proceed with the control of the robot, any other value returned will not start the endless loop, causing the program to end. The control loop will obtain the sensor readings from the hardware once the readings are available. The hardware state will be conveyed to the variable named "robot"; this information will be used to make decisions. After that, the control loop will call the function "FollowLine()." This function will compare the values of the left and right infrared sensors to follow the line; based on these comparisons, a motor command will be selected. Motor commands are used by the function "SetMotorCmd()" to be translated into hardware signals and finally to be sent to the motors of the robot. Functions "ReadSensors()," "SetMotorCmd()," and "Init()" are hardware dependent; therefore, we did not explicitly show them here. Another point about the example code is that it needs to be tuned for the certain values of the sensors for comparison. For these, we prefer to use names that provide a better understanding of what this value should be; in the actual implementation, these names correspond to the readings from the sensors.

The robot is expected to follow the line; when it reaches the end of the line, it will stop. The robot will detect the difference between the left and right infrared sensors to determine where the line is located and produce the appropriate command. Since the robot is reacting to this difference and the processing of this value is fast, the reaction of the robot to these stimuli is almost instantaneous.

Once the Follow-Line behavior has been tested and properly debugged, then it is time to write the higher-level behavior to avoid the collision and to go around the object. Usually higher-level behavior would be more complex than those in the lower levels. The proposed Round-Obstacle behavior may require a few steps, and it may not be possible to achieve on a single iteration of the program. In this case, a finite-state machine (FSM) is used to represent a behavior that is composed of a fixed number of states and transitions. An FSM consists of states and transitions; a state would represent a state of the environment or the robot, while the transitions would represent the actions that the robot need to execute to reach a particular state. The proposed behavior is illustrated as FSM in Figure 12.8.

Note that the transition among the states may take several iterations, and it is important to monitor the sensors to perceive the changes in the states. Figure 12.9 presents the MATLAB® code implementation of the Round-Obstacle behavior as described by the FSM in Figure 12.8. The names written in uppercase letters correspond to the values that need to be tuned for actual system. For example, when using the infrared sensors to check for a black line, the real value returned by the sensor should be replaced with the name "LINE" in the code. The value of the LINE should be greater than the background color value. The final remark is the function "Init()"

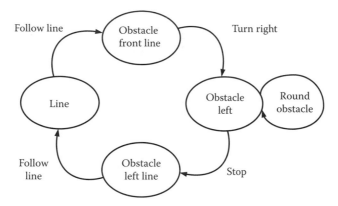

FIGURE 12.8 The finite-state machine representation for the Round-Obstacle behavior.

will now also initialize the value of the robot.state variable to NO_STATE, and this will ensure that the FSM will start with the state determined by the sensors.

Introducing this behavior as a higher-level behavior would require modifying the main function to give the priority to the Round-Obstacle behavior when an obstacle is detected and thus suppress the output of the Follow-Line behavior. Figure 12.10 presents the main function after adding the suppression of the Follow-Line behavior by the Round-Obstacle behavior.

In Figure 12.10, the function "RoundObstacle()" is called; if the function returns true, then the "FollowLine()" function will not be called. This may seem similar to suppress the output of the "FollowLine()" function. The advantage is that the performance of the system will not be affected by increasing computations. The emerging behavior of the system can be seen as the robot follows the line and avoids the obstacle.

12.3.3 HYBRID DECISION-MAKING

Reactive control has very a fast response when compared to the classic sense–plan–act control paradigm. On the other hand, it is not a flexible method when it comes to deliberating a plan. The hybrid control mechanism involves a combination of reactive and deliberative control paradigms to get the best of both the worlds. A hybrid control mechanism aims for fast response to sensory inputs as well as for producing a flexible plan to attain the goal (Mataric 2007). A common way of implementing these systems consists of three layers as shown in Figure 12.11.

Signals from the sensors will usually input to the reactive and middle layer. Actuators of the robot are also connected to the reactive layer. The role of the middle layer is to generate environment models that will be used by the planning layer, and to keep these updated regularly. In addition, the middle layer receives the actions to execute from the planning layer. These actions are passed down to the reactive layer for execution and the success or failure of them is monitored by the middle layer. In the case of failure, the middle layer avoids replanning by trying other actions that could lead to the same goal. The design and implementation of hybrid systems is challenging because they must bring together the components of the reactive and

```
function ret = Init(robot)
% Any necessary setup for the robot...
robot.state = NO_STATE;
ret = 1;

function ret = RoundObstacle(robot)

is_line = 0;

if (robot.ir_sensors(LEFT) >= LINE || robot.ir_sensors(RIGHT) >= LINE)
    is_line = 1;
end

if (robot.state == NO_STATE && robot.ultrasonic_sensors(FRONT) <
OBJECT_NEAR && is_line == 1)
    robot.state = OBSTACLE_FRONT_LINE;
else
    ret = 0;
    return
end

switch robot.state
    case OBSTACLE_FRONT_LINE
        robot.motor_command = TURN_RIGHT;
        if (robot.ultrasonic_sensors(LEFT) < OBJECT_NEAR)
            robot.state = OBSTACLE_LEFT;
        end
    case OBSTACLE_LEFT
        dist = OBJECT_NEAR - robot.ultrasonic_sensors(LEFT);
        if (abs(dist) <= 10)
            robot.motor_command = FORWARD;
        else
            if (dist > 0)
                robot.motor_command = TURN_RIGHT;
            else
                robot.motor_command = TURN_LEFT;
            end
        if (is_line == 1)
            robot.state = OBSTACLE_LEFT_LINE;
        end
        end
    case OBSTACLE_LEFT_LINE
        robot.motor_command = STOP;
        robot.state = NO_STATE;
end

ret = 1;
```

FIGURE 12.9 The implementation of the Round-Obstacle behavior. Note that the state variable is part of the robot structure, and it is used to point at the current state in the FSM.

deliberative systems. All these components should be coupled in such a way that they work seamlessly as a single system.

12.4 ROBOT LEARNING

The challenges put in robotic games are increasing continuously. Robots are expected to accomplish tasks in unknown and dynamic environments. This objective is in line with expectations from mobile robots in practice.

```
function ret = CoSpace2(robot)
% This is the main function of the CoSpace. It will loop reading the
% sensors, following the line, go around the obstacle,
% and sending the motor command.

% Initialized robot
ret = Init(robot);

while ret == 1
    % Read the New Values of the Sensors
    ReadSensors(robot);

    % The output FollowLine will be supressed is there an obstacle
    % RoundObstacle will return a true if there is an object
    % or if robot is going around the object.
    if (RoundObstacle(robot) == 0)
        FollowLine(robot);
    end

    % Write to Motors
    SetMotorCmd(robot);
end
```

FIGURE 12.10 The suppression of the lower-level behavior could be implemented in code with if-else or switch structures. Notice that the higher-level behavior here returns a Boolean (one or zero) value that is used to determine if the lower-level behavior is suppressed or executed.

Nowadays, mobile robots are utilized to perform tasks in remote, dangerous, and unknown environments such as in nuclear disasters, remote planets, or ocean explorations. It is important for a mobile robot to have an understanding of its environment to operate in that environment and to deal with any possible situation. However, it is impossible to write programs that are able to predict every single possible situation that might arise in an environment. This problem becomes even more complex if the robot has to work in an unknown and unpredictable environment. On the other

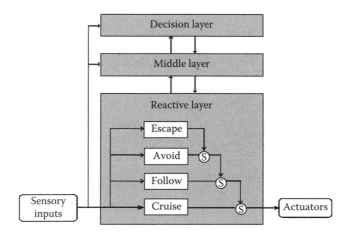

FIGURE 12.11 A hybrid system brings together the classical deliberative system and the reactive systems.

hand, the learning capability provides the robot to learn about its environment and the task it has to accomplish so that it can adapt to the changing environment and achieve its goal. In this section, we will introduce some learning methods that can be implemented for different applications.

12.4.1 ARTIFICIAL NEURAL NETWORKS

An artificial neural network (ANN) is a network that connects artificial neurons that are programmed to behave like biological neurons. A neural network is able to gain knowledge through the process of learning, also referred to as training. The knowledge is stored within the connections of the artificial neurons in the form of weights. These weights play an important role in decision-making. Neural networks are good at generalizing information, and they are able to respond to new situations. As neurons form the basis of a network, we will first discuss these basic elements and behavior of an artificial neuron.

There are four key elements in an artificial neuron; they are: inputs, weights, activation function, and output. As illustrated in Figure 12.12, the inputs (like synapses) are multiplied by weights (strength of the respective signals). All these values are then added and then computed by a mathematical function that determines the activation of the neuron. The amplitude of the output of the artificial neuron depends on the activation function. In mathematical terms, we describe a neuron with a pair of equations:

$$v = \sum_{i=1}^{M} w_i x_i \tag{12.1}$$

$$y = \varphi(v) \tag{12.2}$$

where $[x_1, \ldots, x_i, \ldots, x_M]$ are the input signals, $[w_1, \ldots, w_i, \ldots, w_M]$ are the synaptic weights of the neuron, v is the induced local field or activation potential of neuron,

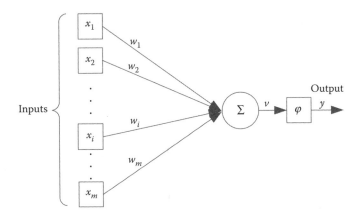

FIGURE 12.12 Basic elements of an artificial neuron.

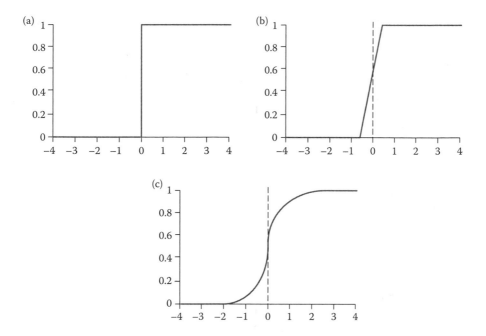

FIGURE 12.13 Three different types of activation functions: (a) threshold, (b) piecewise-linear, and (c) sigmoid functions. According to the required output, an activation function can be used for the neurons in a network.

ϕ is the activation function, and y is the output signal of the neuron. The activation function defines the output of the neuron in terms of the induced local field. The output of the activation function is in the range of 0–1, and this output depends on the type of activation function employed by the neuron. Figure 12.13 shows three types of common activation functions: threshold, piecewise-linear, and sigmoid functions.

There are many different types of neural networks; each type needs to be modeled according to the task in hand and the number of inputs and outputs. An ANN is typically defined by three parameters:

1. The interconnection pattern between different layers of neurons
2. The learning process for updating the weights of the interconnections
3. The activation function that converts a neuron's weighted input to its output activation

12.4.1.1 Perceptron

The perceptron is the simplest neural network. It is designed to solve two-class pattern classification problems. The McCulloch–Pitts neuron model is adopted in the perceptron, and it uses a threshold activation function. All the inputs are connected to a single neuron, and it has only one output. It is considered as a feedforward network. Feedforward neural networks solve problems by input–output functional mappings, which are formed by learning from examples used as training data. Feedforward

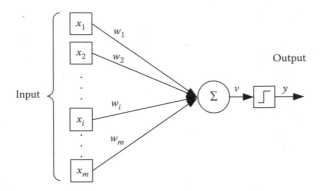

FIGURE 12.14 A perceptron with McCulloch–Pitts neuron.

networks have the good ability to approximate new situations. This is because feed-forward networks use the new inputs and the weights of the networks to interpolate new output data that were not encountered in the training data set.

Figure 12.14 shows a perceptron using the McCulloch–Pitts neuron. The neuron computes the weighted sum v of all M inputs x, as shown in Equation 12.1. As this neuron uses a threshold activation function, according to Equation 12.2, the weighted sum v is then compared to a fix threshold value to produce the output. Equation 12.3 presents the threshold activation function:

$$\phi(v) = \begin{cases} 1 & \text{if} \quad v \geq 0 \\ 0 & \text{if} \quad v < 0 \end{cases} \tag{12.3}$$

The perceptron is capable of computing any function when suitable weights have been given. As mentioned before, the information to solve a given problem or to take a decision resides in the weights. The problem is how to choose those suitable weights. When the number of inputs is small, the values of the weights can be manually adjusted to obtain the desired behavior. However, when the number of inputs is large, this method will not be effective. In those cases, the network can learn by using the backpropagation algorithm which exposes the neural network to training data constantly. This would mean that some data are collected from the inputs and desired output, and this training set of data is used to teach the network to produce similar outputs when similar inputs occur. The training data are then defined as

$$\left[x_1(n), \ldots, x_i(n), \ldots, x_M(n), d(n)\right]_{n=1}^{N} \tag{12.4}$$

where $[x_1, \ldots, x_i, \ldots, x_M]$ are the input signals, and $[d]$ is the desired output for the nth training example of the training dataset. The number of training examples N must not be too small; otherwise, the network will not be flexible enough to react to different situations. If the training set is too large, the learning process will take a very long time. There is also the risk of overtraining, which means making the network specialized to respond to only certain inputs or situations. Overtraining happens when a large number of training samples is biased toward specific inputs or situations. A good training set should have a broad distribution of samples representing most of the possible situations that a robot will encounter.

An important issue is how to obtain the training data. In robotics, we commonly use the sensor values as inputs and the motor drive values as outputs. Thus, it is possible to write a program to log both the sensor values and motor drive values while the robot is performing a task or robot being teleoperated for the task to perform.

In the backpropagation algorithm, the learning consists of two stages, namely forward and backward stages. In the forward stage, an input activates the network to produce an output. The network output is then compared with a desired output to compute the error. During the backward stage, this error is fed back to the network to adjust its weights and to reduce the error. Thus, the model gets closer to the desired output with every iteration, and this is repeated until the error is small. Backpropagation calculates the gradient of the error of the network regarding the network's weights. The error signal at the output of the neuron at iteration n (i.e., presentation of the nth training example) is defined by

$$e(n) = d(n) - y(n) \tag{12.5}$$

where $d(n)$ is the desired output for the nth training example at the neuron. The total error of the network is obtained by summing all the errors of the neurons. Apparently, for a perceptron, this would be only one neuron.

$$\varepsilon(n) = \frac{1}{2} \sum e^2(n) \tag{12.6}$$

The average squared error is obtained by summing the total error of all the data in the training set. The objective of the learning process is to minimize this average error.

$$\varepsilon_{av} = \frac{1}{2} \sum_{n=1}^{N} \varepsilon(n) \tag{12.7}$$

The update rule of the synaptic weights is defined by the correction Δw_i

$$w_i(n + 1) = w_i(n) + \Delta w_i(n) \tag{12.8}$$

$$\Delta w_i(n) = -\eta \delta(n) x_i(n) \tag{12.9}$$

$$\delta(n) = e(n) \varphi'(v(n)) \tag{12.10}$$

The correction Δw_i is defined by the learning parameter η, the local gradient δ, and the current input x_i. The local gradient δ is composed of the neuron error and the derivative of the activation function.

The activation function used by the McCulloch–Pitts model presented in Equation 12.3 is not differentiable; therefore, it is necessary to use an activation function such as a sigmoid or a hyperbolic tangent. As discussed in the previous section, there are a number of activation functions, and many can be modeled according to the desired output. The sigmoid function and its derivative are represented as follows:

$$\varphi(v) = \frac{1}{1 + e^{-av}} \tag{12.11}$$

$$\varphi'(v) = a\varphi(v)(1 - \varphi(v)) \tag{12.12}$$

Parameter a in the equation defines the length of the shape produced by the equation. The training algorithm for the neural network is described with the following steps:

1. Initialize the weights of the neuron to zero.
2. For each training data
    2.1. Calculate the output of the network $y(n)$ for the training data n with Equations 12.1 and 12.2.
    2.2. Calculate the error of the neuron $e(n)$ with Equation 12.5.
    2.3. Calculate the local gradient for the neuron $\delta(n)$ with Equation 12.10.
    2.4. For each input
        2.4.1. Calculate the weight correction $\Delta w_i(n)$ with Equation 12.9.
        2.4.2. Update the weight $w_i(n + 1)$ with Equation 12.8.
    2.5. Calculate the total error of the network $\varepsilon(n)$ for the training data n with Equation 12.6.
3. Calculate the average error of the total errors for all the training data sets, with Equation 12.7.
4. If the average error is greater than a threshold, go back to step 2. If the learning process has been repeated by a maximum number of epochs and the average network does not converge, then stop to avoid an endless loop.

In robotics, input to neural network will be the sensor signals, such as sonar and lasers range sensors, which deliver distance to objects and the output from the network will be translated to a signal for actuators. So, the decision on robot movements will be dependent on the neural network directly. The distance values from the sensors have to be normalized to the 0–1 range before passing them to the network. The output from the network will be in the range of 0–1. Similarly, the output from the network also needs to be translated to proper motor signal.

12.4.1.2 Case Study on Perceptron with Learning

In this case study, we will again study the CoSpace robot presented in Section 12.3.2.1 to illustrate how a robot navigates in an unknown environment after learning with an ANN. We will focus on the scenario where the robot needs to move in random directions and avoid obstacles. Obstacle avoidance is a behavior that could be programmed in any robot; however, using an ANN will provide an advantage when the robot faces new types of obstacles or unknown environments.

We need to start by modeling the ANN. In this case, we need to observe the robot and decide how the neural network will be interfaced with the robot hardware. The robot consists of four ultrasonic sensors, and they are used to determine the distance of the obstacles from the front, the left, and the right directions. A fourth ultrasonic sensor is located at the rear of the robot, and it can only perceive obstacles at the back of the robot. Since the front and the two side sensors are mainly used for avoiding obstacles, only these sensors will be used as inputs to the neural network. The output of the network should then produce the necessary speed for the left and right wheels so that the robot can avoid the obstacle. The neural network in this case will consist of two neurons as shown in Figure 12.15. Since two output signals are needed for the motors, each neuron will be associated to one motor. The output signal to motors can be positive (forward) or negative (backward); we need to have an activation function that can provide us the required range. Here, we will employ the hyperbolic tangent function, which is a nonlinear sigmoid function, and its output range extends from −1 to 1.

$$\varphi(v) = \tanh(v) \tag{12.13}$$

The derivative of the hyperbolic tangent activation function is needed for the learning using the backpropagation method.

$$\varphi'(v) = 1 - \tanh^2(v) \tag{12.14}$$

As we have a model of the neural network, it is necessary to collect the training data. As described previously, the data can be collected from a previous program, or by teleoperating the robot and logging the sensors and motor values. Most of the mobile robotic systems nowadays have an option to be controlled by a human

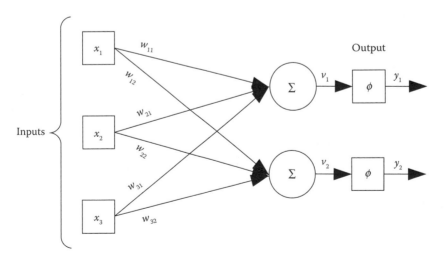

FIGURE 12.15 The proposed neural network for the obstacle avoidance in the CoSpace robot with two perceptrons and three common inputs.

operator. In the case of CoSpace robot, we developed a PC-based program that sends commands, entered by the user via keyboard, directly to the controller board of the robot. The controller board then sends these commands to the wheel motors of the robot. It also reads the three ultrasonic sensors and the speeds of the two motors to record this information into a file. A new set of data is added in every cycle of the controller board; for the CoSpace robot, that means every 100 ms. We stop data recording when the robot has been driven to a few situations which we would like to teach. A sample of the content of this file is presented in Table 12.1. The user controls the robot to avoid obstacles when the obstacles are presented in the front, right, or left side of the robot, speeds are decreased when obstacles are detected, and a simple strategy of turning to the right when an obstacle is faced in front is applied. For the obstacles at the sides, the strategy is simply to turn to the opposite direction. When there is no obstacle in view of the robot, the speed is increased and robot is allowed to move forward.

From Table 12.1, we observe that none of the input values range from 0 to 1, and neither the output range is from −1 to 1. The next thing is to normalize these values by taking into consideration the maximum and minimum range of the sensors and motors. Table 12.2 shows the normalized values of the file presented in Table 12.1.

Once the network is trained and used for controlling the robot, all the inputs must be normalized and the output signal must be converted to suit actual motor commands. After collecting enough training data, the learning process can start. The program presented in Figure 12.16 performs the learning of the network, using the backpropagation method introduced previously.

TABLE 12.1
Parts of the Training Data Set Generated for the CoSpace Robot Generated by Controlling the Robot with a Keyboard

Sensor Front	Sensor Left	Sensor Right	Velocity Left	Velocity Right
1000.00	1000.00	1000.00	16,384	16,384
1000.00	1000.00	1000.00	16,384	16,384
1000.00	1000.00	995.58	4096	4096
1000.00	1000.00	982.52	4096	4096
1000.00	1000.00	795.15	4096	4096
1000.00	1000.00	781.75	4096	8192
988.90	1000.00	767.11	4096	8192
982.73	1000.00	745.75	4096	8192
1000.00	1000.00	445.66	4096	8192
1000.00	1000.00	459.46	4096	8192
1000.00	1000.00	476.16	4096	8192
1000.00	1000.00	769.26	4096	8192
1000.00	1000.00	823.71	4096	8192
1000.00	1000.00	888.03	4096	8192
1000.00	1000.00	956.39	4096	8192
1000.00	1000.00	1000.00	16,384	16,384
1000.00	1000.00	1000.00	16,384	16,384

TABLE 12.2

Normalized Training Data Presented in the File of Table 12.1

Sensor Front	Sensor Left	Sensor Right	Velocity Left	Velocity Right
1	1	1	0.5	0.5
1	1	1	0.5	0.5
1	1	0.99558	0.125	0.125
1	1	0.98252	0.125	0.125
1	1	0.79515	0.125	0.125
1	1	0.78175	0.125	0.25
0.9889	1	0.76711	0.125	0.25
0.98273	1	0.74575	0.125	0.25
1	1	0.44566	0.125	0.25
1	1	0.45946	0.125	0.25
1	1	0.47616	0.125	0.25
1	1	0.76926	0.125	0.25
1	1	0.82371	0.125	0.25
1	1	0.88803	0.125	0.25
1	1	0.95639	0.125	0.25
1	1	1	0.5	0.5
1	1	1	0.5	0.5

Note: The original input values range from 0 to MAXDIST, and now they go from 0 to 1. The original output values range from (−) MAXSPEED to (+) MAXSPEED; after the conversion, they range from −1 to 1.

Once the learning process is completed, the final weight values are saved into a file to be used during the realization of the network on the actual system. It is important to note that neural networks do not always converge; in other words, its learning is not complete. As described in Step 4 of the presented learning algorithm, the learning process will sometimes stop after a number of iterations. This happens when the error of the network is not decreasing during training and thus if the error is not less than a threshold value, the network may not be able to provide a good performance or exhibit the expected behavior.

If a network does not converge, there are a few methods for troubleshooting. The first thing to do is to identify what was the final error of the network. If the error is somehow close to the threshold value, the performance of the network could be acceptable. This happens when the threshold is too low to be reached. If the error is high and the behavior is not what is expected, then we should be looking at the training data. When a few training sets describe a particular situation, and many others describe other possible situations, it is possible that the network has too few information to understand and learn a particular pattern. To approach this situation, it is necessary to collect a new training data that includes a broad distribution of all the possible situations that the network should encounter.

To test the system with the newly learnt weights, connection between sensors, network, and output to motors is needed, and this is done using software. Figure 12.17

```
% Perceptron Error Back-propagation Learning Program
%
% The Learning function perform the learning for a network of perceptrons
% defined by the nInputs and nOutputs, the training data is loaded from a
% file name "TrainingData.csv", after the learning the weights are saved
% into a file name "Weights.csv". The function return 0 if the network does
% not converge or 1 if it does.
%
function ret = Learning()

% Network structure definition
nInputs = 3;
nOutputs = 2;

% Constant parameters
LEARNING_RATE = 0.05;
MAX_EPOCHS = 500;
MIN_ERROR = 0.000005;

% Weights initialisation
weights = rand(nOutputs, nInputs) - 0.5;

% Training samples
samples = csvread('TrainingData.csv');
nSamples = size(samples, 1);

ret = 0;

% Start learning, Repeat for MAX_EPOCHS times
epoch = 1;
while (epoch <= MAX_EPOCHS)
    for (i = 1 : nSamples)
        % Forward calculation
        inputs = samples(i, 1:nInputs)';
        desired_outputs = samples(i, nInputs + 1: nInputs + nOutputs)';
        sum_outputs = weights * inputs;
        outputs = ActivationFunction(sum_outputs);

        % Error calculation
        neuron_error = desired_outputs - outputs;
        total_error(i) = neuron_error' * neuron_error;

        % Backward calculation and weight updating
        local_gradient = neuron_error.* DActivationFunction(outputs);
        dweights = LEARNING_RATE * local_gradient * inputs';
        weights = weights + dweights;
    end
    % Calculate Average Error
    average_error(epoch) = total_error * total_error' / nSamples;

    % Stop if the Average error is less than the minimum error
    if (average_error(epoch) < MIN_ERROR)
        ret = 1;
        break;
    end

    epoch = epoch + 1;
```

FIGURE 12.16 MATLAB code of the learning process for the neural network presented in Figure 12.15. This network uses the backpropagation method to update the weights. The training data used in this code is presented in Table 12.2.

```
end

% Ploting Learning curve
figure(1);
hold on
plot(average_error);
plot([1 size(average_error,2)], [MIN_ERROR MIN_ERROR], 'r:');
xlabel('Epoch');
ylabel('Average Error');
title('Learning Curve');
hold off

% Save the weights in a file
csvwrite('Weights.csv', weights);

end

% The activation function is a Sigmoid Function
function phi = ActivationFunction( v )
phi = tanh( v );
end

% Differential of the activation function
function dphi = DActivationFunction( y )
dphi = (1 - y.^2);
end
```

FIGURE 12.16 (continued) MATLAB code of the learning process for the neural network presented in Figure 12.15. This network uses the backpropagation method to update the weights. The training data used in this code is presented in Table 12.2.

presents the MATLAB code to test the neural network, which controls the robot with the trained weights. If the performance of the robot is up to the expectation, then it is possible to run this program as a controller for the robot. The learning process can be repeated if needed with a new set of training data.

12.4.1.3 Multilayer Perceptron

One of the commonly used neural network model is the multilayer perceptron (MLP). MLP is a popular machine learning solution and finds applications in diverse fields such as speech recognition, image recognition, and stock market forecasting. More recently, there has been some renewed interest in this network due to its success of superior learning capability.

The MLP is a feedforward ANN, consisting of an input layer, one or more layers of hidden neurons, and an output layer as shown in Figure 12.18. The hidden neurons are not directly accessible, but they extract important features contained in the input data. Learning occurs in the perceptron by changing the connection weights after each piece of data is processed and based on the amount of error in the output compared to the expected result.

These networks have found their way into countless applications requiring static pattern classification. Their main advantage is that they are easy to use, and that they can approximate any input/output map. However, a major disadvantage is their training is slow and it requires a large set of training data. Among the various artificial intelligence techniques available in the literature, neural networks offer promising

```
%
% The class DiffRobot the members of the class robot are declared as:
%
% robot.ultrasonic_sensors(3)    - 1 - Front, 2 - Left , 3 - Right
% robot.motor_vel(2);            - 1 - Left, 2 - Right
%
classdef DiffRobot < handle
    properties
        motor_vel = [ 0.0 0.0 ];
        ultrasonic_sensors = [ 0.0 0.0 0.0 ];
    end
    methods
        function obj = DiffRobot() % constructor
        end
    end
end

% This is the main function of the mobile robot. It will
% loop reading the sensors, Processing, and sending the motor command.
%
function MainRobot()

% Initialized robot
robot = DiffRobot();

% Load the learned weights from the file
weights = csvread('Weights.csv');

while 1
    % Read the New Values of the Sensors
    ReadSensors(robot);

    % Perceptron Forward calculation
    Perceptron(robot, weights);

    % Write to Motors
    SetMotorCmd(robot);
end
end

function ReadSensors(robot)
% Obtain the new readings of the sensors
robot.ultrasonic_sensors = [1000 1000 1000];
end

function Perceptron(robot, weights)
% Parameters to normalize the sensor data and the motor data
MAXDIST = 1000;
MAXSPEED = 32768;
% Calculation
inputs = (robot.ultrasonic_sensors / MAXDIST)';
sum_outputs = weights * inputs;
outputs = ActivationFunction(sum_outputs);
robot.motor_vel = outputs' * MAXSPEED;
end

% The activation function is a Sigmoid Function
function phi = ActivationFunction( v )
phi = tanh( v );
```

FIGURE 12.17   MATLAB code for testing the obtained weights from the learning process presented in Figure 12.16.

```
end

function SetMotorCmd(robot)
% Send the motor velocities to the robot
robot.motor_vel
end
```

FIGURE 12.17   (continued) MATLAB code for testing the obtained weights from the learning process presented in Figure 12.16.

solutions to robot navigation problem because of their ability to learn complex non-linear relationships between input sensor values and the output control variables. This ability of neural networks has attracted many researchers across the globe in developing neural network-based controllers for reactive navigation of mobile robots in indoor as well as outdoor environments.

12.4.2   Q-Learning

Q-Learning is a simple algorithm from reinforcement learning that has been used in many applications in robotics such as navigation, manipulation, motion planning, and multirobot systems. As the name implies, in a reinforcement learning algorithm, a reinforcement signal or reward is given to the robot according to the selected action for a particular environment. The robot constantly adjusts its actions and finds the optimal strategy through trial and error. This will maximize the reinforcement function, which is the objective of this method while the robot learns how to deal with specific situations.

In Q-Learning, a matrix named Q represents the "brain" of the robot and the Q matrix represents the learning of the robot. The objective of the method is to obtain this representation. In the Q matrix, each row represents a state of the robot and each column represents an action. In other words, the matrix tells the robot what action should be taken when the robot is in a particular state. The entries of the Q matrix are updated by the reinforcement learning function (Sutton and Barto 1998) defined as

$$Q(s,a) = R(s,a) + \gamma \max(Q(s',a'))$$
(12.15)

where R is a matrix that contains the reward or reinforcement signals for taking the action a in the state s. γ is the learning parameter that multiplies the maximum value of the Q matrix for all actions a' in the next state s'. The next state s' is determined by taking the action a in the current state s. Given a reward matrix for the states and the actions, the Q-Learning algorithm will converge to an optimal solution to reach the goal state defined by the problem. The algorithm for Q-Learning is as follows:

1. Initialize the Q matrix for each state and action with zeros.
2. Repeat for each episode:
    2.1.   Select a random current state s.
    2.2.   Repeat while s is not the goal state.
        −   Select an action a among all the possible action of the state s.
        −   Receive the immediate reward $R(s,a)$.
        −   Observe the next state s'.

Hidden

Inputs

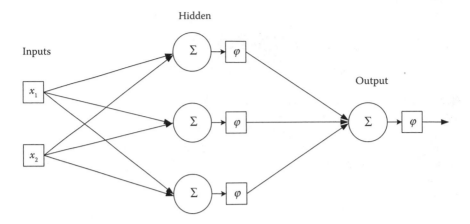

Output

FIGURE 12.18   A multilayer perceptron is a feedforward artificial neural network.

    – Update the entry in the Q matrix with Equation 12.15.
    – Set the next state s' as the current state s.

In the above algorithm, an episode is referring to a scenario where the robot will explore different states in the reward matrix until it reaches the goal state. At the same time, this exploration will update the Q matrix of the robot. More training scenarios will produce further improvements in the values of the Q matrix.

12.4.2.1   Case Study Q-Learning

The Intelligent Robot competition (Intelligent Robot 2013) is a game where a robot must collect objects of different types and colors and sort them into three different baskets (see Figure 12.19). In this case study, we will use Q-Learning to teach the robot about the objects and the actions that the robot must achieve to complete the challenge. Let us discuss a simple scenario to illustrate the application of Q-Learning here.

Let us consider a scenario where there are two different objects (red and blue), and two baskets are provided in the environment. We assume that the robot is equipped with a gripper that allows it to grab one object at a time. Each state in this scenario

FIGURE 12.19   The Intelligent Robot competition is a game where robots must collect and sort colored objects into three different baskets.

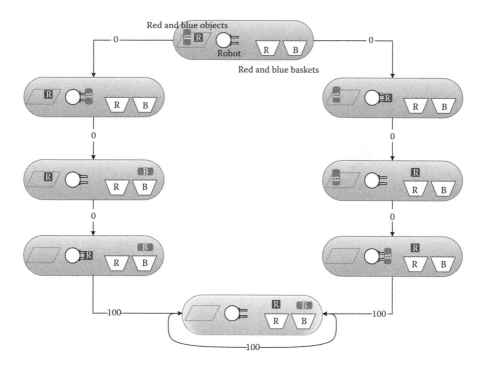

FIGURE 12.20 The finite-state machine for the states in the scenario of two objects for the Intelligent Robot competition. The rewards for taking an action are indicated in the transitions. The goal state has an additional loop that makes the robot to remain in this goal state after it has reached it.

consists of the gripper of the robot, the objects on the field, and the baskets. Let us define the possible states for the Finite State Machine (FSM). Figure 12.20 shows a FSM with all the possible states for this scenario. The transitions between the states in Figure 12.20 represent the reward values for executing that action. The Reward matrix for this FSM is expressed in Table 12.3.

TABLE 12.3
Reward Matrix Obtained from the Finite-State Machine Presented in Figure 12.20

Action State	1	2	3	4	5	6	7	8
1	–	0	–	–	–	0	–	–
2	–	–	0	–	–	–	–	–
3	–	–	–	0	–	–	–	–
4	–	–	–	–	100	–	–	–
5	–	–	–	–	100	–	–	–
6	–	–	–	–	–	–	0	–
7	–	–	–	–	–	–	–	0
8	–	–	–	–	100	–	–	–

```
%%%%%%%%%%%%%%%%%%%%%%%%%%%%%%%%%%%%%%%%%%%%%%%%%%
% Q learning
%%%%%%%%%%%%%%%%%%%%%%%%%%%%%%%%%%%%%%%%%%%%%%%%%%
function Learning

% Immediate reward matrix
R = [-1,    0,   -1,   -1,   -1,    0,   -1,   -1;
     -1,   -1,    0,   -1,   -1,   -1,   -1,   -1;
     -1,   -1,   -1,    0,   -1,   -1,   -1,   -1;
     -1,   -1,   -1,   -1,  100,   -1,   -1,   -1;
     -1,   -1,   -1,   -1,  100,   -1,   -1,   -1;
     -1,   -1,   -1,   -1,   -1,   -1,    0,   -1;
     -1,   -1,   -1,   -1,   -1,   -1,   -1,    0;
     -1,   -1,   -1,   -1,  100,   -1,   -1,   -1;];

% Learning parameter
gamma = 0.80;

Q = QLearning(R, gamma);

% save the QMatrix into a file
csvwrite('QMatrix.csv', Q);


end


% QLearning Function
% Input:
%       R          -    Immediate reward matrix
%       gamma      -    Learning parameter
% Output:
%       Q          -    Q matrix learnt
function Q = QLearning(R, gamma)

% Number of States
nStates = size(R,1);

% 1. Initialize the Q matrix for each state and action with zeros.
Q = zeros(size(R));
% This matrix is used to calculate the error
oldQ = Q;

% 2. Repeat for each episode:
for (episode = 1:500)
    % 2.1. Select a random current state s.
    state = randindex(nStates);

    % 2.2. Repeat while s is not the goal state
    while (1)
        % - Select an action a among all the possible action of the state s
        x = find(R(state,:)>=0);
        action = x(randindex(length(x)));

        % - Update the entry in the Q matrix with the Equation
        Q(state, action)= R(state, action) + gamma * max(Q(action,:));

        % If goal state then break
        if (state == action)
            break;
        end
```

FIGURE 12.21 The MATLAB code for the *Q*-Learning for the Intelligent Robot scenario.

```
      % - Set the next state s' as the current state s
         state = action;
   end

   % Calculate the error of the episode
   err(episode) = sum(sum(abs(oldQ - Q)));
      oldQ = Q;
end

% Plot the error in each episode
plot(err);
end

function x = randindex(n)
px = randperm(n);
x = px(1);
end
```

FIGURE 12.21   (continued) The MATLAB code for the Q-Learning for the Intelligent Robot scenario.

The MATLAB code for the Q-Learning algorithm is presented in Figure 12.21, the QLearning() function receives the reward matrix R and the learning parameter γ and the learned Q matrix is saved in to a file named "QMatrix.csv" so that it can be used later to control the robot. The learning parameter γ ranges between 0 and 1. When it is close to zero, the robot tends to consider the immediate reward; in contrast, if the value is close to one, the robot tends to consider the future reward associated with taking an action.

The code in Figure 12.21 will converge to produce the following matrix:

$$
Q = \begin{bmatrix}
0 & 256 & 0 & 0 & 0 & 256 & 0 & 0 \\
0 & 0 & 320 & 0 & 0 & 0 & 0 & 0 \\
0 & 0 & 0 & 400 & 0 & 0 & 0 & 0 \\
0 & 0 & 0 & 0 & 500 & 0 & 0 & 0 \\
0 & 0 & 0 & 0 & 500 & 0 & 0 & 0 \\
0 & 0 & 0 & 0 & 0 & 0 & 320 & 0 \\
0 & 0 & 0 & 0 & 0 & 0 & 0 & 400 \\
0 & 0 & 0 & 0 & 500 & 0 & 0 & 0
\end{bmatrix}
$$

From this matrix, it is clear that the robot could pick either the red or blue object at the beginning and converge to the goal. Let us now consider the scenario where the robot should collect the objects in a particular order. In that case, the reward matrix should indicate that a particular state like placing the object red in the basket has a reward:

$$R = \begin{bmatrix} - & 0 & - & - & - & 0 & - & - \\ - & - & 20 & - & - & - & - & - \\ - & - & - & 0 & - & - & - & - \\ - & - & - & - & 100 & - & - & - \\ - & - & - & - & 100 & - & - & - \\ - & - & - & - & - & - & 0 & - \\ - & - & - & - & - & - & - & 0 \\ - & - & - & - & 100 & - & - & - \end{bmatrix}$$

With this new reward, the Q-Learning matrix looks like:

$$Q = \begin{bmatrix} 0 & 272 & 0 & 0 & 0 & 256 & 0 & 0 \\ 0 & 0 & 340 & 0 & 0 & 0 & 0 & 0 \\ 0 & 0 & 0 & 400 & 0 & 0 & 0 & 0 \\ 0 & 0 & 0 & 0 & 500 & 0 & 0 & 0 \\ 0 & 0 & 0 & 0 & 500 & 0 & 0 & 0 \\ 0 & 0 & 0 & 0 & 0 & 0 & 320 & 0 \\ 0 & 0 & 0 & 0 & 0 & 0 & 0 & 400 \\ 0 & 0 & 0 & 0 & 500 & 0 & 0 & 0 \end{bmatrix}$$

With these values, the robot will give priority to pick up the red object, represented by the state in the second row of the Q matrix.

To apply the learned Q matrix to control the robot, a hybrid architecture can be employed as presented in Section 12.3.3. Figure 12.22 presents a code for decision-making. The code related to hardware such as the sensor readings and motor drives is not shown here, though it gives an idea of how to use the Q matrix for the robot's decision-making. The decision layer is the important part in this code and shows the application of the learned Q matrix. After each change of state, the decision function will get a new action from the Q matrix, and this action can be broken into primitive actions or actions that can be applied on the robot platform. Once an action is completed and the state changes, a new action must be obtained from the Q matrix. The process will continue until the robot state remains the same as the action, in other words, when the Q matrix reaches the goal state.

12.5 CONCLUSION

Autonomy is a desirable feature in robots, and it is a fundamental part in many robotics competitions. This chapter discussed decision-making and learning for autonomous robots. Autonomous robots act on the environment based on their sensorial

```
%
% The class IRobot the members of the class robot are declared as:
%
%
classdef IRobot < handle
    properties
        state = 0;
        action = -1;
    end
    methods
        function obj = DiffRobot() % constructor
        end
    end
end

% This is the main function of the mobile robot. It will
% loop reading the sensors, Processing, and sending the motor command.
%
function MainRobot()

% Initialized robot
robot = IRobot();

% Load the learned Q-Matrix from the file
Q = csvread('QMatrix.csv');

while (robot.state ~= robot.action)
    % Decide what action to do next according to the state of the robot
    Decision(robot, Q);

    completed = 0;
    while (completed == 0)
        % Read the New Values of the Sensors
        ReadSensors(robot);

        % Obtain a new action, break it into subactions
        % update the state of the robot
        completed = Middle(robot);

        % Navigation, gripping, Obstacle avoidance, etc
        Reactive(robot);

        % Write to Motors
        SetMotorCmd(robot);
    end
end
end

function ReadSensors(robot)
% Obtain the new readings of the sensors
end

function Decision(robot, Q)
% Update the action with the new state of the robot
[val robot.action] = max(Q(robot.state,:));
end
```

FIGURE 12.22 Code for a hybrid architecture taking a high-level decision with the states and goals learned from the *Q*-Learning algorithm.

```
function completed = Middle(robot)
% Check the robot state is still the same...
% or has it change? if so update the state of the robot
robot.state = ...;
completed = 1;
% try a primitive action to achieve the selected robot.action
end

function Reactive(robot)
% use obstacle avoidance, wall following, and other beahviours but give
% priority to those selected by robot.action
end;

function SetMotorCmd(robot)
% Send the motor velocities to the robot
end
```

FIGURE 12.22   (continued) Code for a hybrid architecture taking a high-level decision with the states and goals learned from the *Q*-Learning algorithm.

information. The chapter presented some of the algorithms employed in learning as well as the implementation of the decision in an autonomous robot.

REFERENCES

Autonomous Sumo Robot Competition, Singapore Robotic Games website 2013, http://guppy. mpe.nus.edu.sg/srg/asumo.pdf.

Braitenberg, V. 1984. *Vehicles: Experiments in Synthetic Psychology*. Cambridge, MA: The MIT Press.

Brooks, R.A. 1985. Layered control system for a mobile robot. *IEEE Journal of Robotics and Automation* RA-2:14–23.

Intelligent Robot Contest, Singapore Robotic Games website 2013, http://guppy.mpe.nus.edu. sg/srg/srg13/irc.pdf.

Jones, J.L. and Flynn, A.M. 1993. *Mobile Robots: Inspiration to Implementation*. Wellesley, MA: A K Peters.

Mataric, M.J. 2007. *The Robotics Primer*. Cambridge, MA: The MIT Press.

RC Sumo Robot Competition, Singapore Robotic Games website 2013, http://guppy.mpe.nus. edu.sg/srg/rcsumo.pdf.

RoboCup CoSpace Robot, CoSpace Robot, September 2012, http://www.cospacerobot.org/.

Sutton, R.S. and Barto, A.G. 1998. *Reinforcement Learning: An Introduction*. Cambridge, MA: The MIT Press.

Index

A

AC, *see* Alternating current (AC)
Accelerometer, 57; *see also* Sensors
    application, 54, 57
    mechanical, 190–191, 195
Activation function, 372
    artificial neuron output amplitude, 368
    hyperbolic tangent function, 373
    sigmoid function, 372
    threshold, 370
    types of, 369
Actuators, 103–104; *see also* Electrical actuators;
        Electrical machines
    electrical, 104
    and reactive layer, 365
    springs, 179
Adaptive controllers, 171; *see also* Control
        system
ADC, *see* Analog-to-digital convertor (ADC)
Akermann steering, 34; *see also* Wheel driver
        system
Allegro A132X family Hall effect sensors, 56
Alternate torque equation, 111–112
Alternating current (AC), 104
    induction motor, 113
    motors, 112; *see also* Electrical actuators
    rotating magnetic field, 112
    synchronous motors, 112–113
Analog controllers, 168
    vs. digital controllers, 220
    limitations, 217
    on signal values, 218
Analog plants, 221
Analog-to-digital convertor (ADC), 61
Angular friction estimation, 300, 301
    attenuation factor for oscillation, 303
    example, 304–305
    force balance equation, 301
    frictional torque, 301
    MATLAB program for simulation of pole as
        "normal" pendulum, 305
    pendulum experiment, 301
    simulation result, 305
    time domain solution, 302
ANN, *see* Artificial neural network (ANN)
Arbitration, 360
Ariel robot, 326; *see also* Robot
    frame reference, 327–328
    navigation, 326

Armature, 108–109; *see also* Direct current
        machines
    air-core, 112, 143
    conductor number, 100
    flux density, 111
    winding, 122
Arm parameters, 45; *see also* Robotic arms
    for robot, 148
Artificial neural network (ANN), 368; *see also*
        Perceptron; Robot learning; Training
        data
    activation functions, 369
    in CoSpace robot, 374
    elements of, 368
    MATLAB code for testing weights, 378–379
    modeling for robot, 373
    for obstacle avoidance, 373
    system testing, 375
    troubleshooting methods, 375
Autonomous robot, 14, 317, 321; *see also*
        Micromouse
    decision-making, 358
    fully, 357
    games, 10
    multiple, 5
    navigation problem, 341
    semi, 357
    sumo robot competition, 358
    with USB camera, 66
Autonomous underwater vehicle (AUV), 5
AUV, *see* Autonomous underwater vehicle
        (AUV)
Average squared error, 371

B

Back EMF, 106
    induced in DC motor, 109–110
    pattern sequence, 135
    problems in BLDC, 135
    sensing-based BLDC motor, 134; *see also*
        Brushless DC motors (BLDC motors)
    signal conditioning, 136
Backpropagation algorithm, 370, 371
Behaviors, *see* Reactive rules
Bernoulli's
    equation, 314
    principle, 313
    WCR, 312, 314
Bifilar windings, 132

Biped robot, 170; *see also* Robot
    joints, 171
    state equations, 191
BLDC motors, *see* Brushless DC motors
    (BLDC motors)
BRA, *see* British Robot Association (BRA)
British Robot Association (BRA), 13
Brushless DC motors (BLDC motors), 114;
    see also Back EMF; Drive systems;
    Robotics drives
    bridge circuit to drive, 134
    commutator in, 122
    controllers, 133
    delta-connected, 124–126
    drive, 133
    Hall effect sensor-based control, 136
    phases, 122
    sensor-based switching, 136–137
    star-connected, 122–124

C

Camera systems, 66; *see also* Robot vision
    CMU cam, 66, 67
    in game robotics, 70
    SRV-1 Blackfin, 66, 67
    USB camera, 66
Capacitor, 178
Caster wheels, 29, 31; *see also* Track
    drive system
    robot motion control, 32
    in WCR, 312
CCD, *see* Charge-coupled device (CCD)
Center of rotation (COR), 33
Charge-coupled device (CCD), 69
    image sensors, 71; *see also* Image
    formation
Chow and Kaneko method, 80
Classical decision-making, 359–360; *see
    also* Decision-making; Hybrid
    decision-making
Closed-loop, 169, 180
    computer-controlled system, 231
    controller, 169, 170–171
    MATLAB code for, 237, 279
    open-loop *vs.*, 168
    poles, 278, 284, 289
    root locus plots for pure pole angle
    control, 280
    system equation, 254, 255
    transfer function, 185–186, 204, 231, 232,
    234, 236, 273, 276
    used for simulation, 204
CM, *see* Controllability matrix (CM)
CMOS, *see* Complementary metal–oxide–
    semiconductor (CMOS)
CMU cam, 66, 67

Colony game, 4, 63; *see also* Game robotics
Color detection, 83; *see also* Feature extraction
    MATLAB code for, 85–86
Color sensors, 63; *see also* Sensors
    circuit, 63–64
    MT104Bx, 63
Commutation, 108; *see also* Direct current
    machines
Compact infrared sensors, 61; *see also* Sensors
Compass, digital, 56; *see also* Sensors
Complementary metal–oxide–semiconductor
    (CMOS), 69
    sensors, 71; *see also* Image formation
Computational hardware, 264
Computer vision, 65; *see also* Robot vision
Conflicts solving between reactive rules, 360
Connected component labeling, 92; *see also*
    Symbolic feature extraction methods
    algorithm, 93
    binary image, 94, 97
    centroid of blob, 96, 98
    iterative, 94
    labeled regions in image, 97
    MATLAB code, 95–96
    output in 3D, 97
Continuous state space model, 242; *see also*
    Discrete state space systems
Control system, 166, 171–172; *see also* Plant;
    Control theory; Digital control
    adaptive controllers, 171
    analog *vs.* digital systems, 168
    basic components in, 177
    capacitor, 178
    classification based on, 167
    closed-loop controller, 170–171
    closed-loop systems, 169, 231
    dampers, 178
    DDC, 240
    development setup for robot, 265
    distributed, 171
    electrical components, 177–178
    Faraday's law, 177
    hierarchical, 171
    input to output relationship, 173
    intelligent robot structure, 169–170
    mechanical components, 178–179
    neural network-based controllers, 171
    Ohm's law, 177
    open-loop systems, 168
    PID controller, 169
    springs, 178–179
    trends in control, 171
    variable structure controllers, 171
Control theory, 165, 171–172; *see also* Control
    system; Plant
Controllability, *see* State controllability
Controllability matrix (*CM*), 251

Controller software implementation, 237;
    see also Digital control
    derivative calculations, 240
    digital controller implementation, 240–241
    integral calculations, 240
    PID controller, 239
Controllers, 173, 237, 317, *see* Digital controllers;
        Discrete state feedback controllers;
        Mathematical modeling; Pole
        placement regulator; Servo controller;
        State equation; Transfer function
    design, 215
    elementary, 203
    hardware implementation of, 264, 266
    high-level, 317
    microprocessor as, 219
    popular types of, 215
    state feedback, 251
    transfer functions, 237
Convolution operation, 74; *see also*
        Image-processing
    in MATLAB, 75
    neighborhood operations, 74
    spatial operators, 74
    two-dimensional, 74
Coordinate frame origin, 24
COR, *see* Center of rotation (COR)
Corner frequency, 147
    form, 176
CoSpace robot ANN, 374

D

DAC, *see* Digital-to-analog converter (DAC)
Dampers, 178; *see also* Control system
DDC, *see* Direct digital control (DDC)
Decision-making, 358–359; *see also* Classical
        decision-making
    uncertainty reduction, 318
Degree of freedom (DOF), 13
    six, 14
Degree of mobility (DOM), 13
Delta-connected BLDC motor, 124; *see also*
        Brushless DC motors (BLDC motors)
    advantages of, 126
    winding switching sequence for, 125
Denavit–Hartenberg algorithm (D–H algorithm),
        40; *see also* Arm parameters
    steps, 44–45
D–H algorithm, *see* Denavit–Hartenberg
        algorithm (D–H algorithm)
Differential drive, 33; *see also* Track drive system
    challenge in, 32
    odometry for, 37–39
    parameters, 33, 37
DiffRobot, 39, 40
Digital compass, 56–57

Digital computers, 220, 266; *see also* Digital
        controllers
Digital control, 217, 266; *see also* Controller
        software implementation; Control
        system; Digital systems
    direct, 220
    signal sampler, 218–219
    synchronous sampling system, 218
    system, 218
    zero-order hold, 219
Digital controllers, 217, 219; *see also* Controller;
        Digital control
    advantages, 220
    implementation, 240
    parts of, 220
Digital image-processing, 71; *see also*
        Robot vision
    color and color models, 71
    gray scale, 72
    HSV color model, 73
    RGB color cube, 72
    RGB model, 72
    sampling density, 71
    YUV color space, 73
Digital systems, 221; *see also* Plant; System;
        Z-transform
    continuous signal, 221
    plant representation in, 228
    sampled signal as weighted impulses, 222
    sampler switch, 221
    sampling, 221, 223
    signal Laplace transform, 222
    signal reconstruction, 229
    signal representation in, 221
    switch output impulse approximation, 221
    transfer function of ZOH, 229
    Tustin's approximation, 230–231
    Z-transform of plant fed from ZOH, 230
Digital-to-analog converter (DAC), 219
Dijkstra's shortest path algorithm, 352
Dilation algorithm, 98; *see also* Robot vision
Direct current (DC), 104
Direct current machines, 106; *see also* Electrical
        actuators
    alternate torque equation, 111–112
    armature, 108
    back EMF induced in, 109–110
    commutation, 108
    commutator and armature setup, 109
    DC motor types, 112
    EMF equation alternative form, 110–111
    four-pole winding, 109
    force, 111
    as generator, 106
    induced EMF, 107
    maximum flux linkage, 107
    permanent magnet motors, 114

Direct current machines *(Continued)*
 primitive AC generator, 107
 series winding, 112
 shunt winding, 112
 slip ring generator, 106
 speed, 113–114
 split ring DC machine, 107
 stator, 108
 torque equation, 111
 total peripheral force, 111
Direct current motor
 permanent magnet motors, 114
 types, 112
Direct current motor controller, 126, 127;
   see also Drive systems
 application, 127–128
 commercial, 128
Direct digital control (DDC), 240; *see also*
   Control system
 system, 220
Discrete state feedback controllers, 250; *see also*
   Controller; Pole placement regulator;
   Servo controller; State controllability;
   State observability concept; Steady-
   state quadratic optimal control
Discrete state space systems, 241; *see also*
   System; Transfer function
 analytical method, 242–245
 computer calculations, 246
 from continuous state space model, 242
 discrete time state space representation, 242
 from discrete transfer functions, 241–242
 MATLAB approach, 245
 time domain solution of, 245
 Z-transform approach, 246–250
Distance sensor, 331
Distributed control system, 171
DOF, *see* Degree of freedom (DOF)
DOM, *see* Degree of mobility (DOM)
Drive motor power requirement, 141
 example, 143
 force to overcome friction, 142
 maximum power requirement, 142
 required force, 142
 robot moving on mild slope, 142
 velocity profile, 141
Drive systems, 126; *see also* Brushless DC
   motors (BLDC motors); Electrical
   machines; Stepper motors
 DC motor control, 126
 DC motor controller, 127–128
 H-bridge driver principle, 126–127
Drive wheel speed, 149
Drop-encoder, 293
DSP processors, 264
Dynamic range, 53
Dynamic suction principle, 313

E

Edge detection, 81; *see also* Feature extraction
 filter, 76; *see also* Smoothing
 gradient of image, 81
 Laplacian operator, 81
 MATLAB code, 84
 pixel coordinates and corresponding mask, 83
 Sobel operators, 81, 82
Edutainment robotics, *see* Game robotics
Egocentric grid, 334
 function, 333
 in MATLAB code, 339
 representation of, 340
Electrical actuators, 104; *see also* Alternating
   current (AC)—motors; Actuators;
   Direct current machines; Electrical
   machines
 Faraday's law, 104
 generating and motoring, 104–106
 induced EMF, 105
 Lenz's law, 104
 LHR and force, 105
 RHR and induced voltage, 104–105
Electrical machines, 103; *see also* Electrical
   actuators; Drive systems; Robotics
   drives
Electrical motors, 139; *see also* Brushless DC
   motors (BLDC motors); Direct current
   machines; Electrical actuators; Servo
   motors; Stepper motors
Electromotive force (EMF), 104
 constant, 111
 equation alternative form, 110
EMF, *see* Electromotive force (EMF)
Encoders, 55, 166; *see also* Sensors
 magnetic, 55, 56
 optical, 55
 resolution of, 55
Erosion algorithm, 98; *see also* Robot vision
Error signal, 371

F

Faraday's law, 104, 177
Feature extraction, 319–321; *see also* Color
   detection; Edge detection;
   Image-processing; Robot vision;
   Thresholding
 algorithms, 73, 77
 symbolic, 86
 timing of, 318
Federation of International Robot-Soccer
   Association (FIRA), 6; *see also* Game
   robotics
Feedforward neural networks, 369
Finite-state machine (FSM), 364

FIRA, *see* Federation of International Robot-
    Soccer Association (FIRA)
FIRST, *see* For Inspiration and Recognition of
    Science and Technology (FIRST)
Flipper wall-climbing robot, 306, 309; *see also*
    Wall-climbing robot (WCR)
    climbing action stages, 311
    control of suction pad arms and cruise motor,
        308, 310
    operation sequence of, 310–312
    pneumatic wiring diagram, 307
    P-only controller, 309
    processor, 309
    system configuration of, 307, 308
    transition from wall to base, 312
f-number, 70; *see also* Lens
Focal length, 68; *see also* Lens
Focal point, 68; *see also* Lens
Follow-Line behavior, 362
    MATLAB code for, 363
Force balance equation, 188, 270, 271, 301
For Inspiration and Recognition of Science and
    Technology (FIRST), 9; *see also*
    Game robotics
Forward kinematics, 14; *see also* Robotics
    composite rotations, 17–18
    composite transformations, 20, 21
    coordinates of point *P*, 15
    homogeneous transformation matrix, 19
    mathematical description of objects, 23–28
    matrix multiplication order, 20–23
    mobile robot, 25–28
    objects' center of gravity, 24
    origin of objects' coordinate frame, 24
    rotation along *z* axis, 15
    rotation matrix, 16, 18
    solutions, 45–46
    translation in *xy* plane, 15
    translation matrix, 17
Friction, 146
    of pole support joint, 270
Friction measurement, 146
    example, 147
    MATLAB code, 148
    overall linear, 146
Frictional torque, 301
FSM, *see* Finite-state machine (FSM)
Fully autonomous robot, 14, 357
Fusion, 360

G

Game robotics, 1; *see also* RoboCupJunior
    Rescue; Robotics; Singapore Robotic
    Games (SRG); Robot vision; Sensors
    categories of, 10
    challenges, 10
    classification of, 10
    colony game, 63
    competition platforms, 61
    and engineering education, 1–2
    FIRA, 6
    FIRST, 9
    International Aerial Vehicles Competition, 9
    MATE's ROV competition, 8–9
    micromouse, 6
    RoboCup, 6–8, 9
    Robotic SUMO wrestling, 9
    speed of response, 288
    stipulations, 267
    Sumo robot competition, 358
    around the world, 6–11
    World Robotic Sailing Championship, 9
Game robots, 313
Gateway, *see* Node
Gaussian filter, 76; *see also* Smoothing
    application of, 77
Gear ratio, 139
    on current drawn by motor, 155
    design, 148–153, 156–158
    drive wheel speed, 149
    on efficiency, 156
    equivalent reflected value, 141
    gear-driven robot, 149
    inertia equivalent values, 139
    on maximum velocity, 154, 155
    motor parameters, 149
    motor speed, 150
    power, 140
    speed reduction gear, 140
    system performance, 153–156
    torque, 140, 149, 150
Gear-driven robot, 149
Generator, 104; *see also* Direct current machines;
        Electrical actuators
    DC motor, 106
    modes of operation, 130–131
    sequence, 129
    sinusoidal waveform, 107
    slip ring, 106
    state transit, 130
GP2D15 sensor, 62; *see also* Optical sensors
Grayscale, 72; *see also* Digital image-processing
    conversion, 72–73
Ground LED sensors, 293
Gyroscope, 57; *see also* Sensors

H

Hall effect, 56
    sensor-based control, 136
    sensors, 56; *see also* Sensors
H-bridge driver, 275
    principle, 126–127; *see also* Drive systems

Hierarchical control, 171; *see also* Control system
High-level controller, 317
Hobbywing Pentium-85A, 136; *see also* Brushless DC motors (BLDC motors)
Hokuyo URG-04LX-UG01, 60; *see also* Sensors
Homogeneous transformation matrix, 19; *see also* Forward kinematics
  composite, 20
  example, 22–23
Hough function call, 89
Hough space, 90, 93
  for points in metric map, 325
  and sinusoidal curve, 324
Hough transform, 86, 323–325; *see also* Symbolic feature extraction methods
  computation of, 87
  for detecting circles, 91–92, 93
  equation for circle, 90
  example and result, 90
  Hough function call, 89
  MATLAB code for, 88–89
  polar form of lines, 87
HSV color model, 73; *see also* Digital image-processing
Hue, 83
Humanoid robot, 66
  competition, 4, 5; *see also* Singapore Robotic Games (SRG)
  joints, 13, 148, 171
  soccer-playing, 98
  topple prevention, 56
  torque, 103
Hybrid architecture
  code for, 385–386
  in intelligent robot competition, 384
Hybrid decision-making, 365, 367; *see also* Classical decision-making
Hydraulic actuators, 104; *see also* Actuators

I

Image
  color, 72
  plane, 69
  quality, 73
  sensing unit, 70–71; *see also* Image formation
  sharp, 68
  size of, 69
Image formation, 67; *see also* Lens; Robot vision
  CCD image sensors, 71
  components of, 67
  illumination, 67–68
  image-sensing unit, 70–71
  optical system, 68
  parameters, 70

Image-processing, 73; *see also* Color detection; Convolution operation; Edge detection; Feature extraction; Smoothing; Thresholding
  algorithms, 65
  digital, 71
  in game robotics, 83
Induced EMF, 105
Induction motor, 113; *see also* Alternating current (AC)—motors
Infrared transmitter and receiver, 60
Intelligent robot, 169–170
Intelligent robot competition, 5, 380; *see also* Q-Learning
  finite-state machine for, 381
  hybrid architecture, 384, 385–386
  learned Q matrix, 383–384
  MATLAB code for, 382–383
  reward matrix, 381
  in robotic games, 10
Intermediate-level algorithms, *see* Symbolic feature extraction methods—algorithms
International Aerial Vehicles Competition, 9; *see also* Game robotics
International Standards Organisation (ISO), 13
Inverse kinematics, 46–47
Inverse Z-transforms, 234; *see also* Transfer function; Z-transform
  difference equation techniques, 234–235
  partial fraction technique, 234
  time domain solution by MATLAB, 235–237, 238, 239
ISO, *see* International Standards Organisation (ISO)

L

Laplace transform
  ratio, 173
  technique, 210–211
Laplacian operator, 81; *see also* Edge detection
  pixel coordinates, 83
Laser rangefinders, 59, 326
  advantages, 59–60
  principle of, 59
Learned Q matrix, 383–384
Learning, 366–368; *see also* Q-Learning
Left-hand rule (LHR), 105
  and force, 105
Lens, 68; *see also* Image formation
  f-number, 70
  focal length, 68, 70
  focal point, 68
  magnification ratio, 69, 70
  path of light rays through, 68

selection, 70
    sensor size and focal length, 69
Lenz's law, 104
    induced EMF, 107
LHR, *see* Left-hand rule (LHR)
Light sensing, 359
Linear friction force at mass center of pole, 270
Linear quadratic controllers (LQC), 250
    design results for Q matrix values, 290
    MATLAB code for, 288
    Q matrix change effect, 287–290
    response for, 289, 290
    servo control with integrator, 289
    with servo input, 286
    step input, 289
Load angle, 113
Low-level image processing, *see*
        Image-processing—algorithms
Lower-level behavior suppression code, 367
LQC, *see* Linear quadratic controllers (LQC)

M

Magnetic encoder, 55, 56; *see also* Encoders
    application, 54
Magnetoresistive effect, 57
Magnetoresistivity, 56
Magnification ratio, 69, 70; *see also* Lens
Map representation, 321; *see also* Metric map;
        Perception
    building, 341
    choosing, 321
    metric map, 322
Marine Advance Technology and Education
        Center (MATE), 8
Mass damper and spring assembly, 186–187
Master controllers, 297, 298
MATE, *see* Marine Advance Technology and
        Education Center (MATE)
MATE's ROV competition, 8–9; *see also* Game
        robotics
Mathematical description of objects, 23–28
Mathematical modeling, 174, 269; *see also*
        Controllers; State equation; Transfer
        function
    steps in, 176–177
MATLAB code
    closed-loop system root locus, 279
    color detection, 85–86
    for computing root locus of simple closed-
        loop system, 279
    connected component labeling, 95–96
    convolution operation, 75
    for driving model, 257
    Follow-Line behavior, 363
    friction measurement, 148
    homogeneous transformations, 29–31

Hough transform, 88–89
    intelligent robot competition, 381, 382–383
    in LQC design, 259–260, 261, 262, 263, 288
    obstacles and target, 346–349
    odometry calculation, 41–43, 95–96
    path planning, 352, 353–356
    PID controller simulation, 208
    pole placement design, 285
    for pole placement design, 285
    pole simulation as "normal" pendulum, 305
    P-only controller, 206
    potential field, 346–349
    robot learning, 376–377
    Round-Obstacle behavior, 364, 366
    simulation of pole as "normal" pendulum, 305
    smoothing, 79
    Sobel operators, 84
    testing weights, 378–379
    thresholding, 80
    time domain solution by, 235–237, 238, 239
    of VFH algorithm, 336–339
    wall following, 332
Matrix multiplication order, 20–23
Maze representation, 330
McCulloch–Pitts neuron model, 369
Mean filter, 76; *see also* Smoothing
    image blurring, 77
Mecanum wheels, *see* Omniwheel
Median filter, 77; *see also* Smoothing
    3 × 3 median filter, 78
MEMS, *see* Microelectromechanical systems
        (MEMS)
Metric map, 322, 328; *see also* Ariel robot;
        Two-dimensional Cartesian grid
    cell in, 322
    Hough transform, 323–325
    occupancy grid, 324, 322
    of room, 345
    by ultrasonic readings, 325
    ultrasonic sonar sensor model, 323
Microelectromechanical systems (MEMS), 57
Micromouse, 6, 165, 328, 329; *see also* Game
        robotics; Robot
    infrared sensors, 330
MIMO, *see* Multi input multi output (MIMO)
Mobile robot, 25–28, 164; *see also* Drive motor
        power requirement; Forward
        kinematics; Friction measurement;
        Gear ratio; Motor; Wheel driver system
    Ackermann drive, 34
    design procedures for, 158
    differential drive, 33
    encoders, 55
    motor power and gear ratio, 161–164
    omniwheel drive, 36
    robotic arm joint design, 158–161
    with SRV-1 Blackfin, 66

Motor, 104; *see also* Electrical actuators
    commands, 364
    inertia and friction, 143–144
    motor characteristics data sheet, 145–146
    power, 139, 114; *see also* Robotics drives
Moving average filter, *see* Mean filter
MT104Bx, 63; *see also* Color sensors
Multi input multi output (MIMO), 241, 209
Multilayer perceptron, 377–380

N

Navigation, 330; *see also* Obstacle avoidance; Path
                planning; *Q*-Learning; Wall following
    odometry, 36
    problem, 341
    vector field histogram, 333–334
    visual feedback, 65
Neighborhood connectivity, 341
Neural network
    based controllers, 171; *see also* Control system
    input, 372
Node, 352

O

Observability, 251; *see also* State observability
                concept
    condition, 252
Obstacle avoidance, 333, 372–373; *see also*
                Navigation
    case study of, 335
    egocentric grid, 334, 340
    MATLAB code of VFH algorithm, 336–339
    obstacle looking-ahead, 334
    polar histogram, 335
Occupancy grid, 324
    maps, 322
Odometry, 36
    case study, 37–39
    DiffRobot, 39, 40
    equations for, 36
    MATLAB code, 41–43
    system, 37
Ohm's law, 177
Omniwheel, 32
    drive, 35–36
    mobile robot, 36
    rollers, 35
Open-loop, 180; *see also* Closed-loop
Optical encoder, 55; *see also* Encoders
Optical sensors, 60; *see also* Sensors
    circuit for IR sensor pair, 61, 62
    GD2D02 sensors, 62
    GP2D15, 62
    IR sensor pairs, 62
    IR transmitter and receiver, 60

    limitations, 62
    principle, 61
Overtraining, 370–372; *see also* Training data

P

Path planning, 341; *see also* Path planning;
                Wavefront
    case study, 344, 350
    Dijkstra's shortest path algorithm, 352
    forces in, 343, 344, 351
    MATLAB code for obstacles and target,
                346–349
    MATLAB code for path planning, 352,
                353–356
    metric map of room, 345
    neighborhood connectivity, 341
    potential field method, 343
    primitives for potential field, 343, 344
    shortest path search, 343, 350
    using topological maps, 350
    vector fields, 343, 350
PBR, *see* Pole-balancing robot (PBR)
Pendulum experiment, 301
Perception, 317–318; *see also* Map representation;
                Metric map; Sensor measurement
                processing
    robot's state, 317
Perceptron, 369; *see also* Multilayer perceptron;
                Training data
    average squared error, 371
    backpropagation algorithm, 371
    case study, 372
    error signal, 371
    input to neural network, 372
    with McCulloch–Pitts neuron, 370
    sigmoid function, 372
    synaptic weights, 371
    training data, 370
Permanent magnet motors, 114
Permanent magnet stepper motors, 121–122;
                see also Stepper motors
    four-phase, 132–133
    two-phase, 133
Pitch, 16
Pixel, 70
    coordinates, 83
Planner algorithm, 341–342
Plant, 165; *see also* Control system; Control
                theory; Digital systems
    analog, 221
    with input and output, 166
    linear *vs.* nonlinear, 166, 167
    time-invariant *vs.* time-variant, 167
    transfer functions, 174
    types of, 166
PLC, *see* Programmable logic controller (PLC)

Polar histogram, 333–334, 334
Pole angle control, 276, 277
   closed-loop root locus plots for, 280
   transfer function for, 278
Pole angle measurement, 292
Pole-balancing robot (PBR), 267, 268; *see also*
        Linear quadratic controllers (LQC)—
        with servo input; Pole-balancing
        robot controller implementation;
        Pole placement controllers (PPC);
        Singapore Robotic Games (SRG);
        Wall-climbing robot (WCR)
   additional friction term, 270
   angular velocity of motor and drive wheel, 273
   applied voltage to velocity, 273, 274
   closed-loop root locus plots, 280
   closed-loop transfer function, 276
   connection in H-bridge drivers, 275
   direction pin, 275
   dynamics, 271, 272
   force balance equation, 270, 271
   force effect on, 272
   frictional torque of pole support joint, 270
   linear friction force at pole mass center, 270
   mathematical modeling, 269
   MATLAB code for closed-loop system root
      locus, 279
   MATLAB code for pole placement design, 285
   moment of inertia of pole, 271
   for pendulum experiment, 301
   plant model calculations, 283
   pole angle control and position control, 276, 277
   PWM implementation, 276–276
   race, 2, 3
   reaction due to linear friction, 270
   rotary friction forces, 269
   Simulink model of closed-loop system,
      284–286, 287
   state equation development, 280–282
   state model, 278, 282
   thrust on wheel contact, 273
   torque balance equation, 271
   torque by motor 273
   torque due to pole mass, 271
   torque due to pure linear motion, 271
   torque on drive wheel, 273
   transfer function for pole angle control, 278
   transfer ratio, 272
   two-degree-freedom, 299–300
   vehicle dynamics, 274
   velocity control, 269, 274
Pole-balancing robot controller
      implementation, 291
   avoiding ribbon cables and
      multiconnectors, 292
   cascade velocity controller, 299
   to control motor, 294, 296

   control system, 291
   distance measurement, 292
   drop-encoder and pole support system, 293
   DSP boards, 292
   ground LED sensors, 293
   hardware setup, 291
   interrupt service routine, 296
   master controllers, 297, 298
   pole angle measurement, 292
   pole detached, 292
   sampling and control interval, 294
   software, 293, 295, 297
   timer interrupts, 297
   xcom, 298
Pole inertia, 271
Pole placement controllers (PPC), 250
   with servo input used as offset, 283–286
Pole placement regulator, 253; *see also* Controller;
        Discrete state feedback controllers
   closed-loop system equation, 254
   coefficients method comparison, 254–256
   control law, 254
   controller performance simulation, 257–258
   discrete system, 254
   MATLAB listing for driving model, 257
   MATLAB method, 256
   methods for, 254
   simulation results, 258
   state feedback control setup simulation, 256
Pole/zero form, 175–176
Polynomial form, 175
P-only controller, 309
   computation, 309
   MATLAB code for, 206
   time response from Simulink, 207
Potential field, 343
   MATLAB code, 346–349
   primitives for, 343, 344
   to produce robot path, 345
Power rating of motor, 114
PPC, *see* Pole placement controllers (PPC)
PPR, *see* Pulses per revolution (PPR)
Primitive AC generator, 107; *see also* Direct
        current machines
Prismatic joint, 40, 45
Programmable logic controller (PLC), 219
Proportional integral derivative (PID), 37
Proportional integral derivative controller, 169,
      239; *see also* Control system
   equation, 207
   MATLAB code, 208
   response, 207–209
   terms, 239
   tuning, 208
Pulse width modulation (PWM), 57, 127
   control, 274
Pulses per revolution (PPR), 55

Q

Q-Learning, 379; *see also* Intelligent robot
        competition; Robot learning
    algorithm for, 379–380
    case study, 380
Q matrix
    change effect, 287–290
    design results for new values, 290
    response for modified, 290

R

RC Sumo robot game, 358
Reactive decision-making, 360; *see also*
        RoboCupJunior Rescue
    case study on, 361–365
    subsumption controller, 361
Reactive rules, 360
Remote operative vehicle (ROV), 5, 104
Remotely operated robot, 14
Response time, 54
Revolutions per minute (rpm), 139
RGB color cube, 72
RGB model, 72; *see also* Digital
        image-processing
RHR, *see* Right-hand rule (RHR)
Right-hand rule (RHR), 104
    for induced EMF, 105
    and induced voltage, 104–105
Rigid caster wheels, 31
RoboCup, 6; *see also* Game robotics
    Ariel robot, 326
    competition, 6–7
    event, 6, 9
    micromouse, 328, 329, 330
    RoboCup@Home competition, 48, 325
    RoboCupJunior, 7–8
    RoboCup Opens, 6
    RoboCup senior, 8
    RoboCup Singapore Open, 8
RoboCupJunior Rescue, 361, 362
    CoSpace, 364
    finite-state machine representation for
        Round-Obstacle behavior, 365
    Follow-Line behavior, 362
    MATLAB code for Round-Obstacle
        behavior, 364, 366
    MATLAB code for Follow-Line
        behavior, 363
    motor commands, 364
    robot controller, 362
Robot, 1, 13; *see also* Game robotics; Mobile
        robot; Sensors; Singapore Robotic
        Games (SRG)
    actuators for, 103–104
    architecture selection factor, 357

autonomous, 14
biped, 170
colonies, 3–4
control system design development, 265
heading, 340
joints having maximum load, 159
micromouse, 165
navigation, 326
parameters, 146
position having maximum torque on
        joint, 163
remotely operated, 14
speed, 33
steering angle of, 34
three-joint robot arm, 158
wall-climbing, 162
Robot autonomy, 357, 384; *see also*
        RoboCupJunior Rescue
    classical decision-making, 359–360
    conflicts between reactive rules, 360
    decision-making, 358–359
    hybrid decision-making, 365, 367
    light sensing, 359
    lower-level behavior suppression code, 367
    reactive decision-making, 360
    robot learning, 366
    subsumption controller, 361
Robot learning, 366–368; *see also* Artificial
        neural network (ANN); Q-Learning
    MATLAB code for, 376–377
Robot vision, 65, 100–101; *see also* Camera
        systems; Digital image-processing;
        Feature extraction; Image formation;
        Image-processing; Robotics; Symbolic
        feature extraction methods
    case study, 98–100
Robotic arms, 39; *see also* Denavit–Hartenberg
        algorithm (D–H algorithm); Robotics
    arm parameters, 45
    case study, 48–52
    forward kinematic solutions, 45–46
    frame assignment, 45
    inverse kinematics, 46–47
    prismatic joint, 45
    revolute joint, 45
Robotic games, *see* Game robotics
Robotic SUMO wrestling, 9; *see also* Game
        robotics
Robotics, 1, 13; *see also* Control theory; Forward
        kinematics; Game robotics; Robotic
        arms; Robot vision; Wheel driver
        system
    application fields, 1
    autonomous robots, 14
    DOF, 13, 14
    DOM, 14
    remotely operated robot, 14

terminology, 13–14
work envelope, 14
Robotics drives, 113; *see also* Brushless DC motors
    (BLDC motors); Electrical machines;
    Servo motors; Stepper motors
    DC permanent magnet motors, 114
    issue in, 114
    power rating of motor, 114
Roll, 16
Rotary friction forces, 269
Rotating magnetic field, 112
    by induction motor, 113
Rotation matrix, 16, 18; *see also* Forward
    kinematics
    homogeneous, 20
Round-Obstacle behavior, 364
    finite-state machine representation for, 365
    MATLAB code for, 364, 366
ROV, *see* Remote operative vehicle (ROV)
rpm, *see* Revolutions per minute (rpm)

S

Sampled data systems, 221; *see also* Plant
Sampling and control interval, 294
Sampling density, 71; *see also* Digital
    image-processing
Saturation, 83
Semiautonomous robot, 357
Sense–plan–act, 359–360
Sensor measurement processing; *see also* Map
    representation; Occupancy grid
    feature-extraction, 319–321
    higher level environment representation,
        318–319
    hybrid process, 319
    using raw sensorial information, 318
    uncertainty, 318
Sensors, 53, 317; *see also* Color sensors;
    Encoders; Laser rangefinders; Optical
    sensors; Ultrasonic sensors
    accelerometer, 57
    accuracy, 53
    classification of, 54
    digital compass, 56–57
    error, 53
    frequency, 54
    in game robotics, 55
    gyroscope, 57
    Hall effect, 56
    linearity of, 53
    magnetoresistive effect, 57
    output, 54
    parameters, 53
    precision of, 53–54
    range, 53, 57–62
    resolution, 53

response time, 54
robot heading and inclination measurement,
    56–57
robot speed measurement, 55–56
sensitivity, 53
service robot using multiple, 60
Sequence generator, 129; *see also*
    Stepper motors
    applications, 131
    L297, 129–131
    signals, 129
Serial parallel interface (SPI), 57
Series winding, 112; *see also* Direct current
    machines
Servo control, 260
    with integrator, 289
    methods of, 264
    in PBR, 284
    for special case using reference as
        offset, 265
    velocity reference value, 294
Servo controller, 260; *see also* Controller;
    Discrete state feedback controllers
    discrete state equation, 261
    feedforward input, 260
    manipulated variable, 261
    special case, 261, 264, 265
Servo motors, 114; *see also* Robotics drives
    and assembly, 115
    control signal for, 115
    control with computers, 116
    specifications, 116
Servo system, 263; *see also* System
SGS Thompson L297 IC sequence generator,
    129–130
Shortest path search, 343, 350
Shunt winding, 112; *see also* Direct current
    machines
Sigmoid function, 372
SIMO, *see* Single-input multi-output (SIMO)
Simulink model, 205–207; *see also* Time domain
    solution
    of closed-loop system, 284–286, 287
    for PID implementation, 208
Singapore Robotic Games (SRG), 2, 267; *see also*
    Game robotics
    competitions and open category, 4
    humanoid robot competition, 4, 5
    interesting games in, 5
    pole-balancing robot race, 2, 3
    robot colonies, 3–4
    wall-climbing robot race, 3
Single-input multi-output (SIMO), 278
Slip ring generator, 106
Smoothing, 76; *see also* Image-processing
    application of, 77
    edge-detection filter, 76

Smoothing (*Continued*)
  Gaussian filter, 76
  MATLAB code for, 79
  mean filter, 76
  median filter, 77, 78
Sobel operators, 81, 82; *see also* Edge detection
  MATLAB code, 84
Software
  flowchart, 295, 297
  for robot, 293
Speed reduction gear, 140
SPI, *see* Serial parallel interface (SPI)
Springs, 178–179; *see also* Control system
SRG, *see* Singapore Robotic Games (SRG)
SRV-1 Blackfin, 66, 67
Star-connected BLDC motor, 122; *see also*
        Brushless DC motors (BLDC motors)
  disadvantage of, 123
  winding switching sequence, 124
State controllability, 250; *see also* Discrete state
        feedback controllers
  condition, 251
  control variable, 250
  matrix, 251
  for sampled data systems, 252–253
State equation, 173, 191; *see also* Controllers;
        Mathematical modeling; Time domain
        solution; Transfer function
  basic concepts of, 191–193
  development, 280–282
  mass, spring, and damper system, 193
  mechanical accelerometer, 195
  from plant knowledge, 193–195
  from transfer functions, 195–203
  vehicle inside vehicle, 194
State feedback controllers, 251; *see also* Controller;
        Discrete state feedback controllers
State model for pole-balancing robot, 282
State observability concept, 251; *see also*
        Discrete state feedback controllers
  observability condition, 252
  for sampled data systems, 252–253
  system equations, 251
State transition matrices, 210
Stator, 112, 122; *see also* Direct current
        machines
  in BLDC motor, 123, 124
  drum-like yoke fitted with poles, 108
  field, 113
  poles, 117, 121
  rotating magnetic field, 112
  teeth, 116, 117, 118, 120
  windings, 121, 122
Steady-state quadratic optimal control, 258;
        see also Discrete state feedback
        controllers
  dlqr command, 259, 260

MATLAB in LQC design, 259–260, 261,
        262, 263
performance index, 258–259
steady state optimization design, 258
Steerable caster wheels, 31
Stepper motors, 116; *see also* Drive systems;
        Permanent magnet stepper motors;
        Robotics drives; Sequence generator;
        Variable reluctance motors (VR motors)
  characteristic graph, 157
  classification of, 116
  designs of, 122
  discrete component-based driver, 131
  drivers, 128
  functions needed for, 128
  gear ratio design for, 156–158
  unipolar four-phase, 132
  working principle, 156
Subsumption controller, 361
Subsystem, 166; *see also* Control theory
Sumo robot competition, 358
Swivel caster wheels, 31
Symbolic feature extraction methods, 86, 73;
        see also Connected component
        labeling; Hough transform; Robot
        vision
  algorithms, 65–66
Synaptic weights, 371
Synchronous motors, 112; *see also* Alternating
        current (AC)—motors
  operation of, 113
Synchronous speed, 112
System, 166; *see also* Control theory; Digital
        systems; Discrete state space systems;
        Servo system; Transfer function
  with digital instrumentation, 232–234
  electromechanical system, 183–186
  mass damper and spring assembly, 186–187
  mechanical accelerometer, 190–191
  simple electrical system, 180, 183
  stability of, 174
  vehicle inside vehicle, 187–189

T

Teleoperated robot, 357
Tele-robots, *see* Remotely operated robot
Thresholding, 78, 81; *see also* Feature extraction
  adaptive, 80
  Chow and Kaneko method, 80
  issue in, 79
  MATLAB code for, 80
  multilevel, 78
Time domain solution; *see also* State equation;
        Transfer function
  using analytical methods, 209–214
  analytical solution, 204–205

code for plotting response, 205
of discrete state space systems, 245–250
Laplace transform technique, 210–211
mass, spring, and damper system,
    203–204
MATLAB code for PID controller
    simulation, 208
MATLAB code for P-only controller, 206
PID controller response, 207–209
PID controller tuning, 208
simulation solution, 205–207
Simulink model, 205–207, 208
of state equations, 209
state transition matrices, 210
time response, 206, 207, 209
using transfer function approach, 203
Time of flight (ToF), 59
Time-constant form, 176
ToF, *see* Time of flight (ToF)
Topological map, 328, 329; *see also* Micromouse;
    Path planning
maze representation, 330
searching runs, 328
Torque, 103, 140, 149, 150
alternate torque equation, 111–112
balance equation, 271
constant, 112
equation, 111
due to linear motion of mass, 271
due to pole mass, 271
Track drive system, 34–35; *see also* Wheel driver
    system
Training, *see* Learning
Training data, 370
for CoSpace Robot, 374–375
to obtain, 371
Training set, 371; *see also* Training data
Transfer function, 173, 216; *see also* Controllers;
    Discrete state space systems;
    Inverse Z-transforms; Mathematical
    modeling; State equation; System;
    Time domain solution; Training data
block diagram concepts, 179–180
block diagram reductions, 180, 181–182
closed-loop, 173, 180, 231
controller, 237
corner frequency form, 176
different forms of, 175
feedback, 180
mass damper and spring assembly,
    186–187
mechanical accelerometer, 190–191
models, 174–175
open-loop, 180
of permanent magnet DC motor drive,
    183–186
for plant, 174

pole/zero form, 175–176
polynomial form, 175
simple electrical system, 180, 183
systems with digital instrumentation,
    232–234
time-constant form, 176
vehicle inside vehicle, 187–189
of ZOH, 229
Transfer ratio, 272
Transformations
composite, 20, 21
homogeneous matrix, 19
Translation matrix, 17; *see also* Forward
    kinematics
Tustin's approximation, 230–231; *see also* Digital
    systems
Two-dimensional Cartesian grid, 333

U

Ultrasonic sensors, 58; *see also* Sensors
distance, 59
issues in, 59
on robot body, 58
signals, 58
sonar sensor model, 323
USB camera, 66

V

Variable reluctance motors (VR motors), 116;
    see also Stepper motors
alternative designs, 120
example, 118–119
multistack, 120–121
operation of four phase, 131–132
requirements for operation, 117
single-stack, 116
stack cut section view, 121
step size, 118
Variable structure controllers, 171; *see also*
    Control system
Vector field histogram (VFH), 333
angle and distance, 339–340
arguments in, 339
disadvantages and limitations, 334
robot heading, 340
Vector fields, 343, 350
Velocity control, 269
implementation, 274
Very large scale integration (VLSI), 66
VFH, *see* Vector field histogram (VFH)
Vision process, 73; *see also* Image-processing
Visual scene, 71; *see also* Robot vision
VLSI, *see* Very large scale integration (VLSI)
VR motors, *see* Variable reluctance motors
    (VR motors)

W

Wall following, 331; *see also* Navigation
    distance sensor, 331
    MATLAB code for, 332
    mechanism of, 333
Wall-climbing robot (WCR), 162, 169–170, 267,
            268, 306, 312; *see also* Flipper wall-
            climbing robot; Pole-balancing robot
            (PBR); Robot; Singapore Robotic
            Games (SRG)
    Bernoulli's equation, 314
    using Bernoulli's principle, 315
    competition platform, 306
    using dynamic suction, 312
    dynamic suction principle, 313
    operation of, 314
    race, 3
    system configuration, 313
Wavefront, 341
    planner algorithm, 341–342
Wheel driver system, 28; *see also* Odometry;
            Robotics
    Ackermann steering, 34
    angular speed, 33
    components, 29
    differential drive, 32–33
    MATLAB code for homogeneous
            transformation, 29–31
    omniwheel drive, 35–36
    robot speed, 33
    steering angle, 34
    track drive, 34–35
    types of wheels, 31–32

Winding, 134
    series, 112
    shunt, 112
    switching sequence, 124, 125
Work envelope, 14
World Robotic Sailing Championship, 9; *see also*
            Game robotics

X

Xcom, 298

Y

Yaw, 16
YUV color space, 73; *see also* Digital
            image-processing

Z

Zero-order hold (ZOH), 219
    implementation, 219
    Laplace transform for, 229
    transfer function of, 229
    Z-transform of plant fed from, 230
ZOH, *see* Zero-order hold (ZOH)
Z-transform, 223; *see also* Digital systems;
            Inverse Z-transforms
    approach, 246–250
    of continuous signals, 224–226
    discrete state space systems, 246–250
    and Laplace transform relationship, 224
    of plant fed from ZOH, 230
    possible scenarios, 223
    of signals, 223, 226–228